Roadside Geology of Northern and Central California

David Alt and
Donald W. Hyndman

Mountain Press Publishing Company
Missoula, Montana
2000

Library of Congress Cataloging-in-Publication Data

Alt, David D.

Roadside geology of Northern and Central California / David Alt and Donald W. Hyndman.

p. cm. — (Roadside geology series)

Includes bibliographical references and index.

ISBN 0-87842-409-1 (alk. paper)

1. Geology—California—Guidebooks. 2. California—Guidebooks. I. Hyndman, Donald W. II. Title. III. Series

QE89 .A84 2000
557.94—dc21

99–048554

PRINTED IN THE UNITED STATES OF AMERICA

Mountain Press Publishing Company
P. O. Box 2399 • Missoula, MT 59806
(406) 728-1900

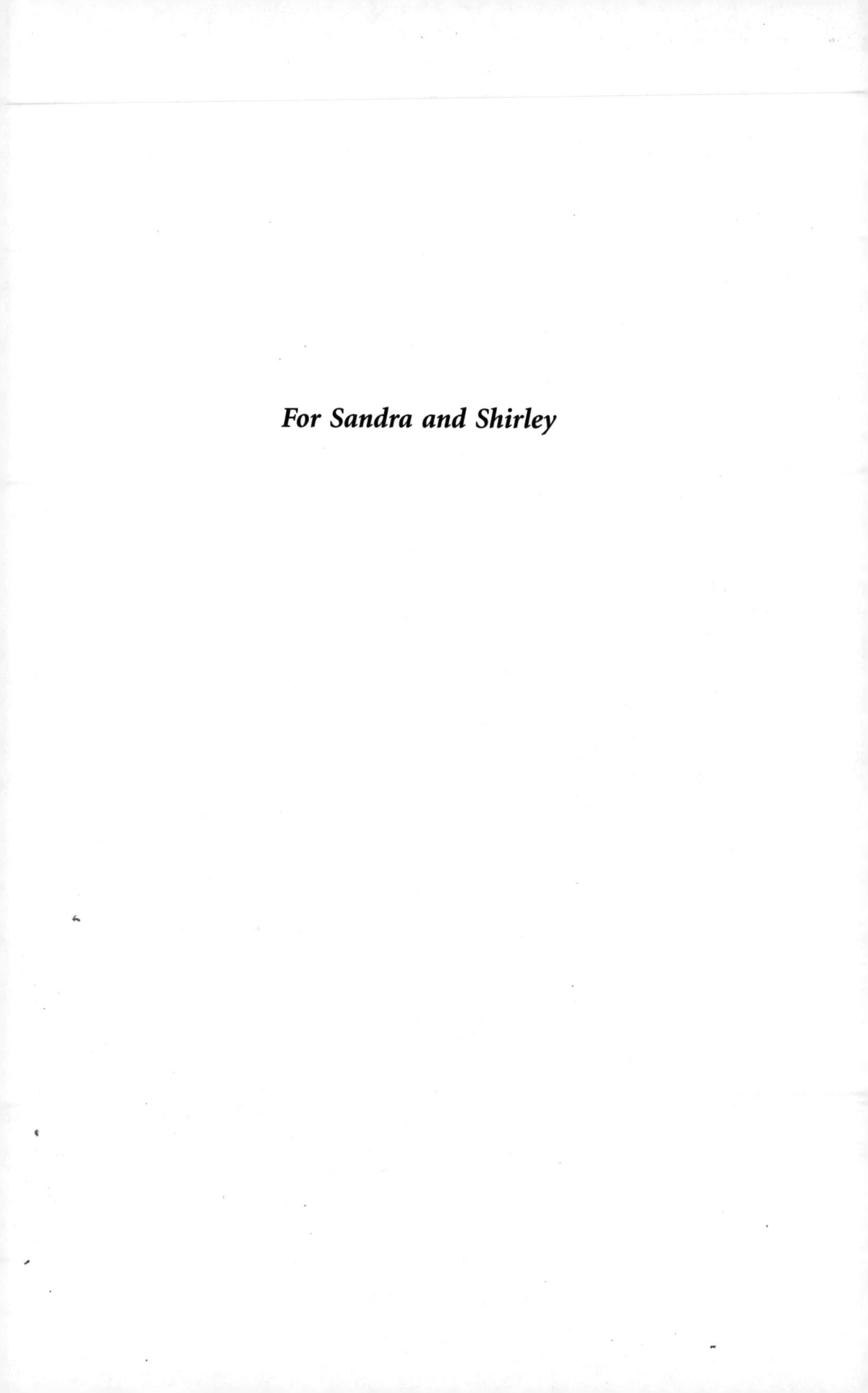

For Sandra and Shirley

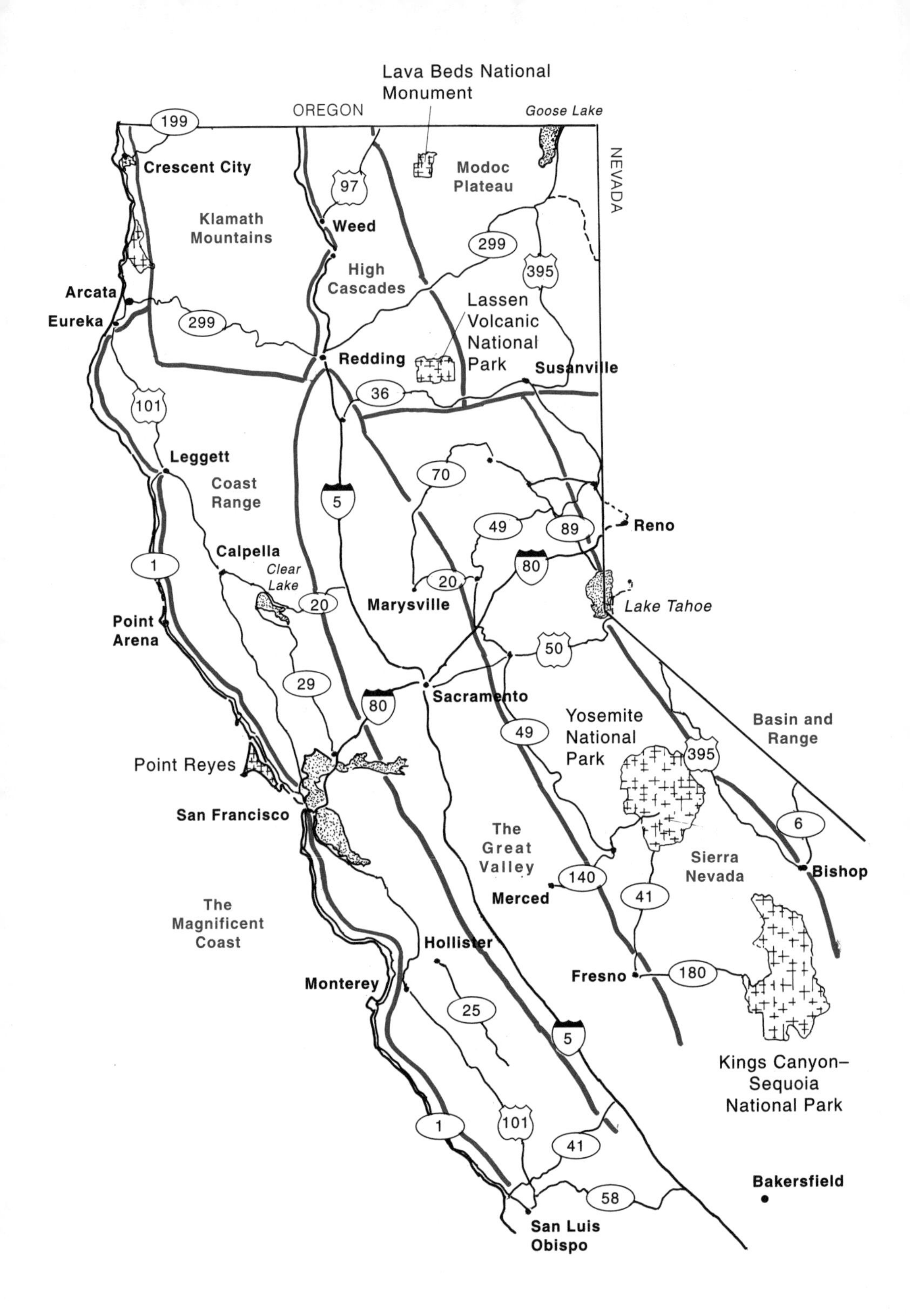

Roads and Sections of
Roadside Geology of Northern and Central California.

Contents

Map Symbols

 alluvium

 glacial deposits

 sediments (lakebeds, marine and continental deposits)

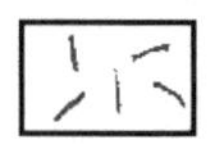 granite

 gabbro

  basalt

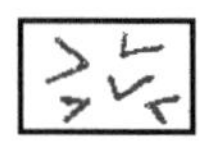 andesite and related volcanic rocks

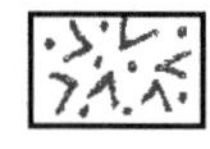 rhyolite and ash

  cinder cones

 volcano

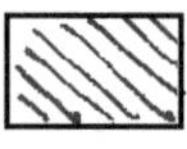 Calaveras complex

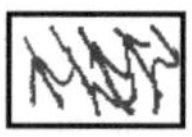 Shoo Fly complex

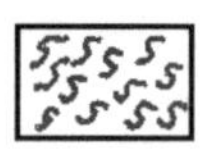 Western Jurassic terrane

 Smartville complex (Sierra Nevada chapter)

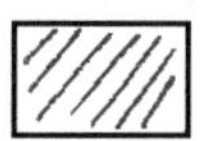 Franciscan complex (coastal chapters)

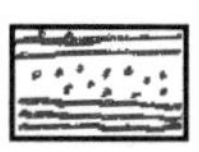 Great Valley sequence

 serpentinite on maps

 serpentinite in cross sections

 fault with sideways movement

 thrust fault

 limit of glaciation

Roadside Geology of Northern and Central California *covers a large and geologically varied area with many different rock types. We used these symbols on roadguide maps throughout the book. We labeled rock units on individual maps as space permitted, but in many cases it was not possible to label every pattern.*

Preface

Some twenty-five years have passed since we wrote *Roadside Geology of Northern California*. Our children have grown up, and our colleagues have grown gray. Only our wives have come through all those years unscathed.

The rocks have suffered most. Somehow, the rocks we knew twenty-five years ago no longer look the same. Nor do they tell the same stories. A lot has happened in California geology during the last twenty-five years; the time has come to reconsider the rocks and write an entirely new book. It seemed so reasonable to include central California that we now wonder why we did not do that the first time.

We wrote this book for people who are not geologists but who would like to know something about the rocks. We tell the big story first—what the rocks are—then fit the more interesting details into their proper places. We skip some rocks for lack of room, others because they are so tedious that only a deeply distracted geologist could love them.

Geology is not one of those subjects that deal with matters of hard scientific fact. Much of what geologists believe consists largely of conjecture, and most geologists love to argue about rocks. In some cases, we tried to present both sides of controversial points. In all cases, we tried to present the major kinds of evidence, to give our readers some idea of how geologists come to their conclusions.

This book begins with an introductory chapter that briefly reviews the geologic history of northern and central California. Then we divide the area into a number of natural geologic regions, consider each one in more detail, and finish each one with a set of roadguides to provide some of the local particulars. As you read the book, be sure to look also at the guides for roads that intersect or come near the road you are on.

Northern and central California contain more rocks than anyone could possibly see within several lifetimes, to say nothing of figuring out what they mean. We relied upon the published literature for all the facts and most of the ideas in this book. It would be nice to acknowledge every one of the geologists who contributed those facts and ideas, but they are too many. It would be nice to refer to their contributions as we come to them, but so many references would make this book unreadable. We did include a few general references to help people who want to read a bit more.

We beg our readers to cherish roadcuts. Most rocks are interesting, and quite a few of them are beautiful. Roadcuts provide our best chances to see and enjoy them. Roadcuts cost taxpayers a wad of money. We think it is an enormous waste for them to spend another wad of money to cover the cuts with soil and plant grass so the rocks are no longer visible. Encourage the highway department to leave roadcuts in their naturally raw, shocking condition, so everyone can enjoy the rocks.

Writing a book like this is fun. We enjoyed reading about the rocks, then getting out and seeing them in person. And most of all we enjoyed getting better acquainted with northern and central California.

Karen Grove, Steve Norwick, and Allen Glazner all read large parts of the manuscript and pointed out errors that would have been embarrassing had they found their way into print. We hope they caught them all. We thank Brady Rhodes for generously contributing to the roadguides for that part of the Coast Range and coast south of Salinas and Monterey. We thank Jennifer Carey of Mountain Press for managing our large and unwieldy manuscript, then neatly compressing it between the covers of a book. Kim Ericsson designed the book and laid out the pages. Andrea Sadler edited the copy to fish out our more flagrant errors of expression. Eva Karau labeled the maps, a task that involved deciphering our scrawled handwriting.

All of the photographs and geologic sections not otherwise credited are ours. We adapted all of the geologic maps not otherwise credited from the state geologic map of California, which the California Division of Mines and Geology has published in sections over a period of many years.

ERA	PERIOD	EPOCH	IMPORTANT GEOLOGIC EVENTS IN CALIFORNIA
	million years ago		
CENOZOIC	Tertiary	Pleistocene	Ice age glaciers carve high areas of the Sierra Nevada and Klamath block. Large lakes flood valleys in Modoc Plateau and Basin and Range. Volcanoes erupt in High Cascades.
		2	
		Pliocene	Sierra Nevada begins to rise; Sutter Buttes volcanoes erupt. Sediments eroded from the Sierra Nevada and Coast Range deposited in Great Valley.
		5	
		Miocene	San Andreas fault lengthens northward as Franciscan trench shortens. Monterey formation accumulates in Coast Range. Flood basalts cover Modoc Plateau. Basin and Range begins to open.
		24	
		Oligocene	Franciscan trench stops jamming oceanic sediments under the Coast Range ophiolite. Cascade volcanoes cover much of Sierra Nevada with ash and lava.
		36	
		Eocene	Streams deposit auriferous gravels in northern Sierra Nevada.
		58	
		Paleocene	
		65	
MESOZOIC	Cretaceous		Klamath block separates from Sierra Nevada and moves 60 miles west. Sediments eroded from Sierra Nevada begin to fill Great Valley. Franciscan trench drags sediments under western edge of Coast Range ophiolite, raising first western margin of Great Valley.
		145	
	Jurassic		Sierran trench jumps 60 miles west to become the Franciscan trench. Oceanic crust of Smartville complex forms about 160 million years ago. Deep continental crust melts into granite magmas that rise into the older rocks of the Sierra Nevada. A chain of volcanoes erupts above the granite batholiths. Sierran trench jams rocks of Calaveras complex onto and under the Shoo Fly complex of eastern Sierra Nevada.
		208	
	Triassic		Ocean floor begins sliding through the Sierran trench, under continental rocks in the Sierra Nevada and Klamath block.
		245	
PALEOZOIC	Permian		Rocks of Calaveras complex deposited on ocean floor and erupted from volcanoes.
		286	
	Pennsylvanian		
		320	
	Mississippian		Antler mountain building event jams older rocks against the margin of the continent, transforming them into tightly folded metamorphic rocks of the Shoo Fly complex.
		360	
	Devonian		
		408	
	Silurian		
		438	
	Ordovician		Oceanic crust of Trinity ophiolite forms and acquires a cover of oceanic sediments.
		505	
	Cambrian		
		570	
PRE-CAMBRIAN	Proterozoic		
		2,500	
	Archean		Ancient North American continental core formed east of California.

1

Introduction

Two hundred million years ago, the western edge of North America was somewhere near the present eastern edge of California. A little more than 200 million years ago, North America began moving west, away from the mid-Atlantic oceanic ridge, where the Atlantic Ocean was beginning to grow. The moving continent collided with the floor of the Pacific Ocean, which sank beneath the continent's oncoming western edge, then disappeared into the hot depths of the earth's mantle. But the sediments and volcanic islands on the oceanic floor were too light to sink. They crumpled in the grinding encounter between continent and ocean floor, to become most of California. And that, briefly told, is the big picture, the basic origin of northern and central California. Now, some details.

PLATE TECTONICS

Like many really powerful scientific theories, plate tectonics is basically simple and easy to understand. It starts with the idea that the earth has a cold and relatively rigid outer rind about 60 miles thick, the lithosphere. The lithosphere consists of segments called plates that cover the earth's surface like the bones in a skull. The earth has about a dozen large plates and an imperfectly known, but small, number of lesser plates. Some plates have only oceanic crust on their surfaces, some have only continental crust, but most have areas of each.

Plates move about the surface of the earth in an apparently random pattern that shows no hint of a grand design. Some pull away from each other, some slide past each other, and some collide with each other, either directly or obliquely.

Oceanic ridges mark the lines where plates pull away from each other. Every ocean has one, a long ridge with a deep rift along its crest that runs the length of the ocean. The opposite sides of the ridge move away from each other at a rate of about 2 inches per year. From time to time, a fissure opens in the rift, and an enormous basalt lava flow erupts from it, forming new oceanic crust. The ocean floor grows flow by flow.

If new oceanic crust forms at the crests of oceanic ridges, then old oceanic crust must disappear somewhere else. That happens where plates collide. One plate slides beneath the other and on down into the earth's mantle. An oceanic trench appears where the sinking plate bends down as it starts its long slide into the depths. The trench appears as a long trough on maps of the ocean floor. A chain of volcanoes rises above the sinking plate parallel to it and 50 to 200 miles away. Continental crust is too light to sink. It stays on the surface, carrying the rocks that contain the record of the earth's history, the planetary archives.

Plates slide horizontally past each other along transform boundaries. The San Andreas fault, which separates the North American and Pacific plates, is the most famous example of such a boundary. All transform boundaries connect oceanic trenches and ridges in some combination. The San Andreas fault connects the East Pacific ridge with a trench off the coast of northern California, Oregon, and Washington. Transform boundaries define the edges of plates moving away from oceanic ridges and oceanic trenches.

THE OLD CONTINENT

A continent of unknown size, shape, and location existed a billion or more years ago. A continent is a floating raft composed mostly of relatively light igneous and metamorphic rocks, mainly granite, gneiss, and schist. Those are the rocks of the continental crust, the rocks geologists often call basement rocks because they lie beneath all younger rocks.

Our best information about the thickness of continental crust comes from the study of earthquake waves. Seismographic records of earthquakes generally show an echo that follows several seconds after the primary shock wave. This echo appears to reflect off the base of the continental crust. Measurements of the timing of the echoing earthquake waves show that, in most areas, the continental crust is about 25 to 30 miles thick.

TALE OF THREE TRENCHES

The story of northern and central California is mainly a tale of three oceanic trenches. The first of our three trenches appeared during late Devonian and Mississippian time, about 350 million years ago. The old continent of North America then collided with the ocean floor in what is known as the Antler mountain building event. That collision crushed the oceanic sediments that had accumulated along the coast of North America into the western edge of the continent. The crushing deformed the sediments into tight folds, broke them along faults, and heated them enough to recrystallize them into metamorphic rocks. Those abused rocks, the Shoo Fly complex, now form most of the eastern Sierra Nevada.

Then the Antler collision ended, its trench disappeared, and the rocks of the Shoo Fly complex became the new western fringe of California. A new generation of sediments quietly accumulated to make a new coastal plain along the new coast. The former edge of the old continent is now somewhere near the eastern border of California.

The second trench, the Sierran trench, appeared about the end of Triassic time, about 200 million years ago. That was when the old continent broke along the line of the mid-Atlantic ridge. North America began moving west as the Atlantic Ocean opened behind it, and Eurasia began to move east, also away from the mid-Atlantic ridge. The new North America again collided with oceanic crust along its western edge. This collision first smashed the sediments that had collected along the new coast against the rocks of the Shoo Fly complex, converting them into another mass of horribly deformed and considerably recrystallized rocks, the Calaveras complex. The Calaveras complex is now a long central strip of the Sierra Nevada.

As the collision continued along the line of the Sierran trench, more sedimentary rocks and many volcanic rocks jammed against the Shoo Fly complex. They included such odds and ends as old volcanic islands and sediments that could only have accumulated on the deepest and most remote reaches of the ocean floor. These rocks were thoroughly deformed and recrystallized. Geologists call them the Western Jurassic terrane.

Then, for reasons that remain unknown, the line of collision between continent and ocean floor jumped west about 60 miles. That probably happened during early Cretaceous time, about 130 million years ago. That was when the northern end of the Sierra Nevada detached and moved 60 miles west to become the Klamath Mountains. The gap between the old and new lines of sinking eventually became the Great Valley. We refer to the new line of sinking as the Franciscan trench. It created the Coast Range.

Sediments that were swept into the Franciscan trench were jammed against the western edge of the Klamath Mountains in the north and against the edge of a strip of oceanic crust some 60 miles wide farther south. It would eventually become the bedrock foundation under the Great Valley.

When the Franciscan trench eventually went out of business, between about 30 and 15 million years ago, the sediments stuffed into it rose to become the Franciscan rocks of the western part of the Coast Range. Some of them had been dragged under the edge of the strip of oceanic crust, and they rose too, jacking the edge of that strip up to a steep angle to become the Great Valley sequence along the eastern side of the Coast Range.

MODOC SEAWAY

No one seems to understand why or how the Klamath Mountains detached from the north end of the Sierra Nevada, but they did. The rocks show that a seaway opened east of that displaced block as it moved west. The seaway was almost landlocked, so sediment entered from all directions; it was filled within a few tens of millions of years, by the end of Cretaceous time. It was a flat plain then, but volcanic rocks have since covered it, and faults have broken it into mountains and basins.

FLOOD BASALT FLOWS

A little more than 17 million years ago, an enormous center of volcanic activity appeared in the southeastern corner of Oregon. Its edges probably lapped into southwestern Idaho and northern Nevada. We think it developed in the crater that opened where an asteroid struck the earth and exploded. Many other geologists argue that a huge mass of especially hot rock rose through the mantle and partly melted to form basalt magma. Whatever the cause, something drastic certainly happened about 17 million years ago that seems to have started a great many other events. The coming of the flood basalts coincided with the establishment of most of the San Andreas fault, the rise of the Coast Range, and the beginning of crustal stretching in the Basin and Range.

That great volcanic center in southeastern Oregon was one of those that erupt flood basalt flows—lava flows so overwhelming that they add an entirely new dimension to our ordinary notions of volcanic catastrophe. They may contain more than 100 cubic miles of lava, hundreds of times more than the volumes of ordinary basalt flows. These basalt flows flooded all the low areas around the great volcano, which was not a mountain but a volcanic plateau.

Some of those floods of molten lava poured southwest across the sediments that filled the Modoc seaway. They covered it with enormous sheets of basalt, making it the southern end of the great Pacific Northwest flood basalt province. The big flood basalt eruptions ended about 15.5 million years ago. The earth has not seen their like since.

California as seen from space.
—G. P. Thelin and R. J. Pike, U.S. Geological Survey

BASIN AND RANGE

The Basin and Range is a vast province of isolated mountain ranges set between broad desert basins. It stretches from eastern Oregon south into Mexico, and from the eastern front of the Sierra Nevada to the Wasatch Front in Utah. The part that embraces most of Nevada and western Utah is the Great Basin. It includes the part of California east of the Sierra Nevada crest and east of the Warner Mountains, in the northeastern corner of the state.

Since middle Miocene time, during the past 16 million or so years, the continental crust in the Basin and Range has stretched from east to west and has greatly thinned. No one knows exactly how far the region has stretched, but most geologists estimate that it is now at least twice as wide as it was.

SAN ANDREAS FAULT

The ocean floor that sank through the Franciscan trench belonged to the Farallon plate, which was growing from an oceanic ridge to the west. New oceanic crust formed at the ridge, then sank beneath the oncoming western edge of North America. As North America moved west, it finally reached the ridge where the Farallon plate was forming. The last of the plate disappeared as the ridge met the trench. The ridge and trench annihilated each other in an act of mutually assured destruction. As the last of the Farallon plate vanished into the earth's mantle, the western edge of North America met the Pacific plate. That new plate boundary is the San Andreas fault, actually a swarm of parallel faults.

All the rocks west of the San Andreas fault belong to the Pacific plate, those east of it to the North American plate. The Pacific plate is moving north at a rate of approximately 2 inches per year and has so far carried the part of California west of the fault at least 350 miles north.

HIGH CASCADES

Volcanoes had been erupting in the Western Cascades before the eruptions of flood basalts. That volcanic chain died when the flood basalt eruptions began. The new chain of volcanoes is the modern High Cascades, parallel to what remains of the Franciscan trench. Shasta and Lassen are the best-known volcanoes of this range in California.

WHITHER CALIFORNIA?

Someone once described history as just "one damned thing after another." That seems equally true of geologic history. The geologic development of California seemed firmly set on a predictable course until the flood basalts, Basin and Range, and extinction of the Western Cascades volcanoes changed the agenda about 17 million years ago. Now the situation again seems fairly stable and the future fairly predictable.

If present trends continue, the slice of California west of the San Andreas fault will continue to move north and will part company with the rest of California sometime around 15 million years from now. It should eventually arrive in Alaska. Other members of the San Andreas system of faults will continue to slice the Coast Range into long slivers. Meanwhile, new faults in the western edge of the Basin and Range will take great slices off the eastern Sierra Nevada, creating new ranges like the White Mountains and new valleys like the Owens Valley.

All that will happen right on schedule if no unpredictable happening again changes the course of events. Given the great length of geologic time and the number of unpredictable events that could happen, one will happen, eventually.

The Sierra Nevada.

El Capitan.

2

Sierra Nevada

The Sierra Nevada is the high eastern backbone of California, its defining ridge. Geographically, it is a high wedge lying on its side, with a steep eastern slope and a long western slope that eases gently down to the Great Valley. Geologically, it is extremely complex, reflecting a long sequence of geologic events.

SUBDUCTION AND OBDUCTION

The normal fate of an oceanic plate is to sink through a trench and down into the earth's mantle, a process geologists broadly call subduction. They use the parallel term obduction to refer to what happens when big pieces of oceanic plates somehow avoid sinking into the trench and instead become incorporated into a trench filling, or perhaps slide right over the top of the trench filling. Exactly how that happens is one of the great unclarities of modern geology. Do not suppose that their having invented the technical term obduction means that geologists understand what happens. Most attempts at explanation posit the leading edge of the oncoming continent somehow trying to dive under the ocean floor, instead of riding over it.

Ophiolites—Oceanic Crust on Land

For more than a century, geologists in many countries have recognized and studied great slabs of rocks they called ophiolites. It came as a revelation when geologists finally realized during the late 1960s that ophiolites are slabs of oceanic crust now on land. Before then, the oceanic crust had seemed as impossibly inaccessible as the far side of the moon. Since then, anyone can see it exposed on sunny hill slopes during a Sunday picnic in the Sierra Nevada, Klamath Mountains, or Coast Range.

The upper part of the oceanic crust is made up of basalt lava flows, piled to a thickness of thousands of feet. These flows erupted at the crest of an oceanic ridge, in the rift where two plates pull away from each other. The eruptions happen quite visibly in Iceland, where the mid-Atlantic oceanic ridge stands above sea level. There, the rift valley cuts a dogleg course across the country, between the Eurasian plate on the east and the North American plate on the west. Surveyors measure the valley growing wider as those two plates draw away from each other at a rate of approximately 2 inches per year. Every few hundred years, a long fissure opens in the valley floor, and a large lava flow erupts from it, producing new oceanic crust.

When basalt lava erupts on the ocean floor, the outer surface of the flow quickly chills against the cold seawater to form a hard shell. Meanwhile, the molten, flowing basalt within that shell swells it, bursts it, and then pours onto the seafloor, where it again chills to form another hard shell. As that sequence of events repeats, the flow converts itself into a pile of basalt blobs and cylinders that range from buckets to barrels in diameter. A cross section of such blobs of basalt exposed in a cliff or roadcut looks like a pile of greenish black pillows—ugly stuff. Geologists call them pillow basalts.

Beneath the basalt lava flows, thousands of feet of vertical basalt dikes are stacked next to each other like books on a shelf. Each fills a fissure that

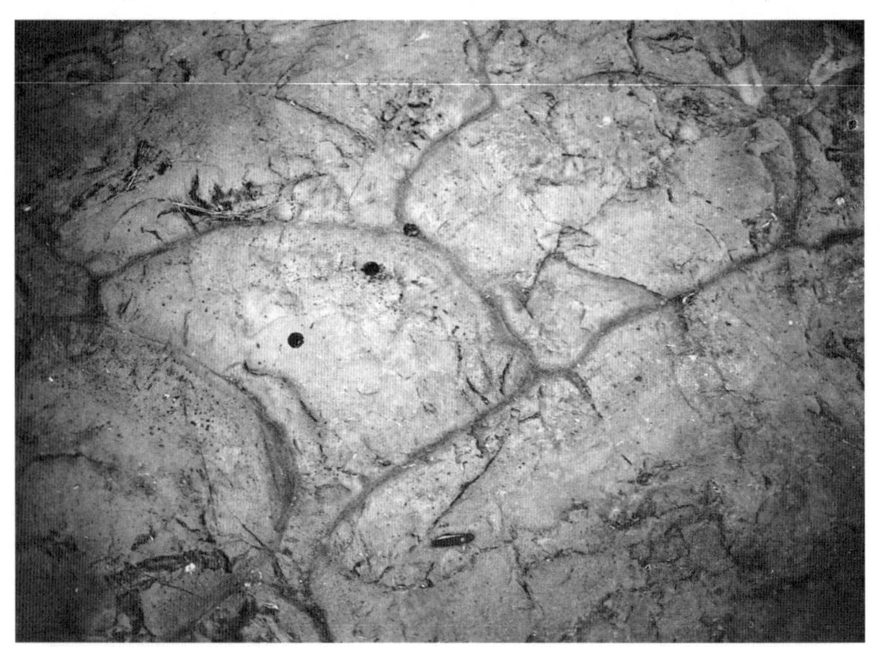

Pillow basalts in the Smartville complex, at the Yuba River, east of Marysville.

Layered gabbro in the Smartville complex east of Marysville.

opened as two plates pulled away from each other. Each is the plumbing that carried basalt magma from pools at the base of the new crust into one of the lava flows above. Most are a few feet thick. They measure the growth of an ocean, dike by dike, in a time sequence that runs horizontally, instead of from bottom to top as geologists usually read their rocks.

Geologists call a collection of those dikes a sheeted dike complex. Careful examination reveals that nearly every individual dike has one margin of extremely fine, almost glassy rock. This margin chilled against the adjacent solid basalt, while the other margin could not cool quickly because it was in contact with magma.

Below the dikes lie thousands of feet of dark gabbro, which would have been basalt had it erupted. Gabbro differs from basalt only in consisting of larger mineral grains, which grew very slowly in the reservoir of molten magma beneath an oceanic ridge. The heavier pyroxene and olivine crystals sank as they grew and accumulated in layers on the floor of the magma chamber. They make the gabbro look like a sedimentary rock, which is exactly what it is—a product of igneous sedimentation within a molten magma chamber.

Mantle Rocks

Some ophiolite complexes continue below the layered gabbro at the base of the oceanic crust and into the peridotite of the upper mantle.

Peridotite is a dark, notably dense rock composed mostly of green olivine and black pyroxene. Some varieties also contain crystals of red garnet. It is the main rock of the earth's mantle and is, therefore, the most abundant rock in the planet. But we rarely see peridotite because continental or oceanic crust covers the mantle nearly everywhere. These deepest levels of ophiolite complexes offer geologists their only opportunities to study the mantle directly, in bedrock outcrops.

The mantle peridotite in most ophiolite complexes is at least partially altered to a strange, darkly greenish rock called serpentinite. The rock feels heavy in the hand, and its surface feels vaguely soapy, which explains why many people call it soapstone. Most specimens are soft enough to carve with a knife. Many people who have never glanced at a roadcut in serpentinite have browsed through gift shops, admiring the little greenish figurines and boxes carved from it.

Serpentinite forms at the crests of oceanic ridges, where seawater sinks into fractures that open as the two plates separate. The tremendous pressure of the depths forces the water down into the extremely hot mantle rocks below the ridge crest. The water reacts with hot peridotite to make serpentinite, which constitutes a large proportion of the uppermost mantle.

Serpentinite really is as slippery as it feels. Large masses of it rise from the slab of ocean floor that is sinking into the mantle at an oceanic trench and intrude the rocks above. The Sierra Nevada and Coast Range are full of intrusive masses of serpentinite, most of which rose through the shattered rocks along fault zones.

When the slab of ocean floor sinking through an oceanic trench reaches a trench depth of about 50 miles, its temperature rises to about 550 degrees centigrade. At that point, the serpentinite loses its water and reverts to peridotite. Under those conditions, the water is technically a supercritical fluid, neither water nor steam in the ordinary sense of the terms. The red-hot fluid rises into the rocks above, melting them to make magmas. The magmas feed the volcanic chain that invariably rises parallel to an oceanic trench, 50 to 200 miles away from it.

STACKED TERRANES

Even in these modern days, geologists often speak of terranes, and that really is the way they spell it. The word looks quaint, but the idea makes sense. A terrane, in the geologic sense, is simply a discrete package of rocks enclosed within faults or other natural boundaries. Rocks within terranes may be simple or complex, and they may differ greatly, but they appear to share a common origin and history. Geologists also use other words in the same sense: complex, belt, slab, formation, plate.

Thinking of complex rocks such as those in the northern Sierra Nevada as an assemblage of terranes, under whatever name, greatly simplifies the problem of figuring out what happened. First, you deal broadly with the terranes, then you tackle the complexities within them, some of which may rank as subterranes.

Geologists tend to differ in their ideas of how to divide the rocks into terranes, but in the end they usually arrive at the same broad picture. An argument still rages over the terranes in the Sierra Nevada, and this is not the book, nor are we the authors, to resolve it. We will stick to the basics, trying not to get too deeply involved in the more contentious details.

The basic bedrock framework of the Sierra Nevada is a series of major terranes, each a long belt of metamorphic rocks and granite intrusions that trends generally north. Each is basically a great and internally complex slab of rock that tilts down to the east and lies on a fault that separates it from the next slab to the west. The arrangement broadly resembles that of shingles on a roof.

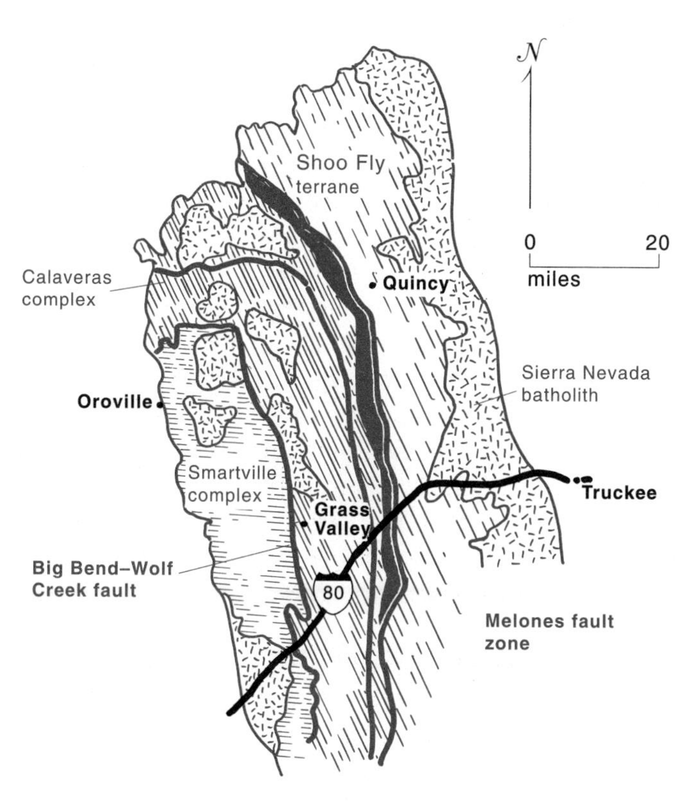

Major terranes and their bounding faults in the Sierra Nevada.

The original sedimentary formations in the terranes of the Sierra Nevada are oldest in the east and become younger westward. In general, those formations were deposited closest to land in the east and become more clearly oceanic westward. The older terranes consistently lie on younger terranes. That is the reverse of the usual stacking, which has younger rocks lying on older rocks.

Obviously, the terranes of the Sierra Nevada did not form where we now see them. They were somehow shuffled into their present positions as North America collided with the floor of the Pacific Ocean. Some may have moved long distances north from their original homes. Crushed sedimentary formations, stray islands in the ocean, and volcanic chains built the western edge of North America seaward as they collided with it. So the continental margin established some 800 to 700 million years ago is now in the mountains east of the Sierra Nevada, dismembered in the Basin and Range.

Shoo Fly Terrane

The oldest sedimentary formations in California, which were deposited along the old continental margin 800 to 700 million years ago, are in the White and Inyo Mountains. Both ranges are slices detached from the Sierra Nevada. The oldest rocks in the Sierra Nevada were deposited during early Paleozoic time, before the Antler mountain building event began about 350 million years ago. They were squashed against the old continental margin during the Antler event. Most are in the Shoo Fly complex of the high Sierra Nevada and in some of the mountain ranges farther east. They include sandstone, shale, limestone, and chert that were deposited on the ocean floor, scrambled with serpentinite from the upper mantle. Sierra Buttes, just north of Sierra City, are the remains of a volcanic chain that erupted along the continental margin during the Antler event.

Those squashed oldest rocks added enough to the western edge of North America to move it west to a line high in the Sierra Nevada. North of U.S. 50, those rocks are recrystallized into slabby schists composed mostly of very fine crystals. Farther south, where the granite magmas of the Sierra Nevada batholith cooked them to high temperatures, they are now coarsely crystalline schists and gneisses.

Rocks of the Shoo Fly terrane lie on the Calaveras and Shoo Fly fault zone south of Interstate 80 and on the northern Melones, Rich Bar, and Goodyear faults in the northern Sierra Nevada. The various names really are confusing.

Calaveras Complex

Rocks immediately west of the tangled assortment of faults that make up the Shoo Fly terrane belong to the Calaveras complex—actually a terrane masquerading under its old name. They were stuffed into the Sierran trench and added to North America during Jurassic time. The contact between the Shoo Fly terrane and Calaveras complex, the approximate line of all those faults, is probably as close as we can conveniently come to finding the coast that existed after the Antler event ended.

After the Antler event ended, another generation of sedimentary rocks accumulated along the newly established continental margin. Then, North America again began to move west against the floor of the Pacific Ocean. That happened as Triassic time ended and Jurassic time began, about 200 million years ago. It happened because the Atlantic Ocean had begun to open, separating North America from northern Africa and southern Europe. The floor of the Pacific Ocean began to sink beneath the advancing western edge of North America. The Sierran trench appeared where the ocean floor flexed down to begin its long slide beneath the western edge of the continent and deep into the earth's mantle.

As the sinking ocean floor slid out from under the mud and other sediments that were deposited on it, the continent crushed them into the Sierran trench and against its western edge. These late Paleozoic to early Triassic rocks recrystallized in the heat of the trench to become metamorphic rocks, mainly greenish slates, glistening phyllites, and dark green serpentinite. Masses of granite magma intruded them. The result was the broad belt of metamorphic rocks and granite that geologists call the

Slates in the gold country.

Calaveras complex. It is a large terrane that occupies most of the core of the Sierra Nevada.

The best that can be said of the Calaveras complex rocks is that they are an unholy mess, difficult to decipher in the field and impossible to describe adequately on maps or in words. The rocks include a scrambled assortment of pieces of old oceanic crust and the sediments that were deposited on them, all torn up along faults, jammed into tight folds, and recrystallized into metamorphic rocks in the heat of the deep trench. Even geologists experienced in working with such rocks find it hard to stop at an outcrop and understand what they see.

Melones Fault Zone

South of Interstate 80, the Melones fault zone separates the Calaveras complex from the younger and radically different assemblage of Mesozoic rocks in the Western Jurassic terrane to the west. Many geologists think the Melones fault zone was the actual surface along which the sinking slab of ocean floor slipped under the western edge of the Calaveras complex and beneath North America. If so, then it is an old plate boundary. The sinking ocean floor dragged the younger rocks of the Western Jurassic terrane beneath the Calaveras complex.

The sinking oceanic crust regurgitated large masses of slippery serpentinite as though they were blobs of grease. Many rose through the fractures of the Melones fault zone, which evidently provided easy passageways. Nearly everywhere you see it, the fault zone is full of serpentinite. Watch for roadcuts in darkly greenish rock broken into chunks with polished surfaces that glitter brightly in the sun.

Odd scraps of evidence keep emerging to suggest that some parts of the Sierra Nevada may have begun their careers in places far to the south: some sedimentary rock formations contain fossils with distinctly southern affinities. The magnetization directions of some bodies of rock are easiest to understand if we assume that they are far north, perhaps thousands of miles north, of where they started. Although none of the individual items of evidence seems quite compelling, their variety and number are highly suggestive. Parts of the Sierra Nevada may indeed have arrived from points far south. If so, they must have moved north along a fault that slips horizontally, like the San Andreas fault.

Many geologists now consider the Melones fault zone a candidate for at least part of the honor. It is obviously a major structure, and it is remarkably straight. Faults that move horizontally normally slice vertically through the earth's crust, and that gives them a straight trace on the map. Slice a loaf of bread vertically, and the cut will look straight from above;

slice it at an angle, and the cut, seen from above, will trace a curving line across the loaf. So the straightness of the Melones fault zone across hills and valleys suggests that it is vertical and moved horizontally. That does not in any way contradict the idea that it is the surface along which the ocean floor sank beneath North America early in Mesozoic time. The ocean floor could have slid obliquely to its fate.

Western Jurassic Terrane

The Western Jurassic terrane is west of the Calaveras complex. The Big Bend, Wolf Creek, and Bear Mountains faults bound it on the east. It is an extraordinary collection of rocks that mainly features old oceanic crust, now metamorphosed to greenstones, and the slates and graywacke that were deposited on them. The Smartville ophiolite, northeast of Sacramento, is a fairly well preserved slab of Jurassic oceanic crust. Its lower part is serpentinite, above that the usual sequence of gabbro, sheeted dikes, and pillow basalt. The base of the Smartville ophiolite dips gently down to the west, under the sediments that fill the Great Valley. The base and the sediments that were deposited on it were tightly folded, broken along faults, and slightly recrystallized as they jammed into the Sierran trench and against the western edge of the Calaveras complex before Jurassic time ended.

SIERRA NEVADA GRANITE BATHOLITH

The word *granite* is an old miner's term that originally referred to almost any massive granular rock composed mostly of feldspar and quartz in crystals large enough to see without a magnifier. A batholith is simply a large expanse of granite, enough to cover more than approximately 40 square miles.

A chain of volcanoes rises parallel to every oceanic trench, for reasons that go back to the oceanic ridge, and hinge mainly on water. The volcanoes depend upon the extremely hot fluid released as serpentinite reverts to peridotite. Add water in any form to any kind of rock, and you lower its melting temperature. In this case, that happens directly above the depth where the sinking slab of lithosphere gets hot enough to break down its serpentinite. The released water rises into the overlying mantle, partially melting it to make basalt magma, which in turn rises into the overlying crustal rocks and partially melts them to make granite magma.

The magmas finally erupt through a chain of volcanoes aligned above the zone where the sinking slab loses its water. The volcanoes may rise anywhere from 50 to 200 miles from the oceanic trench, depending upon how steeply the descending oceanic crust is sinking. Long after the volcanoes die, erosion reveals enormous masses of granite that crystallized

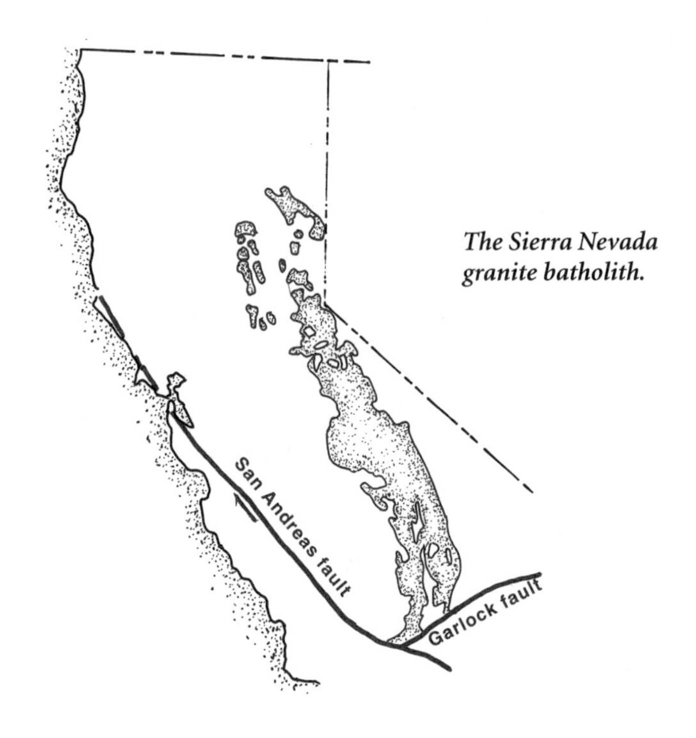

The Sierra Nevada granite batholith.

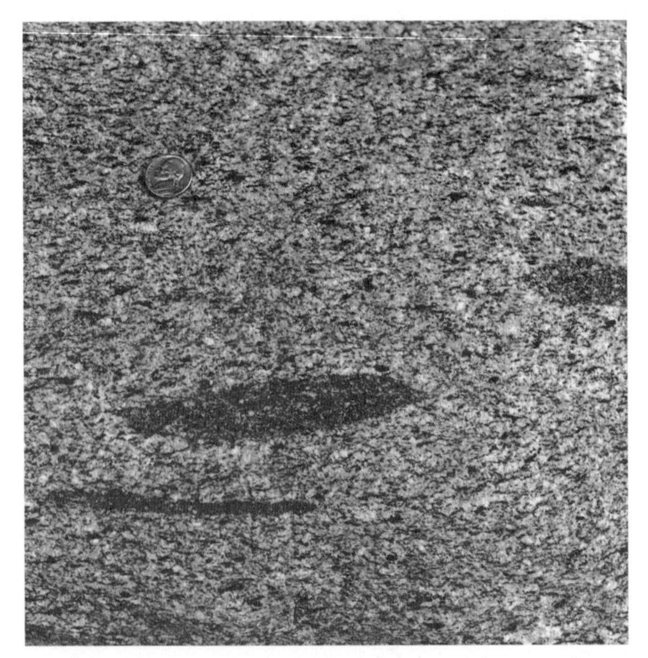

Stretched inclusion of dark rock in pale granite north of Kings Canyon National Park.

from unerupted magma in their roots. The basalt magma that helped provide the heat to melt the granite magma leaves dark inclusions in the granite, big freckles of black rock a few inches across.

The Sierra Nevada batholith was one of the first in which geologists recognized numerous huge masses of slightly different kinds of granite packed together like marshmallows in a bag. At first glance, the varieties of granite in those masses, known as plutons, look alike, in the same way that all house cats look alike. In either case, a closer look reveals small variations on the common theme. Granites differ in color, grain size, texture, and the proportions of the different minerals. The various plutons are distinctive enough that you can recognize them in the field and draw their boundaries on a map.

Most of the plutons in the Sierra Nevada batholith range from a few miles to a few tens of miles across. According to some estimates, the Sierra Nevada batholith contains more than one hundred separate granite plutons. Age dates show that they were emplaced between about 225 and 80 million years ago, as big blobs of magma rose from the depths.

The Klamath block is the original northern end of the Sierra Nevada, now detached and moved about 60 miles west. It contains just a few masses of granite, each one quite separate from the others. Granite becomes progressively more abundant from the northern Sierra Nevada southward, and the individual masses become more closely packed. The southern end of the range consists almost entirely of granite.

Many geologists now contend that the traverse from north to south down the Sierra Nevada is the equivalent of a vertical section through the crust. They believe that the rocks you see at the surface in the southern part of the range are like those that exist at depth in the northern part. Come back in 50 million years to see if deeper erosion of the northern Sierra Nevada has indeed exposed a nearly continuous mass of granite.

Weathering Granite

People who wander the Sierra Nevada see enormous amounts of weathered granite, which normally forms bouldery outcrops and vast amounts of sandy debris. The first step in the weathering process is that water penetrates fractures in the granite, reacts with the rock, and weathers it to soil. Water weathers angular blocks most rapidly at corners, where it attacks from three directions; less rapidly at edges, where it attacks from two directions; and least rapidly on the flat surfaces. So weathering preferentially rounds off the corners and edges, leaving a round mass of unweathered granite within the soil, a core stone. Next, something destroys the plant cover—a fire, or perhaps a period of very dry climate.

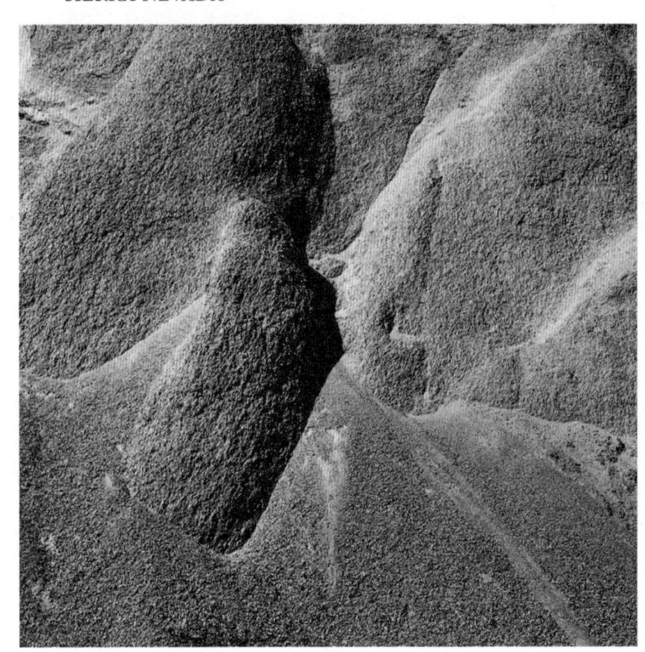

Granite weathering to sandy grus near entrance to D. L. Bliss State Park.

Then erosion by rain splash and surface runoff removes the soil, leaving the rounded core stones at the surface. Granite core stones are typically large because the fractures in the rock tend to be widely spaced. Rocks with more closely spaced fractures weather into smaller and much less conspicuous core stones.

The grains of feldspar and mica in granite swell as they react with water and turn into clay. Then the rock falls apart, for the same reason that a building would fall apart if its bricks were to swell, some more than others. Geologists call the piles of loose and partly weathered mineral grains *grus*, a German word more or less domesticated into the geologic variant of English. Grus is common and conspicuous in the lower elevations of the Sierra Nevada, where no glaciers scraped it off.

Roof Pendants

The great expanses of granite in the southern parts of the Sierra Nevada batholith contain occasional masses of older metamorphic rocks. Early geologists interpreted them as pieces of the older rocks that hung down into a sea of molten magma beneath them, so they called them roof pendants. Later fieldwork showed that most are actually remnants of the older rocks caught between the separate granite intrusions that make the batholith, screens of metamorphic rocks that separate the intrusions. But geologists still call them roof pendants, the demands of tradition being what they are.

Most of the roof pendants were strongly heated in their proximity to the masses of granite magma and are now metamorphic rocks recrystallized

The steep east front of the Sierra Nevada near South Lake Tahoe.

almost beyond recognition. But it is possible to determine, in at least a general way, what those rocks were before they recrystallized and to reconstruct the bedrock that existed before the granite magmas arrived.

Rise of the Sierra Nevada

The Sierra Nevada was a broadly rolling lowland until a few million years ago, when movements on the faults along its eastern face began to raise it, evidently because the westward development of the Basin and Range finally established the Sierra Nevada fault. Gently tilted remnants of the old rolling lowland conspicuously survive in the broad upland flats of the western slope.

The steep eastern front of the Sierra Nevada rises immediately west of the Owens Valley, the westernmost basin of the Basin and Range. As the Sierra Nevada moves west on a fault, it emerges from beneath the burden of the slab to the east, which includes the Owens Valley and the White and Inyo Mountains. That unloads the earth's crust just east of the Sierra Nevada, permitting the range to rise because it floats up. Meanwhile, volcanic rocks erupt along the western edge of the Owens Valley, probably because the drop in pressure on the hot rocks at depth permits them to partly melt.

In time, a new fault will probably develop farther west. The eastern part of the Sierra Nevada will shear off along it and move east to become part of the Basin and Range, and a new version of the Owens Valley will

A deep canyon cut into granite of the Sierra Nevada batholith northeast of Fresno.

open west of the newly detached slice. As that happens, what is now the gentle western slope of the Sierra Nevada will continue to rise, and a new crest will appear west of the present crest.

GLACIERS

Pleistocene time, the past 2 million or so years, saw an unknown number of ice ages come and go—eight according to some authorities, twelve according to others, perhaps as many as twenty according to a few. It seems clear that ice ages were times of cold weather and heavy rain and snowfall. Although theories abound, no one actually knows what caused them. Nor does anyone know whether the earth will see more ice ages, why the last one suddenly ended, or when the next one may come.

Ice age glaciers grew all along the crest of the high Sierra Nevada and flowed down the valleys, especially to the west, but to a lesser extent to the east. So the upper parts of major Sierran valleys are glacially eroded, and the lowest parts are filled with glacial sediment. The largest glaciers grew where the crest is highest and caught the most moisture, and they eroded the deepest canyons.

Glaciers polished large areas of granite bedrock in the high Sierra Nevada and left them covered with patterns of parallel scratches. Glacial ice is not really the clear blue stuff so often featured in advertisements for gin or vodka. It is actually full of rocks, sand, and mud—so much in places that it is hard to tell where the glacial ice ends and the frozen mud begins.

So the glacial ice that scrapes across bedrock outcrops is full of abrasive particles in widely assorted sizes. The finer particles polish the bedrock, while the larger objects gouge grooves into the polished surfaces. The grooves exactly record the direction of ice movement. If you find sets of grooves that cross, they record a change in the direction of ice movement.

Actually, the grooves record the direction of ice movement with an ambiguity of 180 degrees. Did the ice move from left to right, or from right to left? Geologists resolve that question by gently stroking the surface with their hands. It feels much smoother if you move your hand in the direction the ice was moving. People who have not already done so might try fondling rock outcrops, quite possibly the least famous of all outdoor sports.

Looking at glacially polished and striated surfaces is sobering if you consider that those surfaces have been there since the last ice melted, about 10,000 years ago. The processes of weathering that break rock down into soil operate very slowly indeed. Soil is not a renewable resource within any humanly meaningful span of time.

Glacial ice also freezes tight to the bedrock and plucks out blocks broken along the fractures that all rocks naturally contain. The combination of glacial rasping and plucking straightens the original stream valley and steepens its walls. The ice scours hills into rounded knobs. It shapes rock outcrops into streamlined forms with smoothly rasped surfaces that face

Glacially polished granite surface along Tioga Pass Road near Cathedral Lake.

into the direction of ice flow and raggedly plucked surfaces that face down the direction of ice flow. Each glaciated valley heads in a broad basin called a cirque, with a gnawed peak rising above it.

Glacial ice dumps its load of debris where it melts. Most of it is in deposits of till, a disorderly mixture of all sizes of material with little internal layering. Till looks like something a bulldozer might have scraped together. Melting ice also drops scattered boulders, which are called erratics because most of them are unlike the bedrock beneath. You see them here and there on glaciated surfaces throughout the high Sierra Nevada.

Anything made of till is a moraine. Many moraines are ridges made of debris that melting ice dropped along the edge of a glacier. Plot them on a map, and you have a precise picture of an ice age glacier. Geologists who do that generally find two sets of glacial moraines, one much farther down the valley than the other. The outer moraines apparently formed during an ice age of approximately 100,000 years ago—the exact date is a matter of much dispute. The inner set of moraines date from the last ice age, a lesser event that reached its maximum about 15,000 years ago and ended about 12,000 years ago. No one has clearly recognized in the landscape the mark of whatever ice ages preceded those two.

Melting ice sent torrents of water heavily laden with outwash sediment down glaciated valleys, many miles beyond the farthest moraine. Most outwash deposits survive as stream terraces within the lower valleys and as smooth alluvial fans and plains beyond the valleys, mainly on the floors of the Owens Valley and the Great Valley.

GOLD

The northern Sierra Nevada produced enormous amounts of gold. Almost all of the primary bedrock deposits are within a narrow belt close to the Melones fault. Somehow, the juxtaposition of the fault, serpentinite, intrusions of granite, and circulating hot water created large numbers of quartz veins that contain gold, the fabled mother lode.

Once the bedrock deposits were exposed to the processes of erosion, the gold found its way into the streams, which concentrated it into deposits of heavy minerals called placers. The northern Sierra Nevada contains two generations of gold placers, one in ancient streams that flowed during Eocene time, the other in the modern streams. The problems of mining the bedrock deposits and the two generations of stream placers were so different that they led to three quite different versions of gold mining.

Stream Placer Mining

After gold was discovered at Coloma in 1848, all sorts of unlikely people suddenly became prospectors. Within a few months, that motley horde was prospecting the western slopes of the Sierra Nevada, finding gold in many places. The news traveled east as fast as news could, and the miners rushed west to stage the tremendous gold boom of the 1850s.

The debris of weathering gold quartz veins erodes down slopes and into streams. Every stream is a natural sluice box, a channel through which running water flushes sand and gravel. Gold settles to the bottom of the streambed and catches against irregularities in the bedrock, just as it catches behind the cleats across the bed of a sluice box. That secondary concentration makes stream placer deposits much richer than the original bedrock ores.

Pebbles moving down the streambed during a flood pound most of the soft gold into flakes that range in size from the merest specks to corn flakes, the merest specks being by far the most abundant. Meanwhile, all the flakes of gold, being equally dense regardless of their size, settle through the moving mass of sediment. The gold that emerges from a pan of washed surface gravel gives only a hint of what may or may not exist at depth. The only way to find out is to mine the gravel down to bedrock.

The early placer miners who worked the shallow stream gravels in the northern Sierra Nevada skimmed the cream off the gold belt within a few years, with little capital investment. The first step in working a stream placer was to clear the trees, brush, and top layer of gravel off the mining claim. Then it was time to dig ditches to bring water and build sluice boxes or rockers to wash the gold out of the gravel. All of that required a tremendous investment in labor.

When the spring runoff came, the claim holder and his crew spent weeks standing waist deep in cold water shoveling gravel out of the streambed and into a sluice box or rocker. In most cases, they worked the claim to shovel depth in one season. So the first phase of placer mining generally ended in two years, the typical boom period of the early mining camps.

The number of miners who got rich probably amounted to less than 1 percent. Many more went broke.

Dredging

Invariably, the best placer deposits lie at the base of the stream channel, where gravel lies directly on bedrock. The bedrock surface typically lies at shovel depth only in small streams, the kind you can jump across. So the first wave of miners were not able to dig deep enough to reach many of the best deposits. They left those for the big dredges, which appeared during the 1890s.

Gold dredges are barges that float in ponds of their own making, slowly navigating through the floodplain by scooping gravel from the bottom ahead with a chain of buckets, washing it through sluice boxes, and dumping

Windrows of dredge gravels.

the tailings from a conveyor belt at the back. They filled the pond behind them as they dug it in front. Some dredging operations became terribly nautical: the captain and his mates supervised an able-bodied crew that worked on deck, leaned on bulkheads when no one was watching, and took shore liberty when their shift ended.

The early hand workings of placer mining left small heaps of gravel, each the former site of a sluice box or rocker. You can still see them in places where the dredges never ventured. Dredge tailings lie in long ridges with small cross ridges that record the swings of the conveyor belt as it stacked the tailings aft. Meanwhile, the soil flushed down the river. Only truly catastrophic floods will enable the streams to rework the placer mine tailings and restore their floodplains to a natural condition. That may not happen for thousands of years. In the meantime, people may partially reclaim those ravaged floodplains by using the old tailings for road metal and construction aggregate.

Mercury

Old-time placer miners washed gravel down to a small residue of black sand, then stirred that concentrate into mercury. The gold dissolved in the mercury, and the worthless heavy minerals floated on top, where the miner could skim them off with a scrap of cardboard. So placer gold created a tremendous demand for mercury, most of which came from the Coast

Range. After enough gold dissolved in the mercury to make an amalgam too stiff to run, the next problem was to separate the two metals.

Most early smelters could roast the mercury off the amalgam, condense it, and return pure gold and reconstituted mercury in separate containers. But smelters charge for their services, and many early miners would not trust anyone with their gold. They did their own separating.

The simplest procedure was to roast the amalgam in a frying pan, driving the mercury off in a cloud of poisonous fumes that everyone could inhale. Some miners attempted to trap the mercury vapor, probably motivated more by the idea of saving mercury than by environmental concerns. One widely recommended method was to cut a large potato in half, scoop a hole in the cut side, and put that over the amalgam, hoping the mercury vapor would condense in the moist interior of the potato. In fact, the miner rarely recovered more than a fraction of his mercury from the ruins of the potato. The rest wafted off in the breeze. Many early miners inhaled far more mercury vapor than their central nervous systems could handle. The proverbial crazy prospector was probably suffering from mercury poisoning.

Auriferous Gravels and Hydraulic Mining

Early prospectors were amazed to find that most of the gold in the streams was coming from older stream gravels, which they called the auriferous gravels. The streams that deposited them flowed in Eocene time, about 50 million years ago. They faltered as the climate dried after Eocene time. Then volcanoes filled their valleys. Hardly a trace of those old valleys existed in the landscape to influence the directions of the modern streams when they began to flow a few million years ago. So the modern streams chose their own courses across the paths of their ancient predecessors.

Gold in the auriferous gravels posed the problem of washing it from dry gravel. The solution was hydraulic mining, which began in 1852. Hydraulic miners started a ditch far enough upstream and gently enough inclined to bring water to a point more than 100 feet above the gravels they planned to wash. That provided the pressure. Then they piped water from the ditch to huge nozzles that washed the gravel down and flushed it through sluice boxes to catch the gold. The washed gravel went down the slope and into the stream.

The net effect was to add a tremendous load of sediment to the streams of the northern Sierra Nevada with no corresponding increase in their flow. The streams responded by dumping great sheets of gravel on their floodplains. Sediment raised the level of the Sacramento River as much

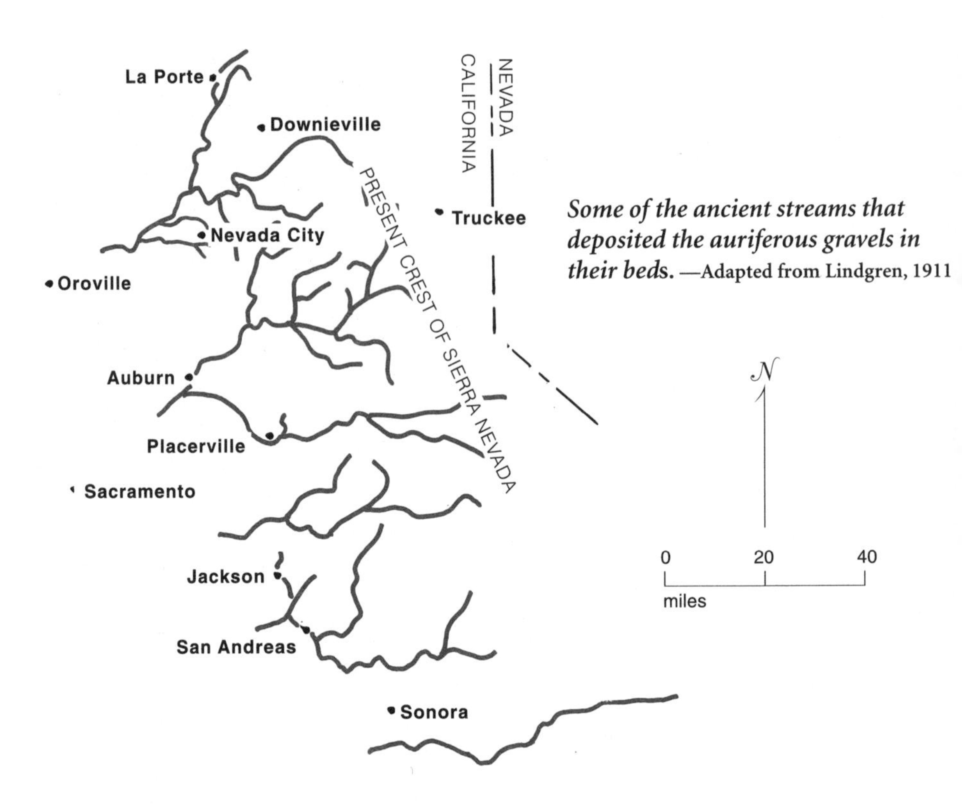

Some of the ancient streams that deposited the auriferous gravels in their beds. —Adapted from Lindgren, 1911

as 8 feet. The Sacramento Delta grew visibly, and muddy water billowed through the Golden Gate into the Pacific Ocean. In the 1860s and 1870s, farmers in the Sacramento Valley saw spring floods beyond any that anyone could remember. Their fields turned into wetlands more suitable for bullfrogs, ducks, and catfish than for crops. The farmers sued, asking the court to enjoin the miners from dumping gravel into the streams. Their suit ground its tedious way to the California Supreme Court, which finally issued its Sawyer Decision in 1884, forbidding the hydraulic miners to dump their gravel into streams. That stopped hydraulic mining in the Sierra Nevada.

Of course, the mining interests appealed the Sawyer Decision to the United States Supreme Court, which asked the U.S. Geological Survey to investigate the relationship between hydraulic mining in the Sierra Nevada and the turn of events in the stream valleys. The Survey assigned the job to G. K. Gilbert, its most distinguished geologist.

Gilbert showed how deposits of gravel in streambeds steepened their gradients and so speeded their flow. That explained the sudden floods. He also showed how deposits of mud along the banks of streams in the

Sacramento Valley created new wetlands by raising the levels of the stream channels, making it impossible for them to drain their floodplains. While Gilbert was working out his theoretical explanations, the rivers of the Sierra Nevada, freed of their unseemly loads of gravel, were rapidly eroding their channels back to the levels of 1852.

The ugliness of the raw scars the hydraulic jets carved in the hills appalled even the miners who made them. But the bare gravel freshly exposed in hydraulic mine scars made an excellent seedbed for some kinds of trees. The seedlings that sprouted soon after mining ceased have long since covered parts of the old mine scars with beautiful forests of mature trees. Other parts are still as ugly as ever.

Bedrock Mining

The ultimate source of all the gold was in the bedrock vein deposits, the prospector's mother lode. Most are gold quartz veins in metamorphosed sedimentary and volcanic rocks near the Melones fault zone and in the large bodies of serpentinite in the ophiolites west of the fault. They lie within a belt a few miles wide that trends north along the low western slope of the northern Sierra Nevada.

Watch in the northern Sierra Nevada for quartz veins cutting their milky white seams across the other rocks. Vein quartz is normally white instead of transparent because it is full of minute inclusions filled with water or gas that reflect and scatter light. White quartz veins rarely contain interesting

A quartz vein about 1 foot across fills a fracture in metamorphosed Jurassic volcanic rocks in the southern gold country.

minerals. Mineralized quartz veins typically weather to a rusty mess because they generally contain the iron sulfide mineral pyrite, along with various other minerals. Pyrite weathers to iron oxide, which stains the vein in rusty shades of yellow or brown. Weathering pyrite also produces rather strongly acidic water, basically a solution of sulfuric acid, which dissolves other metallic minerals and redeposits their constituents in new minerals, some of which are colorful.

The early prospectors, like many of those who followed, lived in the belief that the bedrock sources of placer gold, the mother lode, would surely make them rich. In fact, gold quartz veins bankrupt far more people than they enrich. The problems hinge on a disorderly combination of geology, psychology, and accounting.

Most quartz veins contain no gold. In the few that do, the gold occurs mainly in small patches of rich ore, bonanzas set within much larger expanses of barren quartz. Some bonanzas are fantastic; some of the fabulous legends about the old gold mines are true. Geologists have no theoretical way to predict where miners will find those bonanzas, and no indirect method to locate them. Gold bonanzas are too small to make likely targets for core drilling. The only way to find them is to mine through the barren quartz, hoping to hit one before you go broke.

Nothing inspires optimism quite like a good run of bonanza ore. But their wide, apparently random scattering makes it impossible to estimate in advance how much gold a quartz vein contains. Accountants have no way of knowing how much investment the property can amortize and still show a profit. A good many of the old quartz gold mines invested far more in equipment and development than the deposit could ever pay off. Once they did that, eventual bankruptcy became as inevitable as next Tuesday morning. A rare few of the early mines managed to amortize their investment, stay in business for a long time, and make a profit.

SIERRA NEVADA ROADGUIDES

Most of the roads on the low western slopes of the Sierra Nevada pass few bedrock exposures, except in roadcuts and stream banks. Much of the landscape existed before the Sierra Nevada rose and gently slopes to the west with the modern stream valleys deeply entrenched into it. Rocks are much better exposed in the higher, glaciated parts of the range. The landscape on the eastern slope is basically the steep scarp of the Sierra Nevada fault.

Most of the geologic maps look complicated, mainly due to an overlay of younger and mostly unconsolidated sediment on much older hard bedrock. The older rocks are the terranes of sedimentary and volcanic rocks, now mostly recrystallized into metamorphic rocks, and the masses of granite that intrude them. They trend generally north. The younger rocks consist largely of volcanic ash and gravels that filled stream valleys eroded into the older rocks before the Sierra Nevada rose. Their map pattern does indeed look like segments of old valleys that trend west. Remembering that the maps show two sets of unrelated rocks, one on top of the other, makes them easier to read.

Interstate 80
Sacramento—Reno
136 miles

Interstate 80 crosses the gentle western slope of the Sierra Nevada between Sacramento and Reno, following large remnants of the softly rolling lowland that existed before the Sierra Nevada rose. Watch for red soils and dry stream gravels, both relics of the old landscape. The route across the steep eastern slope traverses the Sierra Nevada fault scarp, passing outcrops of pale gray granite but no red soil. The geologic map shows the usual Sierran picture of old bedrock, thinly overlain in many places with deposits of volcanic rocks and stream gravels that fill old valleys.

Boulders in a moraine west of Emigrant Gap. View is about 25 feet across.

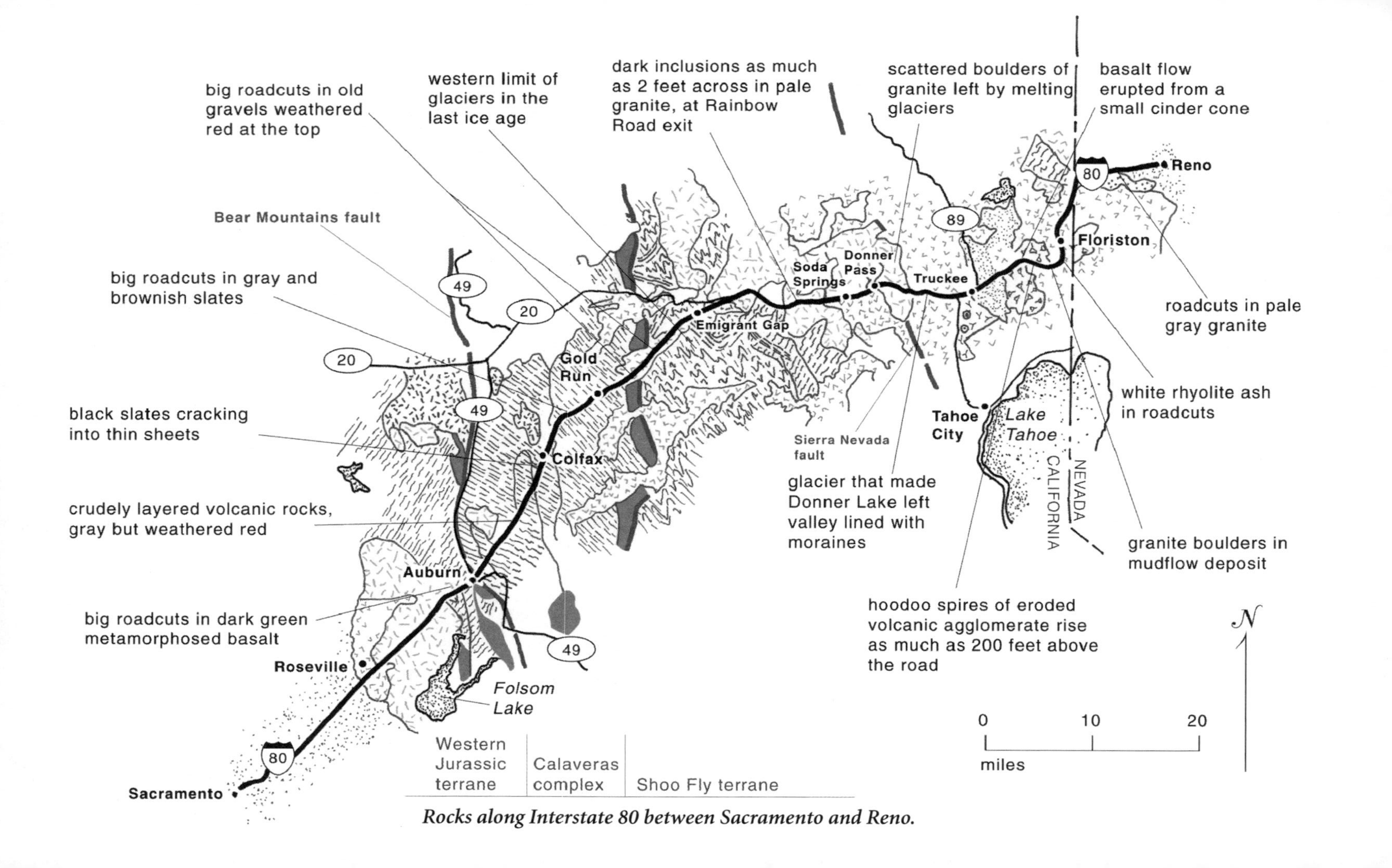

Rocks along Interstate 80 between Sacramento and Reno.

During the ice ages, great glaciers carved the high crest of the range, deeply exposing its granite core in bold outcrops of massive gray rock, still nearly bare of soil. Their melting left dramatic alpine landscapes.

Granite at Auburn

Interstate 80 crosses a mass of granite between Roseville and Auburn, on the low western edge of the Sierra Nevada. The granite is a western outpost of the enormous Sierra Nevada granite batholith farther east. Watch for the great rounded boulders littering the fields. Nothing moved them; they were weathered where they lay. Geologists call them core stones.

The granite crystallized from a mass of magma that rose into the Smartville complex, a large slab of oceanic crust in the western Sierra Nevada that belongs to the Western Jurassic terrane. The Smartville complex is exposed for about 1 mile on either side of Auburn but widens to about 20 miles a short distance to the north. It includes oceanic basalts, the basalt dikes that fed them from below, and the gabbro magma chambers that stored the basalt magma before it erupted.

Remnant of the Old Landscape

The long stretch of highway between Auburn and Emigrant Gap follows a ridge between the valleys of the Bear River to the north and the North Fork of the American River to the south. The ridge crest is a remnant of the gentle landscape that was raised and tilted to become the western slope of the Sierra Nevada. Erosion left this bit of the old landscape almost untouched as the rivers carved their canyons deep into the bedrock on either side. Stream sediments, volcanic rocks, and a deep mantle of red soil blanket the older landscape.

Auriferous Gravels

Coarse stream gravels exposed in spectacular cuts near Gold Run are the famous auriferous gravels that the miners of the nineteenth century so eagerly sought. The highway passes through a long valley that the hydraulic miners hosed out of the gravels. Accounts from the 1860s describe it as a horrifying wasteland of muddy water and gravel. Trees now cover most of the scars.

The gravels fill the channels of streams that flowed during Eocene time, when the Sierra Nevada was a humid lowland. Then volcanic eruptions completely buried the old stream valleys under sheets of andesite and rhyolite ash. Now modern streams are carving their own valleys right across the buried valleys of the Eocene streams. The long pit at Dutch Flat essentially exhumes one of the old stream channels.

Reddish auriferous gravels of Eocene time along Interstate 80 near Dutch Flat.

Western Jurassic Terrane

Bedrock a few miles west of Auburn is in the Western Jurassic terrane, sedimentary and volcanic rocks and old bedrock ocean floor that were jammed against the Calaveras complex to the east in early Jurassic time, starting about 200 million years ago. Intense deformation and prolonged heating metamorphosed the original rocks almost beyond recognition. Now they are slabby rocks that break easily into slaty plates. Many are phyllites whose surfaces glisten with minute flakes of mica. Their colors range from black through various shades of tan and green to pale gray.

Sierra Nevada Batholith

Most of the rock exposed between Emigrant Gap and Reno is granite. It surfaces in the distinctive way of granite, in massive knobs of smooth gray rock seamed with fractures. Big core stones in softly bulging forms litter the countryside.

Granite is the dominant rock near Donner Pass. Ragged peaks of knobby gray rock rise above deep valleys that felt the slow scrape of ice age glaciers. The glaciers left their distinctive mark in peaks bitten away to masses of roughly quarried rock that embrace sparkling little lakes. Trees now green

Sierra Nevada granite west of Donner Pass.

Horizontally layered agglomerate eroded into hoodoos north of Interstate 80, about 13 miles east of Truckee.

these regions so lately inhabited by groaning rivers of ice. But they do not quite hide the bleak landscape the ice left when it finally melted, some 12,000 or so years ago.

Donner Pass was an ice divide during the ice ages, just as it is now a water divide. Ice flowed west and east. Watch for the bouldery deposits of glacial till between the west end of Donner Lake and Truckee. Watch, too, for occasional outcrops of volcanic rocks almost buried beneath the glacial deposits. These rocks erupted onto the valley floor as it dropped while the Sierra Nevada rose.

Northeast of Floriston, the Truckee River and Interstate 80 cut through exposures of andesite, the volcanic rocks that covered the Sierra Nevada before it rose. Volcanic rocks are thicker in the ranges east of the Sierra Nevada.

U.S. 50
Sacramento—Carson City
133 miles

U.S. 50 crosses the flat floor of the Sacramento Valley and the gentle foothills of the Sierra Nevada between Sacramento and Placerville. A few miles east of Placerville, it meets the South Fork of the American River, which it follows all the way to the crest of the range at Echo Summit.

The river eroded its valley deep into the bedrock core of the Sierra Nevada since the range began to rise along the faults at its eastern margin. Rocks exposed along the valley provide a section through the bedrock that is quite different from what you see along Interstate 80. That road follows a parallel route high along the drainage divide just a few miles to the north, across remnants of the lowland landscape that existed before the range rose.

U.S. 50 crosses rocks of the Western Jurassic terrane west of Shingle Springs. They include gabbro and serpentinite, as well as volcanic rocks that erupted during Jurassic time and were then transformed into slates. Several masses of granite intrude them. The road crosses the Melones fault zone at Placerville. All the metamorphic rocks around Placerville belong to the Calaveras complex. Volcanic and sedimentary rocks that fill old stream valleys cover them in places. Most of the rocks in the eastern part of the route are granite, part of the Sierra Nevada batholith.

Sacramento Valley

Nearly 10 miles of road through the eastern part of Sacramento crosses old placer mine tailings, a vast ugliness of mounded heaps of gravel. The highway follows the broad valley of the American River for about 20 miles,

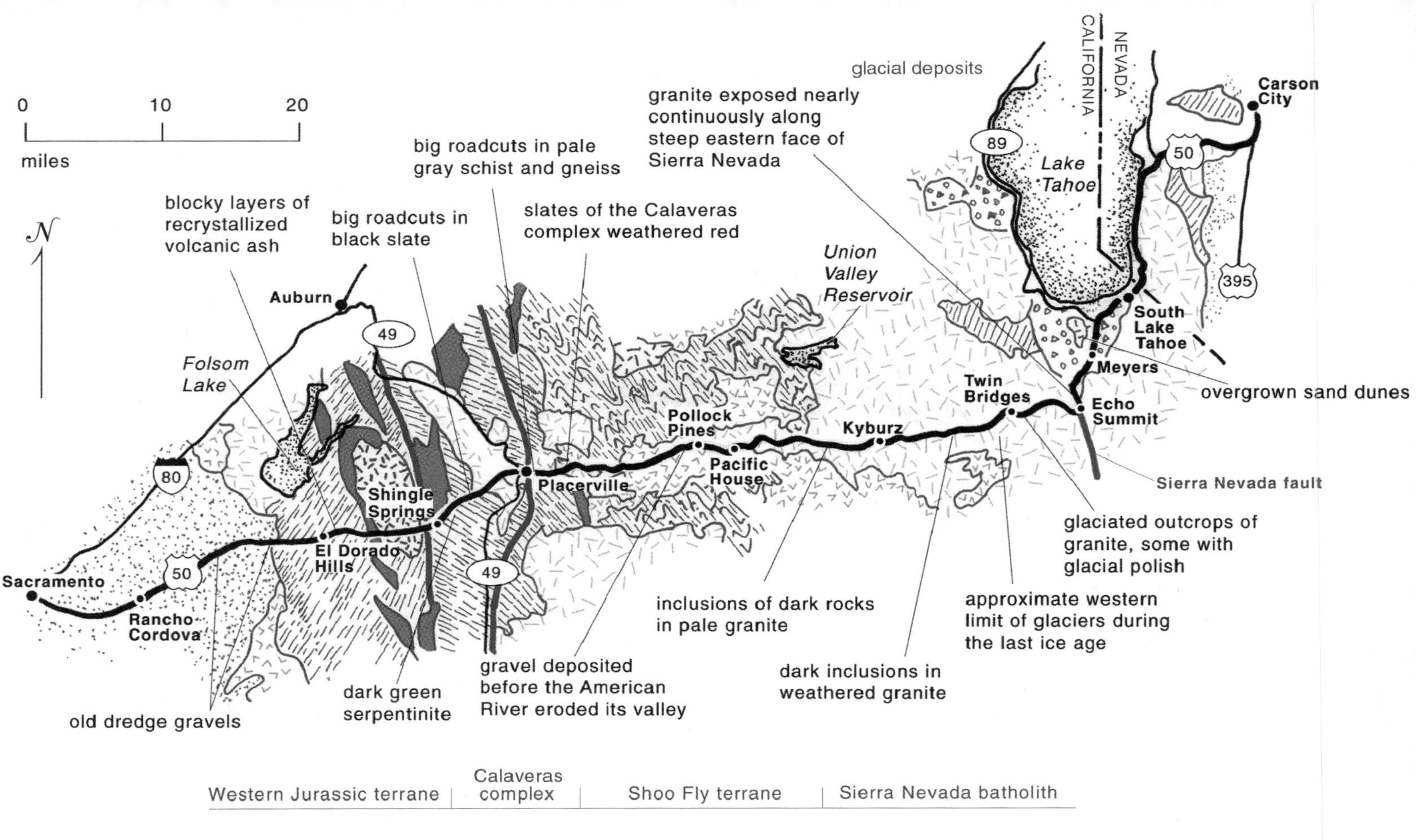

Rocks along U.S. 50 between Sacramento and Carson City.

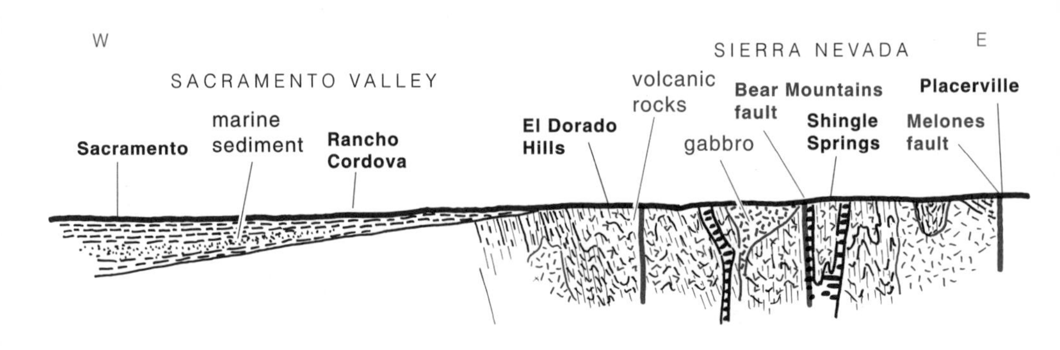

Geologic section along the line of U.S. 50 between Sacramento and Placerville. All the hard bedrock belongs to the Western Jurassic terrane. —Adapted from Wagner and others, 1981

east from Sacramento to where the river leaves the Sacramento Valley and crosses into the foothills of the Sierra Nevada just south of Folsom Reservoir. Floodplain deposits underlie the road. The river cut its low valley walls into deposits of sediment washed from the Sierra Nevada onto the flat valley floor within the past few million years.

Terranes

Between the floodplain of the American River and Placerville, U.S. 50 crosses the old and deeply weathered landscape eroded before the Sierra Nevada rose. Here, the bedrock belongs to the Calaveras complex, a broad belt of sedimentary and volcanic rocks that were laid down during Permian and Jurassic time, then jammed into the Sierran trench during Jurassic time. Now they are metamorphosed into slates and schists. Quartz veins fill some of their fractures. Nice outcrops are scarce.

The highway crosses the main branch of the Bear Mountains fault zone at Shingle Springs. The fault contains serpentinite, which probably squeezed up from below the trench, following the easy passage through fractured rock. The dark green to greenish gray rocks exposed in the 5 miles east of Shingle Springs are metamorphosed oceanic basalt that was stuffed into the Sierran trench.

Placerville is just west of the Melones fault, which separates the Mesozoic rocks of the western Mesozoic belt to the west from the older, mostly Paleozoic, rocks to the east. The red rocks at the east edge of Placerville are gray slates of the Calaveras complex weathered beneath the soil of the old landscape on the west slope of the Sierra Nevada. Most of the rocks near the road between Placerville and Fresh Pond are much younger gravels and volcanic rocks laid on the older rocks.

Most of the route between Placerville and just east of Pacific House crosses volcanic rocks that erupted onto the old landscape of the Sierra Nevada during the past 10 million or so years. They belong to the Mehrten formation, sandstone and pale volcanic ash deposited on the old erosion surface. These are rocks that weather too easily to make prominent outcrops. They perch on a drainage divide, a remnant of the old landscape between modern stream valleys.

The road between Pollock Pines and Pacific House descends into the valley that the South Fork of the American River eroded into rocks of the

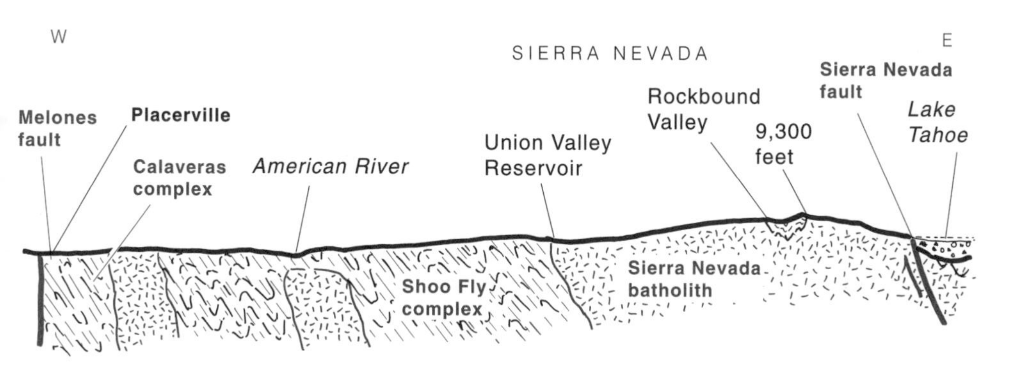

Geologic section drawn north of the line of U.S. 50 between Placerville and Lake Tahoe. —Adapted from Wagner and others, 1981

Dark gray slates east of Shingle Springs.

Calaveras complex. They were deposited as sediments on the ocean floor between 300 and 200 million years ago, before it began sinking through the Sierran trench. Like the younger sedimentary rocks to the west, which they resemble, these rocks hold numerous quartz veins, some of which contain gold. They survive mainly as roof pendants caught between masses of granite.

U.S. 50 follows the valley of the South Fork of the American River east of Pacific House. Volcanic ash and sands of the Mehrten formation cap remnants of the gently sloping west flank of the Sierra Nevada, 1,000 to 2,000 feet above.

Granite

Almost all of the bedrock exposed between Pacific House and Echo Summit is granite, the Sierra Nevada batholith. It makes bold outcrops of great core stones that suggest swollen pillows, welcoming landscapes that invite a walk.

Close examination of the rock reveals that it consists mainly of blocky grains of milky white or pink orthoclase feldspar. The occasional white feldspars with a slightly greenish cast are plagioclase. Quartz crystallizes into irregular grains that look both dark and glassy. Black flakes of biotite mica or stubby needles of hornblende pepper most granites.

Look for the rounded inclusions of dark rock scattered through the white granite. They range from a few inches to a few feet across. Most are

Dark inclusions in granite 2 miles west of Kyburz.

blobs of the black basalt that rose above the descending Pacific plate in Mesozoic time, importing heat to help melt the granite magma. Some are stray pieces of the dark metamorphic rocks that the granite intruded.

Glaciers gnawed the granite crest of the Sierra Nevada during the great ice ages, as recently as about 12,000 years ago. They left their signature in craggy peaks carved into bare rock, with numerous small lakes huddling in rocky basins between them. They also left polished outcrops and scattered erratic boulders stranded where the melting ice abandoned them. Echo Lake, northeast of Twin Bridges, fills a glacially carved bedrock basin.

Sierra Nevada Fault

The highway crosses the high fault scarp that defines the steep eastern face of the Sierra Nevada in the few miles between Echo Summit and Meyers. The bedrock is pale gray granite, part of the Sierra Nevada batholith. The Carson Range in the distance, east of Lake Tahoe, is the high end of a slice of the batholith that detached along a fault. Its bedrock is the same granite that makes the high backbone of the Sierra Nevada.

Occasional earthquakes leave no doubt that the Sierra Nevada fault is still moving. With every strong jolt, the crest of the Sierra Nevada rises a bit and moves north a bit. Meanwhile, the valley to the east drops a bit and grows a bit wider. The bits may be as much as several feet.

California 20
Marysville—Emigrant Gap
67 miles

Sutter Buttes rise jagged from the flat floor of the Sacramento Valley just west of Marysville. The road east of Marysville passes rocks of the Western Jurassic terrane. Between Smartville and Grass Valley, California 20 crosses a large mass of dark gabbro, part of the Smartville ophiolite complex. At Grass Valley, the road crosses the Bear Mountains fault zone and onto rocks of the Calaveras complex.

The road crosses granite and slightly recrystallized slates in its passage across the Calaveras complex. Unfortunately for people interested in the older rocks of the Sierra Nevada, many of those are not visible. The road follows an old valley filled with volcanic and sedimentary rocks of Tertiary time along most of the route between Nevada City and Emigrant Gap. Those younger rocks also bury all but a short stretch of rocks of the Shoo Fly complex just west of Emigrant Gap.

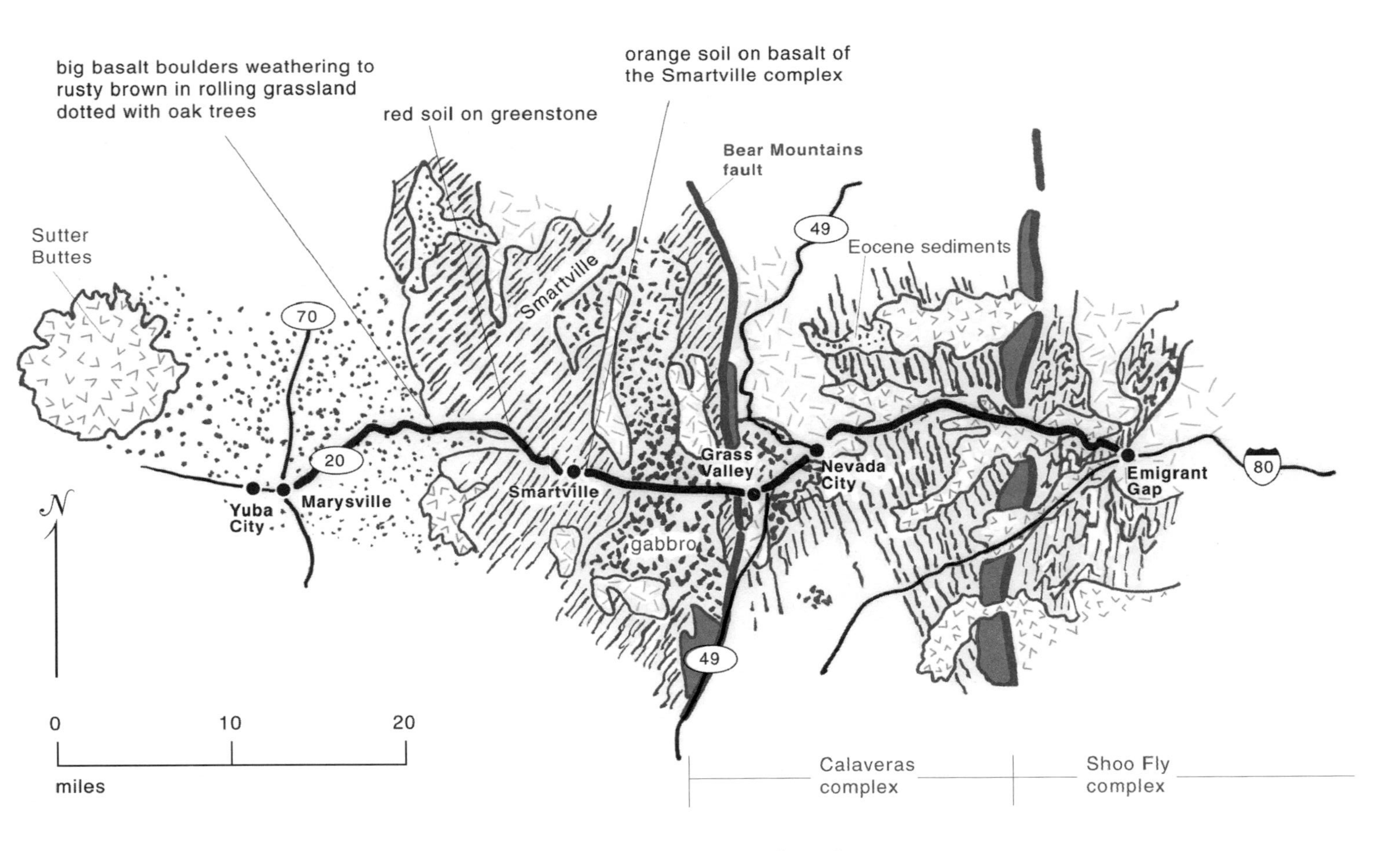

Rocks along California 20 between Marysville and Emigrant Gap.

Dredge Fields

The Hammonton Dredge Field amounts to about 7,000 acres of gravel tailings along the Yuba River floodplain east of Yuba City. It was dredged for gold between 1903 and 1968. You can glimpse a few piles of gravel in the distance southeast of the highway, from the area between 7 and 13 miles east of town.

The bleak expanse of dredge tailings in the Yuba Gold Field stretches for about 8 miles along the Yuba River, just east of Marysville and south of California 20. It is almost 3 miles wide in places. The gold probably came from the bedrock veins around Grass Valley and Nevada City.

Smartville Complex

Huge boulders of basalt that weather to rusty shades of brown litter an area of gently rolling grassland with scattered oak trees east of Marysville. The basalt erupted during Jurassic time. It is in the Smartville complex, part of the Western Jurassic terrane. It is one of the nicest slabs of old ocean floor anywhere.

Massive greenstones partly covered with red soils are exposed about a half mile on either side of the bridge across the Yuba River. They are basalt lava flows in the upper part of the Smartville complex, part of the old bedrock ocean floor. They extend east about 7 miles to the south end of a belt composed wholly of basalt dikes, which are not exposed along the highway. The dikes are the part of the oceanic crust between the basalt lava flows above and the gabbro beneath. Rocks exposed in the 8 miles east of Grass Valley are black gabbro, the lower part of the old ocean floor. Gabbro is like basalt except that it consists of much larger mineral grains. Unlike basalt, gabbro cools slowly, which allows time for mineral crystals to form.

The road between the Yuba River bridge and Grass Valley crosses a section through the oceanic crust that is now standing on end, with its top to the west. It was tilted when it collided with the older rocks of the western Sierra Nevada. Oceanic crust is typically 4 or 5 miles thick. It seems thicker here because the section is folded, so the road crosses parts of it more than once.

Prominent red soils along the way are on the old surface of the Sierra Nevada that was gently tilted as the range rose during the past few million years. The streams have been eroding their valleys into the old landscape as the range rises, so you see the red soils only on the uplands.

Grass Valley Gold

Prospectors discovered placer gold in the Grass Valley district in 1848 and gold quartz veins in 1850. Underground mining began almost immediately after the veins were found and continued for more than a

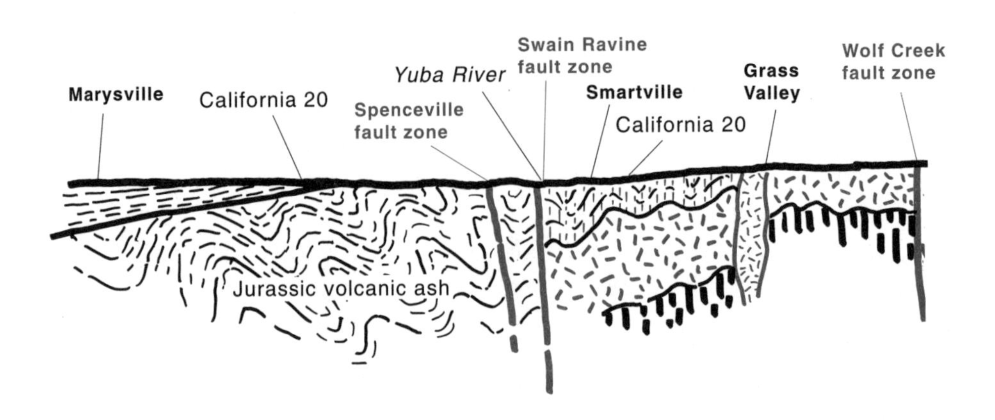

This geologic section illustrates the extreme complexity of the assemblage of metamorphic rocks and granite in the northern Sierra Nevada.

century, with the usual ups and downs. Gold mines thrive best in hard times.

Gold mines flourished nearly everywhere during the Great Depression of the 1930s, when more than 4,000 miners worked in the Grass Valley mines at pathetically low wages. The mines closed during the Second World War, then resumed operations and continued until the late 1950s. One of the mines at Grass Valley reached a vertical depth of slightly more than 1 mile and had more than 200 miles of underground workings. According to some estimates, it produced more than $120 million in gold bullion.

It is hard to estimate production from the early mining districts because the government did not require gold mines to report their production until 1893, and some of those reports have been more fanciful than accurate. The best attempts place the total production of the Grass Valley district at around $300 million, the most of any district in California. Remember, though, that gold production figures represent total cash flow over all the years of production, not profits. Many supermarkets have more cash flow than good gold mines.

Nevada City Gold

The Nevada City mining district, immediately northeast of the Grass Valley district, was one of the largest and most productive in the Sierra Nevada. It began producing placer gold from Deer Creek in 1849. Then

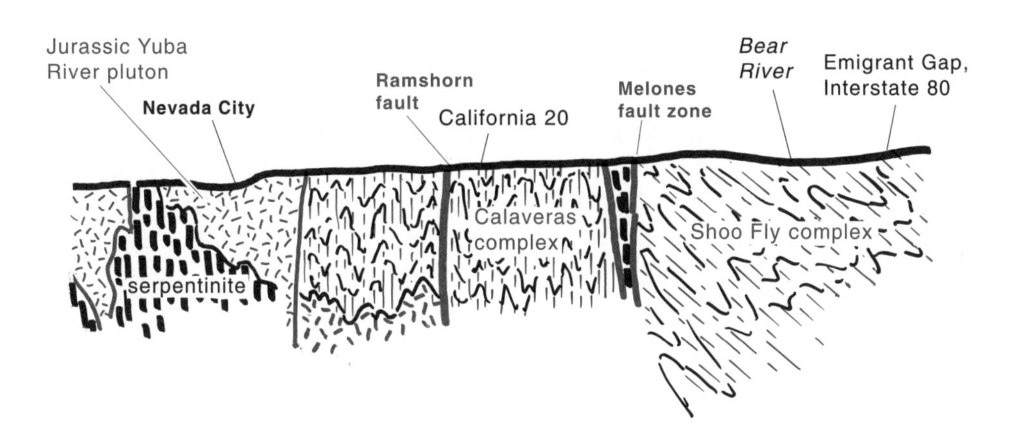

the first hydraulic mine in California started washing down the side of a mountain in 1852. Hydraulic mining continued until about 1880, greatly to the detriment of the creek.

Prospectors located gold quartz veins by 1850, but peculiarities of the ore made it hard to separate the gold from the rock. More than ten years passed before the district saw much underground mining. Then the two largest underground mines produced something like a half million dollars worth of gold annually until the early years of the twentieth century. As in most gold districts, heavy-duty mining resumed during the Great Depression of the 1930s, then stopped in 1942 when the government declared gold mining not essential to the war effort. The district has seen only sporadic activity since. According to some estimates, it produced between $50 and $70 million in gold.

California 41

Fresno—Yosemite National Park

80 miles

North of Fresno, California 41 crosses buff sands and gravels that stand in bluffs 50 feet high bordering the valley of the San Joaquin River. These are glacial outwash that came out of the Sierra Nevada during the ice ages. The same sands and gravels underlie the almost flat countryside to the

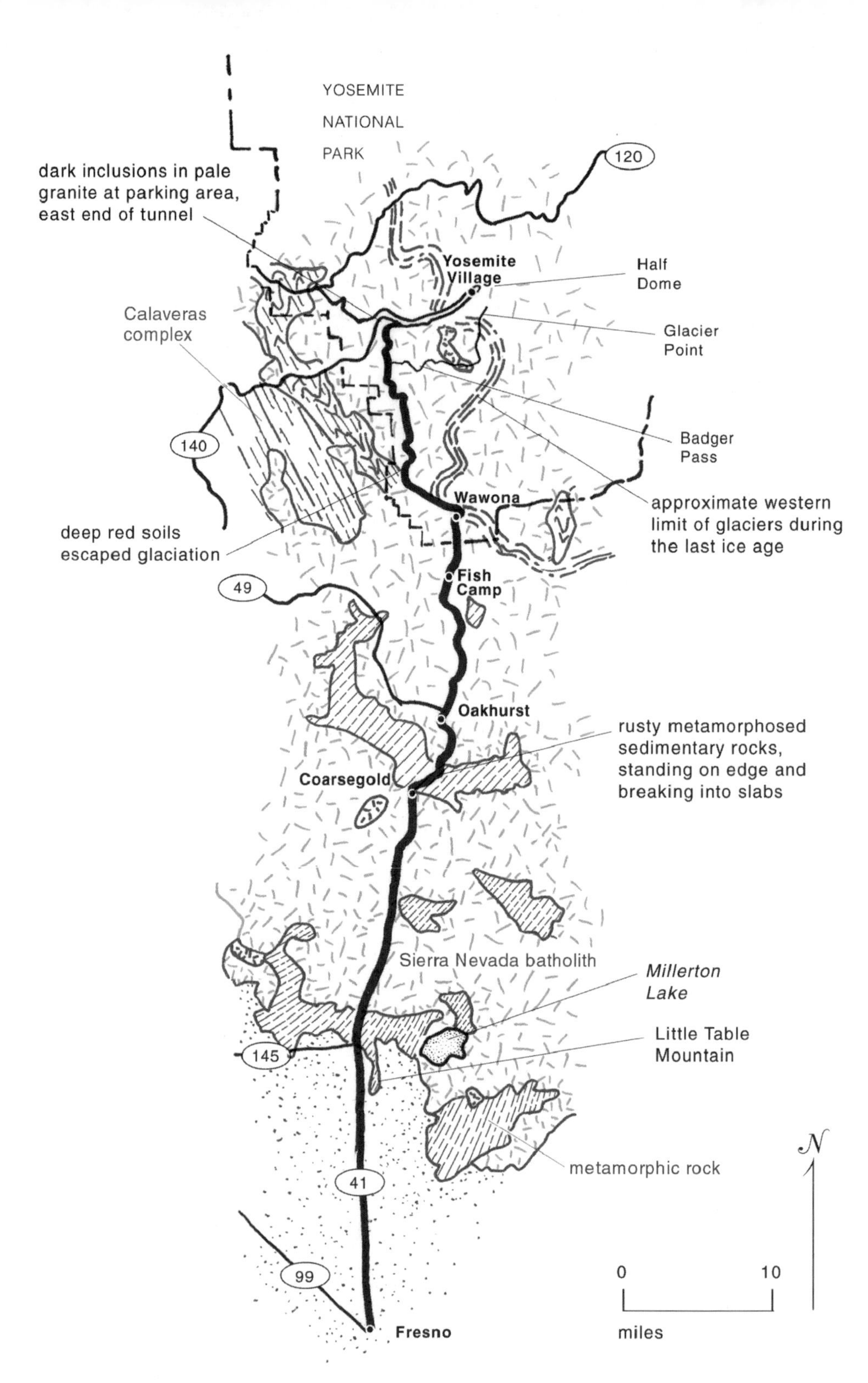

Rocks along California 41 between Fresno and Yosemite Valley.

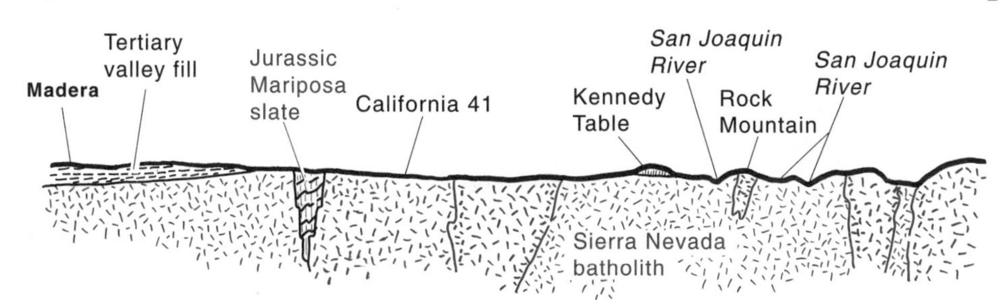

Geologic section between the San Joaquin Valley and Yosemite National Park. The granite bedrock is exposed except at the western end, where it is buried under sediments in the San Joaquin Valley. —Adapted from Bateman, 1978

junction with California 145. To the north the highway climbs gently into open rolling grassland with scattered bouldery outcrops of granite. Beyond, the Sierra Nevada.

The geologic map shows that almost all of the rocks the highway crosses in the Sierra Nevada are granite. It also crosses small areas of metamorphic rocks—roof pendants caught between masses of granite, the rocks in them probably belonging to the Calaveras complex.

Little Table Mountain

Little Table Mountain stands 1 or 2 miles east of California 41, south of its junction with California 145. Its flat top is a layer of gravel that resists erosion and preserves beneath it a nice remnant of the Ione formation. A cap of gravel resists erosion because it absorbs water and so prevents the surface runoff that would otherwise gather into streams.

The Ione formation exists here and there all along the low western front of the Sierra Nevada. It was deposited as a coastal plain during Eocene time, when the Great Valley was still under seawater, the Franciscan trench was still swallowing ocean floor, and the western margin of the Sierra Nevada was the west coast of California. In this area, the eastern part of the Ione formation is mostly gravel, which grades westward into sandstone. It all washed out of the Sierra Nevada and onto the coast landward of the shore.

Sierra Nevada Batholith

The road north of California 145 climbs through rolling grassy hills dotted with oak trees and digger pines. Bouldery outcrops of granite are the Sierra Nevada batholith that underlies most of the Sierra Nevada, except its westernmost foothills. Big roadcuts south of Coarsegold expose pale granite.

A dike of basalt heated and squashed in sheared granite southeast of Coarsegold.

In the 3 miles northeast of Coarsegold, California 41 crosses a remnant of old metamorphosed sedimentary rocks enclosed within the granite batholith—a roof pendant. The old sedimentary rocks break along steeply dipping platy fractures and weather to a rusty color.

The highway in the area a few miles south of Oakhurst and Wawona crosses hills with pine trees growing between outcrops of weathered granite. Watch in the high areas about 3 miles north of Wawona for layered sedimentary rocks that belong to the Calaveras complex and for weathered granite in the low areas. The South Fork of the Merced River, just west of the road, eroded through a thin cover of sedimentary rocks and into the granite.

Yosemite Valley

Outcrops of granite abound along the descent into Yosemite Valley. Many of them are full of dark inclusions. They are probably all that remains of globs of basalt that rose as molten magma into the continental crust, helped melt it, and were then incorporated into the magma that became the Sierra Nevada batholith. Think of the dark inclusions as relics of the source of heat that formed all of that granite.

The viewpoint at the east end of the long tunnel about 6 miles east of Yosemite Village provides one of the best views of Yosemite Valley. The

Yosemite Valley, east from the tunnel overlook on its south side.

granite there is full of dark inclusions. The view east up the valley emphasizes its vertical walls, left as the glacier plucked chunks of granite along a strong set of vertical fractures.

California 49
Mariposa—Placerville
139 miles

California 49 winds among the western foothills of the Sierra Nevada, passing through one historic mining community after another, all relics of the great gold rush. A large proportion of all the gold ever mined in California was recovered within sight of this narrow road through what was once the busiest part of the state. It now seems as though nothing ever happened in this quiet countryside.

The road follows a route within 1 or 2 miles of the Melones fault. The gold belt also lies close to the Melones fault and to the serpentinite along it. Rocks of the Calaveras complex, containing many large masses of granite, lie east of the road; rocks west of the road belong mostly to the Western Jurassic terrane.

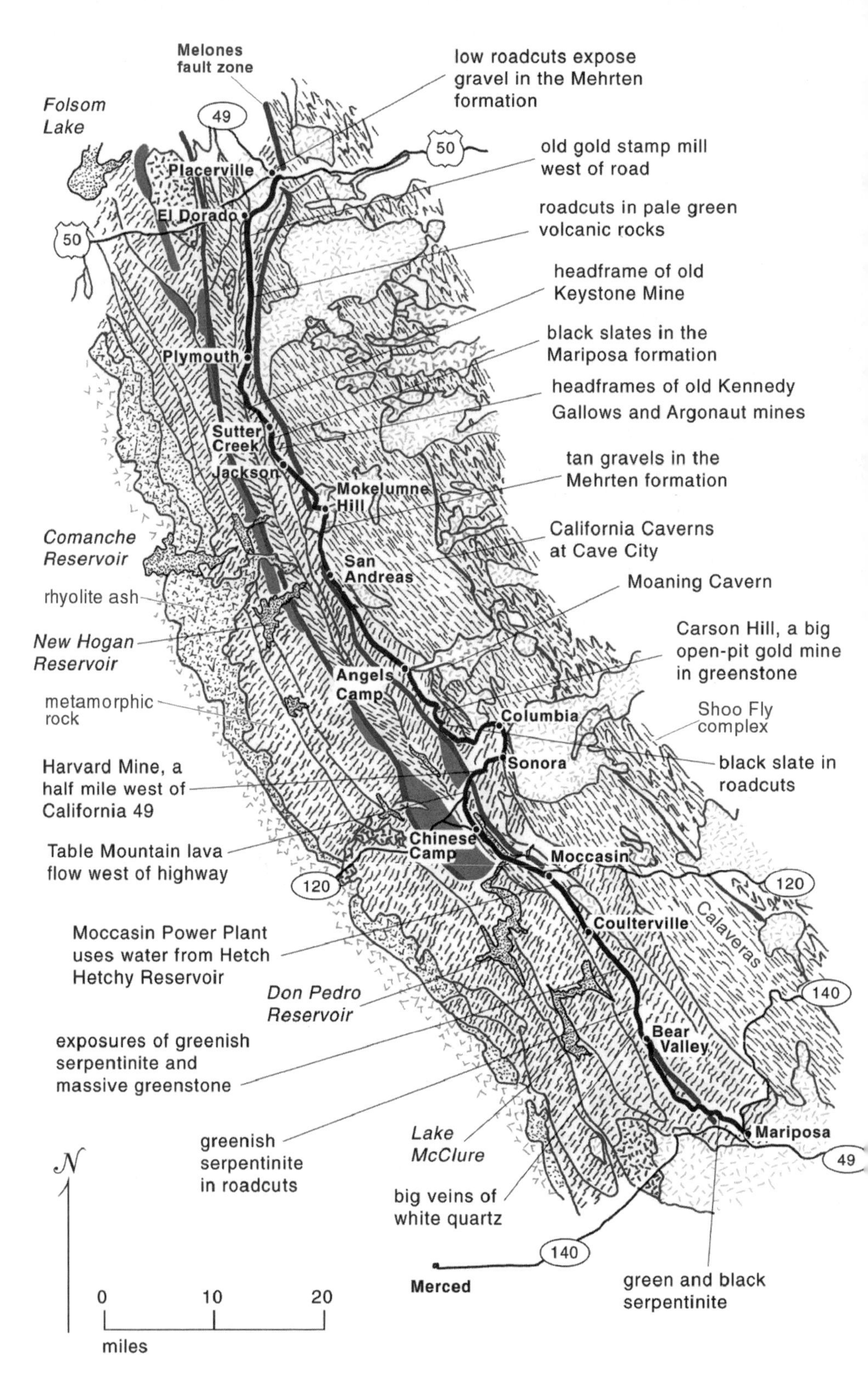

Rocks along California 49 between Mariposa and Placerville.

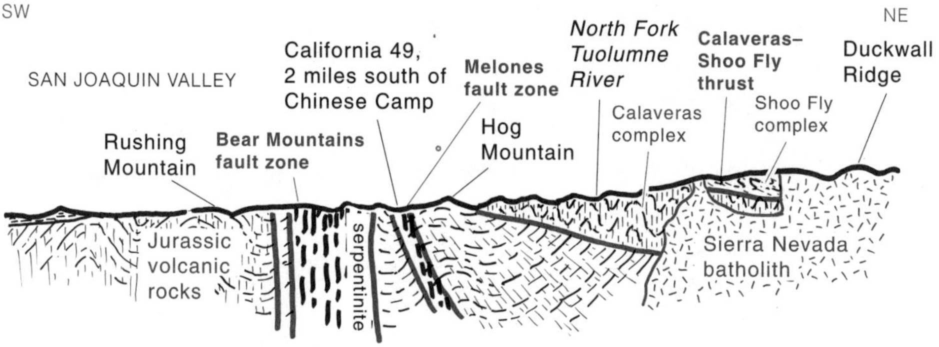

Geologic section across the line of California 49 a few miles south of Chinese Camp. —Adapted from Wagner and others, 1991

Melones Fault

Mariposa lies in a belt of gray volcanic rocks that erupted during Jurassic time and were metamorphosed as they were stuffed into the Sierran trench. They are part of the Western Jurassic terrane. Watch for thin belts of green and dark green serpentinite along the Melones fault between Mariposa and Mount Bullion, 5 miles to the northwest. Rocks east of the Melones fault are metamorphosed sedimentary and volcanic rocks of the Calaveras complex, Paleozoic in age. Granite of the Sierra Nevada batholith intruded them during Mesozoic time.

Mariposa Formation

The highway north of Mount Bullion crosses a rolling and grassy upland dotted with oak trees. A few low roadcuts expose greenish gray or rusty slates that began as mudstone deposited during Jurassic time. These slates belong to the Mariposa formation, part of the Western Jurassic terrane that lies west of the Melones fault zone along its entire length of more than 100 miles.

The road between Bear Valley and just north of Chinese Camp crosses serpentinite and Jurassic sediments of the Mariposa formation west of the Melones fault and crosses Jurassic volcanic rocks east of it near Coulterville. The serpentinite varies from pale green to black and weathers to shades of brown and orange. It is full of shiny surfaces, polished by slippage within the rock. The Mariposa formation just north of Bear Valley is platy black slates that weather to shades of brown. Big veins of white quartz cut them farther north.

Quartz veins in slaty sheared greenstone, just west of the Carson Hill Mine.

Between the area 2 miles north of Coulterville and Chinese Camp, California 49 crosses the Mariposa formation and occasional exposures of serpentinite in the Melones fault zone.

Gold

Watch about 4 miles north of Coulterville for a brilliant white quartz vein in a brushy hill east of the road. It consists of pure quartz deposited from hot water circulating through a fracture. Other quartz veins, many rusty from weathered iron sulfide, were the source of much of the gold produced from the bedrock of the Sierra Nevada.

The long walls of boulders along Moccasin Creek just south of Don Pedro Reservoir were stacked by hand as Chinese miners worked the creek for placer gold. Those scenes were common a century or more ago, before gold dredges began to rework the stream gravels to get the deeper gold.

Table Mountain

The mesa northwest of Chinese Camp is Table Mountain, one of several with that name along the western base of the Sierra Nevada. In the right light, its basalt caprock looks like long palisades of vertical fence posts. The fractures that outline those columns opened as the lava flow shrank when it crystallized, then shrank a bit more as it cooled.

Table Mountain and the Stanislaus River, 3 miles east of Knights Ferry.

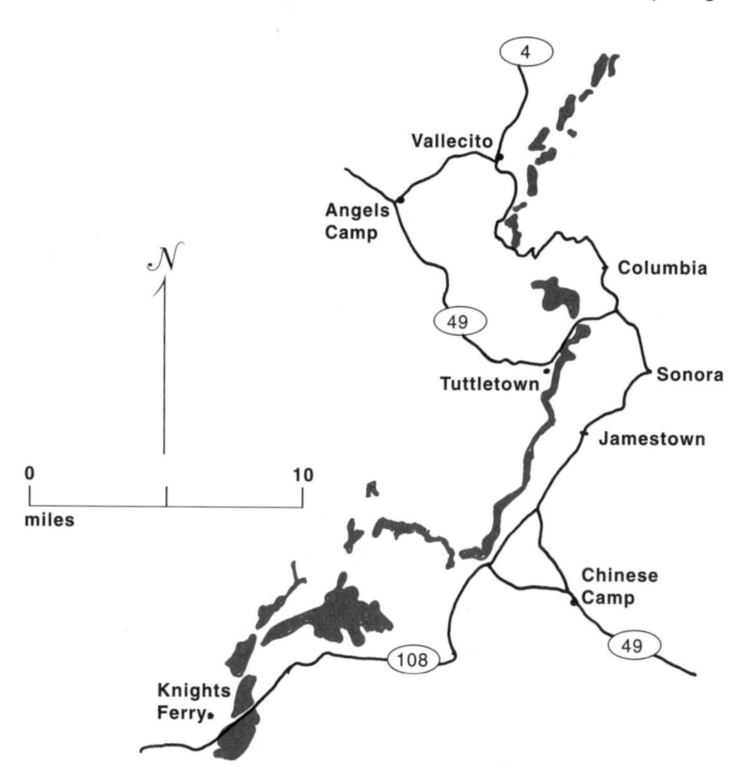

The Table Mountain lava flow traces the ancestral valley of the Stanislaus River. —Adapted from Wagner and others, 1981; 1991

Solution weathering of limestone at Columbia Historic Park.

The lava flow erupted near the crest of the Sierra Nevada about 9 million years ago and flowed down the former bed of the Stanislaus River. The basalt now stands high because it resists weathering and erosion better than the older rocks around it. The basalt also caps some high ridges north and south of the river to its headwaters, more than 40 miles upstream. The Stanislaus River is now in a new course.

Limestone Landscapes

Broad expanses of gray and white limestone, now fairly well recrystallized into marble, exist in the Calaveras complex around Sonora and Columbia. The rock outcrops in a sea of vertical tusks, a landscape suitable for a vision of another planet.

Rainwater invariably becomes slightly acidic as it absorbs carbon dioxide from the atmosphere, and that makes it capable of dissolving limestone and marble. This rock probably weathered beneath a cover of soil. Then erosion stripped off the soil, exposing the weird natural sculptures. Such landscapes commonly develop on limestone and marble, but they are rare in California.

Calaveras Complex

Watch just north of the bridge across New Melones Lake, south of Angels Camp, for slate in shades that range from almost black to a pale greenish color. Rusty stains cover the weathered surfaces. It is metamorphosed volcanic rock of the Calaveras complex, late Paleozoic to early Triassic in age. Slates that began as volcanic ash are typically green because they contain enough magnesium and iron to make green minerals, most abundantly the mineral chlorite. The highway passes similar rocks between the lake and San Andreas.

Oak trees grow on rocks of the Calaveras complex in the grassy hills between Angels Camp and San Andreas. Watch for roadcuts in pale greenish slates, some of which flake into thin chips. This slaty cleavage trends slightly west of north and stands almost vertically, parallel to the Melones fault 1 or 2 miles to the west and to the regional grain of the rocks.

Caves

Mercer Caverns are north of California 4, about 8 miles east of Angels Camp. Moaning Cavern is south of California 4, about 5 miles east of Angels Camp. California Caverns are at Cave City, about 12 miles east of San Andreas. Groundwater eroded all of them in limestone or marble within the Calaveras complex—in these matters, the difference between limestone and marble hardly matters. California Caverns has especially intricate stalactites and stalagmites—all the usual dripstone formations.

Slaty cleavage in slate north of the bridge across New Melones Lake.

Oligocene Gravel

Watch about halfway between San Andreas and Mokelumne Hill for a big exposure of tan gravel. It is part of the Mehrten formation, which covered much of the upland surface of the western Sierra Nevada during Tertiary time. It includes the auriferous gravels the early placer and hydraulic miners so ardently admired.

Bedrock in the area around Mokelumne Hill is an isolated intrusion of granite, an outpost of the Sierra Nevada batholith. Roadcuts near the Mokelumne River expose deeply weathered granite. The road passes greenish black basalt in the Calaveras complex about 1 mile north of the Mokelumne River. Two miles north of the river, the road crosses the Melones fault zone into an area of green slates.

The black slates around Jackson are part of the Mariposa formation of Jurassic age. Similar slates around Sutter Creek also belong to the Mariposa formation. The flaky cleavage is what defines slate as slate. It develops parallel to the direction the rocks flattened as they were compressed, at a right angle to the direction of the force that flattened them.

California 49 follows the bank of Big Indian Creek in the 8 miles north of Plymouth, passing outcrops of black slates and green volcanic rocks in the Mariposa formation. Roadcuts between the area north of the North Fork of the Cosumnes River and El Dorado expose pale green volcanic rocks of Jurassic age that weather to shades of brown.

Metamorphosed Jurassic volcanic rocks north of Jackson.

California 49
Placerville—Sierraville
124 miles

It is easy to cast a romantic spell over the memory of gold mining in California, now that the ravaged streams have cleared their channels and trees have covered the earth's wounds. It is hard to remember in the quaint stillness of the old towns that they thrived in times of economic desperation.

Like most geologic maps of the Sierra Nevada, that of the gold belt seems complex at first glance, but it does make sense. The oldest rocks are in the Shoo Fly complex, which lies next to the younger Calaveras complex along the Melones fault. The still younger rocks of the Western Jurassic terrane, along the western edge of the map, were dragged beneath the Calaveras complex on another large fault. Intrusions of granite complete the assemblage of older rocks. The superficial cover of much younger volcanic rocks that fill old stream channels complicates the map.

Old and New Landscapes

Between Placerville and Grass Valley, California 49 trends north, cutting across the topographic grain of the landscape as it rises over drainage divides and winds in and out of the major valleys. The crests of the divides are remnants of the old landscape that existed before the Sierra Nevada rose. Red soils that mantle this old landscape hide most of the bedrock.

Placerville is on a remnant of that old landscape and so is Auburn. The highway between them winds through the deep and rugged valleys the American River and its tributaries eroded as the Sierra Nevada rose. The route between Auburn and Grass Valley crosses large remnants of the old land surface, a remarkably flat landscape low on the western flank of the Sierra Nevada.

Western Jurassic Terrane

Placerville perches on the Melones fault zone, which separates the Western Jurassic terrane to the west from the Calaveras complex to the east. Both terranes consist largely of muddy sandstones that were deposited on the ocean floor and then jammed into the Sierran trench.

The 11 miles of highway between Placerville and the South Fork of the American River cross a large granite intrusion in the Western Jurassic terrane. The highway crosses onto old ocean floor of the Smartville complex about 11 miles south of Auburn, but outcrops are too scarce to provide a good view.

The highway north of Cool winds through the deep canyon of the Middle Fork of the American River. Roadcuts in the north canyon wall

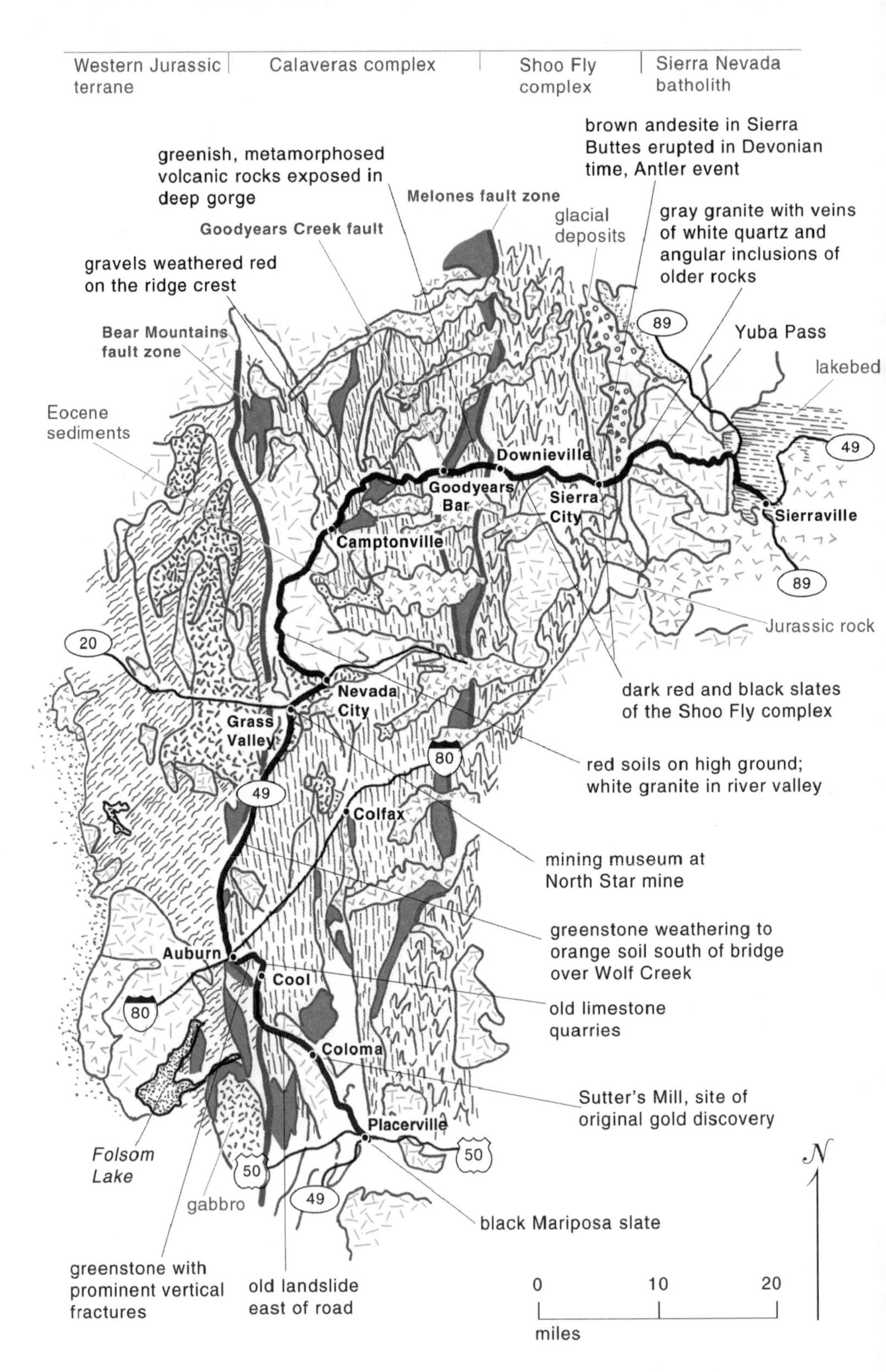

Rocks along California 49 between Placerville and Sierraville.

expose a greenstone mélange of the Western Jurassic terrane. The fresh rocks are intensely green, but their iron pyrite stains them rusty colors as they weather. Near the top of the canyon wall, just south of Auburn, the weathered surfaces become red as the road crosses onto a remnant of the old landscape.

California 49 north of Auburn almost exactly follows the line of the Wolf Creek fault zone. To the west lies the old oceanic crust of the Smartville complex, part of the Western Jurassic terrane. To the east lie late Paleozoic rocks of the Calaveras complex.

About 9 miles north of Auburn, the highway crosses the deep valley the Bear River eroded into the old landscape. Lovely outcrops of green volcanic rocks add interest to a walk along the river in either direction from the bridge. Watch for the offsets along faults and for the small intrusions of darker green serpentinite.

Ancient red laterite soils of the old landscape cover most of the bedrock between Auburn and Grass Valley. Occasional outcrops expose green and gray volcanic rocks and layers of muddy sediments that were scraped off the seafloor into the Sierran trench about 150 million years ago. They are part of the Western Jurassic terrane.

Gold at Coloma

James Marshall's discovery of gold at Coloma in 1848, in gravels along the South Fork of the American River, triggered the great gold rush of 1849.

Rounded outcrops of granite in the picnic area at Coloma.

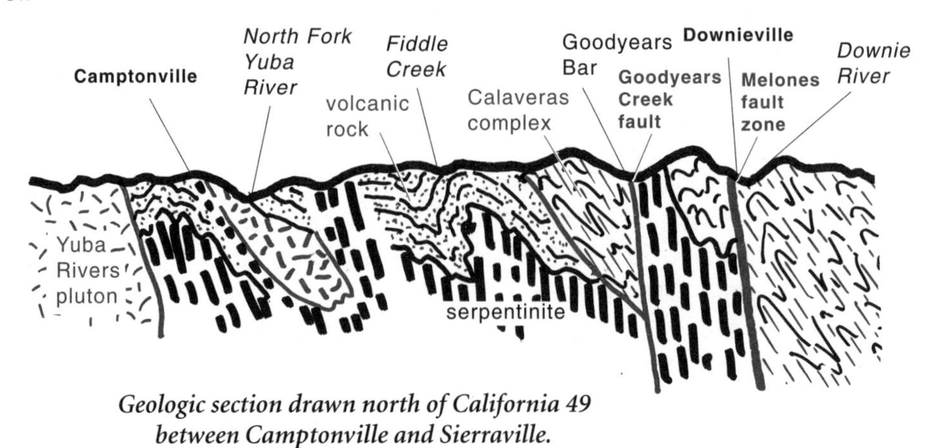

Geologic section drawn north of California 49 between Camptonville and Sierraville.

The Coloma district never produced much gold; nevertheless, some historians consider it the most important of all gold discoveries, in the sense that it brought the greatest consequences.

Grass Valley and Granite Hills

The route between Grass Valley and Downieville winds through the forested hills and valleys of the western slope of the Sierra Nevada. It crosses the South Fork of the Yuba River about 6 miles north of Grass Valley and the Middle Fork south of Camptonville. About 5 miles and several old hydraulic pits north of Camptonville, the road winds down into the deep gorge of the North Fork of the Yuba River, which it follows east to Downieville and to Yuba Pass. The rivers entrenched their valleys as the Sierra Nevada rose. Remnants of the old landscape on which they started flowing survive on drainage divides as much as 1,000 feet above stream level. The road provides a few glimpses.

Bedrock is well exposed along most of the route. It is mostly old sedimentary and volcanic rocks, now metamorphosed and intruded by large masses of granite. Excellent outcrops of serpentinite exist a few hundred yards east of Goodyears Bar.

Most of the route between Grass Valley and Camptonville crosses a large mass of granite that extends to within 1 mile of both towns. Older metamorphic rocks enclosed within the granite appear here and there. They include dark sedimentary rocks, green volcanic rocks, and green serpentinite.

Roadcuts about 1 mile west of Nevada City provide good exposures of the granite. Numerous dark patches are all that remain of chunks of sedimentary rock that got into the molten magma and nearly melted before

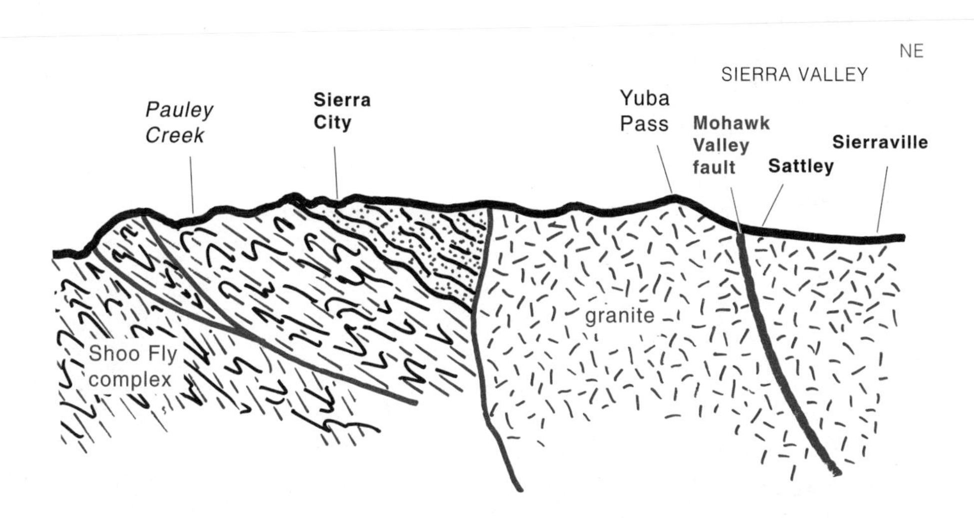

the entire mass solidified. Such inclusions are especially common near the margins of granite intrusions, where the older wall rock is close at hand.

Midway between Nevada City and North San Juan, the highway winds down the valley walls in a series of sharp curves to cross the South Fork of the Yuba River. Watch for spectacular outcrops of granite near the bridge. A few fractures and veins seam the forbidding gray faces.

Many huge blocks of granite fill the river channel below the bridge. It is not clear whether they arrived in a mudflow, or simply tumbled down the valley walls. In any case, they are much too large for the river to move. Torrents of water, muddy with particles of hard sediment, swirl across those boulders when the river is in flood. They abrade the granite as though the stream were a giant sheet of fluid sandpaper. That is how it carved the boulders into their rakish forms—natural abstract sculpture.

Folded Ocean Floor

Bedrock north of Camptonville includes metamorphosed sedimentary rocks and a folded slab of old oceanic crust, which extends to the bottom of the deep canyon of the North Fork of the Yuba River. That oceanic crust comes complete with a generous slice of serpentinized peridotite from the earth's mantle and its overlying gabbro and basalt.

Feather River Peridotite Belt

California 49 passes the Goodyears Creek fault about 3 miles west of Downieville. It is a strand of the Melones fault zone that contains a mass of serpentinite. Between it and the main Melones fault at Downieville, the road crosses metamorphosed sedimentary and volcanic rocks of the Feather River peridotite belt.

Sierra Buttes, a remnant of Devonian volcanic rocks. They stand high because they resist erosion.

The Feather River peridotite belt is a strip of old oceanic rocks a few miles wide that trends north between the northern Melones fault and the Goodyear fault. It is a slab of oceanic crust that was jammed deep into an oceanic trench during the Antler mountain building event and dragged under the Shoo Fly complex to the east. Most of the slab is serpentinite that sheared off the mantle rocks beneath the ocean floor. But it also contains gabbro, dark amphibolite gneiss, and a small amount of blueschist.

Shoo Fly Complex

Sierra Buttes rise west of the highway in the area north of Sierra City. They contain volcanic rocks that erupted during Devonian time, while the Antler mountain building event was in progress. They probably erupted above a sinking slab of ocean floor just as the Cascade Range volcanoes do today. Sierra Buttes are good evidence that the Antler event involved a plate collision and an oceanic trench.

Rocks exposed along the 13 miles between Downieville and Sierra City are part of the Shoo Fly complex. Most are dark slates, but some are more colorful. All began as sediments deposited along the continental margin and on the ocean floor during early Paleozoic time. Then the sinking ocean floor swept them into the oceanic trench about 350 million years ago, during

Early Paleozoic slates of the Shoo Fly complex, east of Downieville.

the Antler event. All were intensely deformed and considerably recrystallized in the heat. Veins of white quartz streak through a few of the roadcuts and through the outcrops along the river.

Gold

The Sierra City mining district includes a number of mines in a belt of slate and metamorphosed volcanic rocks that trends north from town. The volcanic rocks, mostly andesite and rhyolite, are probably part of an old volcanic chain that rose parallel to an oceanic trench during the Antler event. All the rocks are part of the Shoo Fly complex.

The district went into production in 1850, after prospectors found placer deposits that yielded some uncommonly large nuggets. As usual, bedrock lode discoveries followed almost immediately, some with the discovery of bonanza ore actually exposed in outcrops—money lying loose on the ground. One of those, at the Four Hills Mine about 12 miles north of Sierra City, was said to have yielded between a quarter and a half million dollars worth of gold, in 1850 dollars. A number of mines opened within a few years, and some continued to produce until about the time of the First World War. Some authorities estimate total gold production during those years at between $20 and $30 million.

Downieville was founded in 1850 as the center of another mining district, which also began with placer mines, then went into underground mining. The early placer claims were notable both for the amount of gold they produced and the size of some of the nuggets. One nugget, found in the river just above Downieville, weighed about 25 pounds. The quartz veins are in slate and metamorphosed volcanic rocks similar to those around Sierra City. The mines operated fairly continuously until the Second World War and sporadically since then.

Watch along the North Fork of the Yuba River east of Downieville for occasional piles of gravel nearly hidden among the trees. The early miners shoveled them up as they washed the gravel through sluices and rockers. The gravel piles probably date from the gold rush of the early 1850s. Later gold dredges left great corrugated ridges of gravel, not isolated little heaps.

Sierra Nevada Batholith

The road immediately east of Sierra City crosses a narrow belt of pink granite that contains large crystals of quartz and feldspar in a finely crystalline matrix. About 8 miles northeast of Sierra City, at Bassetts, the gold country ends at the western margin of the Sierra Nevada granite batholith. Very little gold has ever been found within granite, but a great deal has been found in veins in the older rocks around the margins of granite intrusions.

Sierra Nevada Fault

Yuba Pass is at the crest and eastern margin of the Sierra Nevada. East of the pass, California 49 steeply negotiates the abrupt fault scarp that defines the eastern face of the range. Sierraville is on the floor of the Sierra Valley, another large block of the continental crust that dropped as the Sierra Nevada rose.

California 49 crosses the remarkably flat floor of the Sierra Valley between Sattley and Sierraville. Such flat ground is likely to be an old lakebed, as this is. The road skirts the northern margin of a patch of low, wooded hills—eroded volcanic rocks that erupted onto the floor of the Sierra Valley after it dropped. A few outcrops of granite that poke through the volcanic rocks show that the Sierra Nevada batholith lies beneath the valley floor, the same rock as that in the high mountain crest outlined on the skyline to the west.

Ice Age Glaciers

The highway between Yuba Pass and the area about 10 miles to the west crosses glacial moraines and passes many large roadcuts in glacial till. Watch for the rubbly exposures that feature large, rounded boulders embedded in a mass of mud and gravel. That is glacial till, the stuff of moraines. The glaciers eroded the lake basins north of the highway, along the Gold Lake Road.

Sierra City is at the lowest limit of former glaciation in this valley. Watch for the large glacial moraines conspicuously visible in roadcuts near town, hummocky heaps and ridges of debris. Those are moraines made of till dumped directly from the ice along the margin of a glacier. Roadcuts through them reveal a disorderly mix of boulders, gravel, and sand, left in a heap as though by a giant bulldozer. Glaciers really do function as giant bulldozers. They scoured the valley between Sierra City and Yuba Pass into a broad and rather straight trough, in striking contrast to the narrow, winding canyon farther downstream.

California 70
Feather River Road
Marysville—Hallelujah Junction
164 miles

The geologic map of the western part of this route shows the usual Sierran picture of older bedrock with a partial cover of much younger volcanic rocks that filled old stream valleys, mainly between 40 and 17 million years ago, during Oligocene and early Miocene time. The older rocks belong mostly to the Calaveras and Shoo Fly complexes. Large masses of granite invaded both.

East of the Sierra Nevada and the town of Graeagle, the road enters the Basin and Range. That part of the route crosses old lakebed sediments deposited during the last ice age, when the climate was wet enough to flood the valleys. Old volcanoes rise south of the road.

Sutter Buttes

The prominent range of rocky hills a few miles northwest of Yuba City is Sutter Buttes. The cluster of hills, about 10 miles in diameter, rises about 2,000 feet above the floor of the Sacramento Valley. It consists of andesite and rhyolite erupted in early Pleistocene time, perhaps about 1.5 million years ago. Sutter Buttes stand on the floor of the Sacramento Valley. They are not part of the Sierra Nevada.

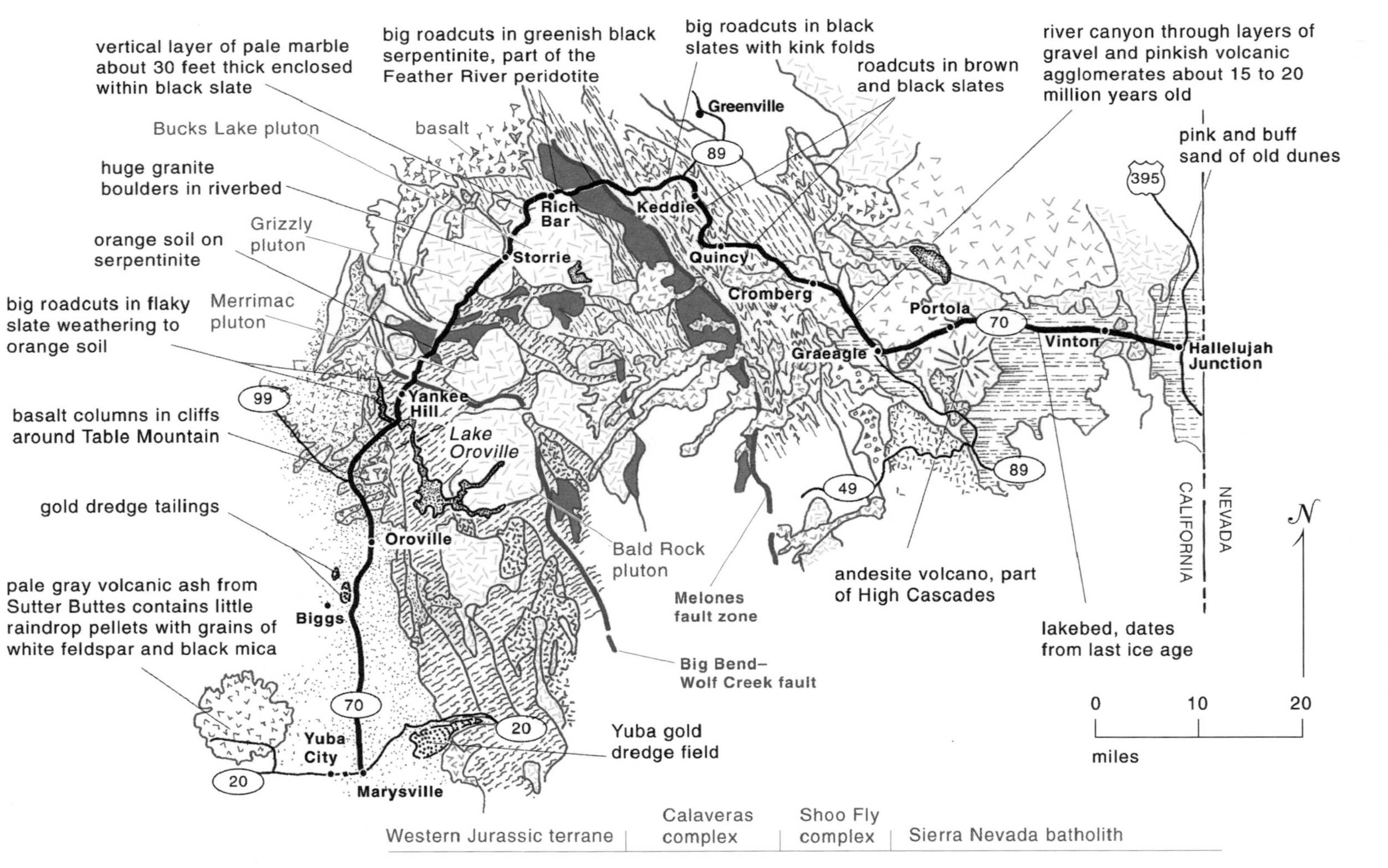

Rocks along California 70 between Marysville and Hallelujah Junction.

Sutter Buttes from the southwest.

Oroville Dredge Field

Oroville is on the thin feather edge of the young sedimentary deposits that fill the Sacramento Valley and lap onto the older rocks of the Sierra Nevada. Prospectors found placer gold in the Feather River in 1849, and they washed some of the shallow gravel then. Most of the gold was far below shovel depth.

Serious mining began with the arrival of gold dredges in 1898. By 1908, at least thirty-five dredges were operating between Oroville and Biggs, about 10 miles to the south. Activity waned a few years after that but resumed between 1936 and 1942 and again between 1946 and 1952. The dredges worked the streambed to a depth of as much as 55 feet, recovered almost 2 million ounces of gold, and left the ghastly desolation of gravel south of Oroville.

The dredges converted almost 9 linear miles of floodplain, a width of 1 or 2 miles, into a sea of raw gravel stacked in long ridges. Without a truly phenomenal flood, the Feather River will not restore this floodplain to anything resembling its natural state for many thousands of years. The cost of artificially restoring it would greatly exceed the value of the gold it produced. Eventually, so will the cost of lost agricultural production. With luck, the gravel may be crushed and used for road aggregate and concrete.

Dredge working gravels near Oroville in 1905.
—G. K. Gilbert photo, U.S. Geological Survey

Table Mountain

About 3 miles northwest of Oroville, the road passes through a gap in Table Mountain, one of California's most renowned geologic landmarks. Its black rimrock is a remnant of a basalt lava flow that poured down a stream that flowed out of the Sierra Nevada about 30 million years ago. Erosion has since removed the soft sedimentary rocks that made the old valley walls, leaving the more resistant lava in the old valley floor standing high above the surrounding countryside.

Metamorphic Rocks

North of Table Mountain, California 70 turns northeast and heads into the Sierra Nevada. Rocks west of Jarbo Pass began as soft sediments deposited beneath the waters of the Pacific Ocean about 200 million years ago, then metamorphosed as they were jammed into the Sierran trench and heated about 150 million years ago. Now the old sediments are an assortment of rather dark gray slates and buff phyllites that contain enough fine mica to glisten in the sun. They belong to the Western Jurassic terrane. The black slates are exposed just north of the big bridge across Lake Oroville. Roadcuts about 3 miles north of the bridge expose massive greenstone, oceanic basalt that belongs to the Smartville complex.

Near the top of the long grade between Jarbo Pass and the bottom of the Feather River Canyon, California 70 crosses the boundary between the Smartville complex of the Western Jurassic terrane and the older rocks of the Calaveras complex to the east.

Serpentinite

The road passes several large bodies of serpentinite east of Jarbo Pass, along the slope between the pass and the canyon floor. Watch for the rock with a greasy look that comes in various dark shades of green. It is part of a big slice of oceanic rocks that jammed into the continental margin and sheared off the top of the earth's mantle, instead of sinking through the Sierran trench.

The rocks that contain these serpentinites are rather darkly nondescript volcanic debris. The debris was dumped into the ocean, then swept into the Sierran trench, where it got hot enough to recrystallize into slates and schists. Watch for the slabby and flaky outcrops.

Diorite

California 70 passes through a large mass of quartz diorite in the 15 miles between the Poe and Rock Creek power dams. The rock would resemble granite if it were not so dark. It consists basically of crystals of the black pyroxene, augite, set in a matrix of white or greenish white plagioclase feldspar.

Large volcanoes almost certainly fed on the diorite magma while it was still molten and erupted part of it as dark andesites. Now that erosion has removed the volcanoes, the magma that crystallized slowly beneath them is exposed as coarsely crystalline diorite. Cliffs of gray diorite make sheer canyon walls rising dramatically toward the plateau surface far above.

The edge of the Grizzly pluton is just east of the Poe Dam. About 10 miles of road cross it, and another 5 miles cross the Bucks Lake pluton farther east. Both are masses of diorite that contains small amounts of quartz. They rose into the older rocks during Cretaceous time. Neither show any sign of the streaky grain that igneous rocks typically acquire if they are metamorphosed. Evidently, the plutons intruded the older rocks after they were metamorphosed.

A landslide or mudflow must have dumped the monster boulders of granite strewn along the floor of the Feather River Canyon. They owe their size to the wide spacing of the fractures in the granite bedrock. Floods sweeping abrasive sediment over the boulders rasped them into their fluidly sculptured shapes.

Slates

The highway between the area east of the Rock Creek Dam and Quincy crosses dark Paleozoic slates similar to those east of Jarbo Pass. These belong to the Calaveras complex, as does a thick bed of pale gray limestone that borders them on the southwest.

Serpentinite with shiny slip surfaces near Rich Bar.

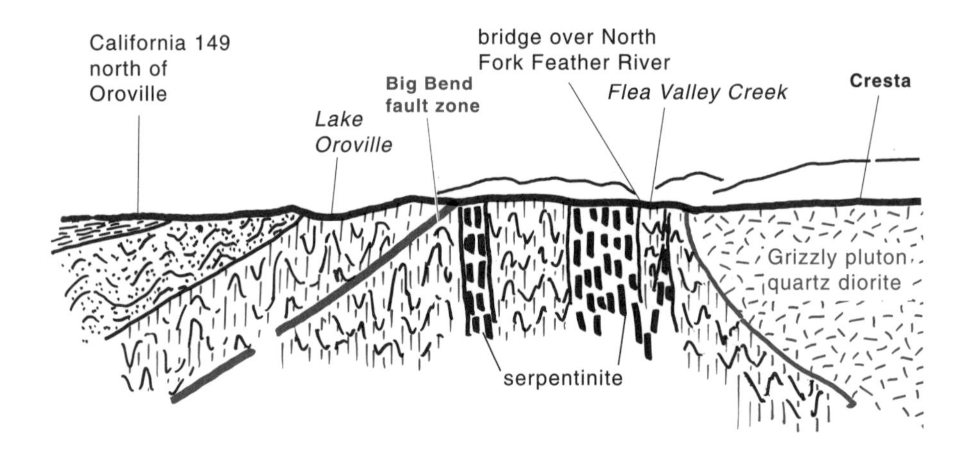

Geologic section approximately along the line of California 70 between Oroville and Greenville, across rocks of the Western Jurassic, Calaveras, and Shoo Fly terranes.

Big Slab of Mantle

Just east of Rich Bar, the road passes nearly continuous big cuts in dark green serpentinites and black peridotites. They are in the Feather River peridotite belt, a slab of mantle rock that somehow found its way into the Sierran trench, along with the sedimentary rocks in the neighborhood. The slab extends for many miles in a belt that parallels the Melones fault zone. Dull greenstones exposed in the eastern part of the belt are part of the old oceanic crust that rested on the mantle serpentinite and peridotite.

Calaveras Complex

About 11 miles east of Rich Bar, California 70 enters a broad belt of Permian and Triassic black slates of the Calaveras complex. The original sedimentary layers stand on edge, upended in the Sierran trench. As always with slates, they break along cleavage that is perpendicular to the direction in which the original sedimentary rocks were compressed.

Big roadcuts about 1 mile west of the junction with California 89 show prominent kink folds. The folds sharply bend the slaty cleavage, so they must be younger than the compression and mild metamorphism that converted the original sedimentary rocks into slates.

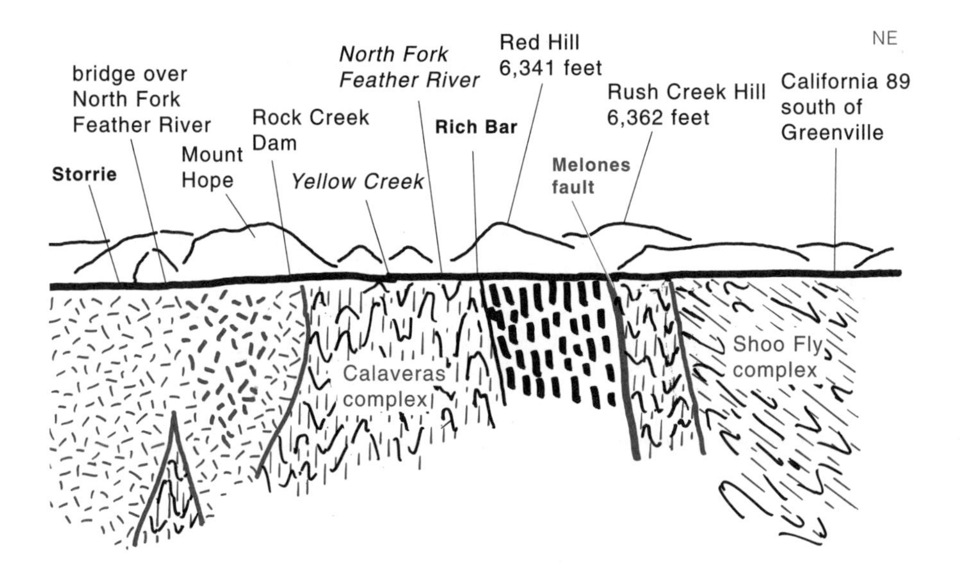

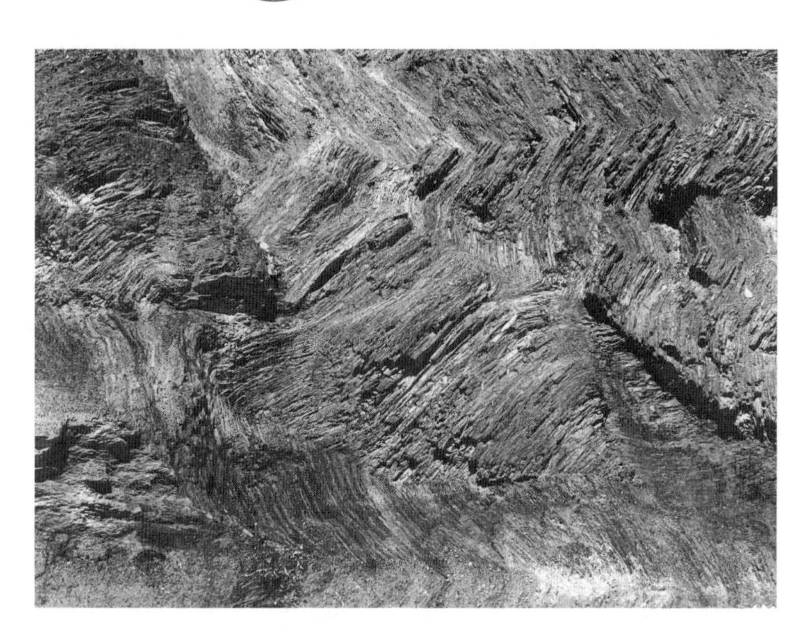

Kink folds in dark slates along California 70.

Quincy

Quincy nestles in a broad valley that apparently formed when a small crustal block subsided along a triangle of faults. Loose sediment deposited on the valley floor washed in during the past several million years. Bedrock in the nearby mountains began as sands and muds deposited on the floor of the Pacific Ocean 300 to 200 million years ago. The heating and tight folding that accompanied their stuffing into the Sierran trench about 150 or so million years ago converted them into a variety of dark gray and brown slates, glistening phyllites, and flaky schists. They are part of the Calaveras complex.

Mohawk Valley Fault

The Mohawk Valley fault is one of several that define the eastern margin of the northern Sierra Nevada. It passes beneath the eastern edge of the valley that contains Quincy. California 70 roughly follows it along most of the distance between Quincy and Graeagle. Near Cromberg the fault consists of several parallel splays. Its exact position is hard to determine farther south because valley fill sediments and young volcanic rocks bury the older bedrock.

Basins

California 70 crosses the floor of a broad basin between about 3 miles east of Portola and Hallelujah Junction. The basin is a block of the crust that subsided east of the Sierra Nevada. Bedrock buried under the basin fill sediments is almost certainly the same granite as that on the high crest of the Sierra Nevada.

Numerous groups of hills within the basin make it hard to see its outlines from the road. They exist because the large block that sank to make the basin is itself broken into smaller blocks along lesser faults, and these rose and sank independently. Some of the bedrock exposed in the hills is granite, the older rock that moved along the faults. Other hills on the basin floor are volcanoes that erupted as the Sierra Nevada rose and the basins subsided.

Volcanic Rocks

The western end of the Feather River valley contains volcanic rocks that erupted in Miocene and Pliocene time over much of the Sierra Nevada. Large exposures appear beside the road where it passes through the canyons north and south of Sloat and from Graeagle to east of Portola. They consist mostly of volcanic ash and mudflow deposits in various pale shades of gray, pink, and lavender. Many are agglomerates of volcanic ash mixed with angular chunks of volcanic rock, probably mudflow deposits.

Granite

Granite appears beside the road in the valley a few miles east of Portola and in the low hills north of there. Watch for it around Beckwourth Pass, about 4 miles west of Hallelujah Junction, where the road passes over the southern tip of a large block of granite that stands high above the valley floor.

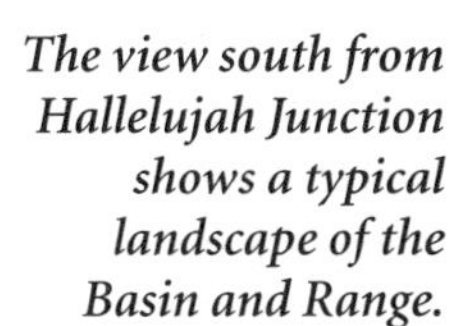

The view south from Hallelujah Junction shows a typical landscape of the Basin and Range.

California 70 west of Hallelujah Junction traverses a sloping surface on gravelly sediments shed from granite in the east face of the Diamond Mountains, a steep fault scarp. A few old sand dunes make yellowish patches low on the slope. North of this area, this fault scarp defines the steep eastern front of the Sierra Nevada block.

Sierra Valley

Sierra Valley from east of Portola to Beckwourth Pass is as flat as the proverbial platter full of milk. Any surface this flat is likely to be lake sediments. Evidently, volcanic rocks impounded a lake here among the volcanoes and granite knobs. The lake filled during the last ice age when rainfall was much greater than it is now, then drained when the outlet stream eroded its channel deep enough to restore drainage into the Feather River.

Feather River Keeps Pace

It is astounding that the Feather River drains this area in the Basin and Range, right through the high rampart of the Sierra Nevada. The river must have been flowing in its valley before the Sierra Nevada rose and eroded its channel fast enough to keep pace. None of the rivers farther south accomplished that feat.

California 89

South Lake Tahoe—Graeagle

96 miles

California 89 stays in view of the steep eastern face of the Sierra Nevada all the way between South Lake Tahoe and Graeagle, following the low country immediately east of the high scarp of the Sierra Nevada fault. The road passes granite of the Sierra Nevada in places along the west side of Lake Tahoe and again just north of Sierraville. Most of the rest of the route crosses volcanic rocks that erupted as the Sierra Nevada rose. The road north of Sierraville passes a flat lake plain, a relic of the last ice age, when the climate was wet and the undrained desert valleys held lakes. The great rivers of ice that gouged the high peaks of the Sierra Nevada left glacial deposits that reach as far east as the road in several places.

Volcanoes

The route between Tahoe City and Graeagle crosses the block that dropped along the Sierra Nevada fault while the mountains rose. A large volcanic pile nearly covers the Sierra Nevada granite. All of the bedrock

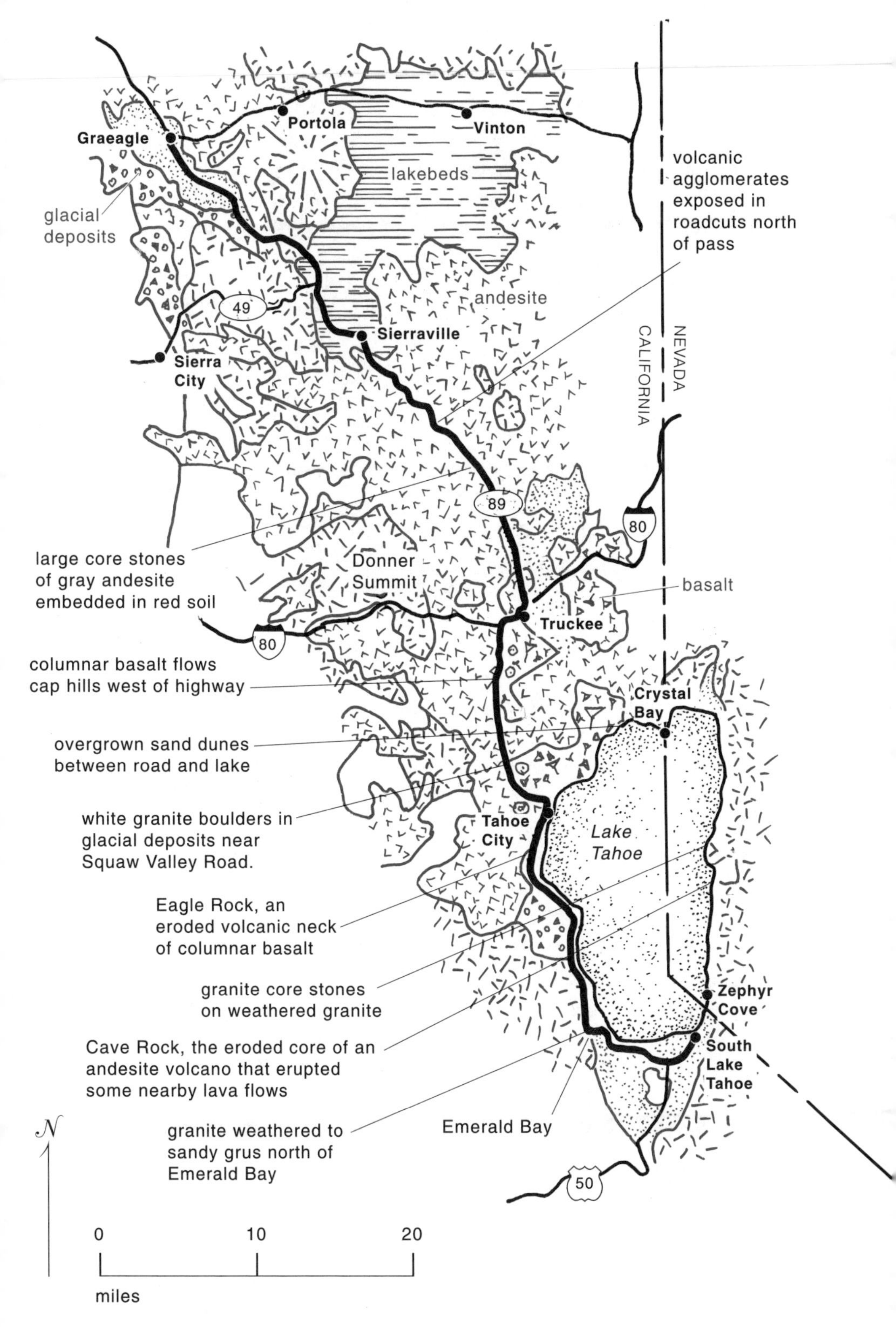

Rocks along California 89 between South Lake Tahoe and Graeagle.

Andesite agglomerate north of Tahoe City.

exposed near the road is volcanic, all erupted within the past 10 million years. Erosion has thoroughly dissected some of the older volcanoes, but the younger ones retain their original form. The rocks are basalt, andesite, and rhyolite.

As in most volcanic piles, these rocks come in disguises as various as the ways volcanoes can erupt. It is easy to recognize the black basalt lava flows, which erupted quietly, and the layers of sparkling white rhyolite ash, which erupted with extreme violence—never mind their look of angelic innocence. Some of the andesites are lava flows, solid rocks that contain a few crystals scattered through a nondescript gray or brown matrix composed of microscopic mineral grains. Most of the andesites are messy agglomerates of ash mixed with angular chunks of lava.

Younger glacial deposits cover most of the volcanic rocks between Tahoe City and Truckee. In some exposures the glacial deposits and volcanic rocks look surprisingly alike. The glaciers scooped up some of the messy agglomerates and converted them into even messier glacial deposits. Both are full of angular chunks of lava in a variety of pastel and reddish colors. Distinguish the glacial deposits by looking for speckled chunks of pale gray granite, which the glaciers imported from the crest of the Sierra Nevada. The volcanic rocks do not contain granite.

The road crosses some of the southern and western part of the Sierra Valley in the few miles around Sierraville. The flat landscape is the floor of a vanished lake. Volcanic rocks dammed the drainage, creating a lake that lasted until the outlet stream could erode its valley deep enough to restore the drainage.

Flowing artesian wells in the Sierra Valley produce water trapped under pressure beneath the lake sediments that cover the flat floor of the valley. Lake sediments typically consist of fine silt and clay. They are generally impermeable, so they make a good cap to confine the pressured artesian water beneath.

Lake Tahoe

Lake Tahoe floods part of a Basin and Range valley. The Carson Range east of Lake Tahoe rose, and the valley that contains the lake sank, as a slice of rock rotated while moving east on a curving fault. Meanwhile, volcanic eruptions built a volcanic pile across the valley, creating the natural dam that impounds Lake Tahoe. Interstate 80 crosses the volcanic dam between Truckee and Reno. The mountains that flank the highway are part of the volcanic pile.

Truckee River

Lake Tahoe drains into the Truckee River, a modest little stream doing its ineffective best to empty the lake. Clear streams like the Truckee River do not carry enough sediment to make them effective eroders of bedrock. It is the abrasive sediment a stream carries that carves bedrock, not the running water, just as sand entrained in a blast of air scours buildings, though clean wind does not.

After running so brightly through the mountains, the Truckee River sets off into the Nevada desert. There, it withers and finally comes to its wretched destiny in the undrained chemical sump of Pyramid Lake.

Ice Age Glaciers

Two of the enormous glaciers that drained from the high snowfields of the Sierra Nevada during the last ice age ground their way down Squaw Valley and Pole Creek to the Truckee River. They dammed the outlet of Lake Tahoe between Tahoe City and Truckee, raising the water level as much as 600 feet. Ice dams float in the lakes they impound for the same reason that ice cubes float in lemonade. The ice dam at Lake Tahoe floated several times, each time breaking up and releasing a catastrophic flood that carried huge boulders downstream past Truckee. Then the glaciers again advanced across the Truckee River, again raising the lake level and setting the stage for another catastrophic flood.

The glaciers of the last ice age finally melted sometime around 12,000 years ago, returning Lake Tahoe to about its present level. Ancient beaches

Waves eroded the caves above the highway tunnel during a time when Lake Tahoe stood at a much higher level.

precisely record the high-water levels of the last ice age. High benches of sediments deposited below those beaches survive as large terraces, generally forested, that rise above long stretches of the lakeshore. The most conspicuous looks like a giant stair step and tread rising to the mountains. Developers can hardly resist the flat top, but construction is hazardous because the soft muds within are likely to slide.

Landslides have indeed been a problem. A large mass began to move slowly above California 89 at the head of Emerald Bay in 1953. It came down all at once during a long rainy spell two years later, obliterating several hundred feet of road. Weak bedrock was the basic cause: granite that had been thoroughly broken in a fault zone, then weathered soft along the fractures. A number of other large landslides warn that the problem will become serious indeed if heedless development continues.

Pollution

It would be easy to pollute Lake Tahoe but hard to restore it, once polluted. The drainage basin of Lake Tahoe is mainly just the mountain slopes that overlook the lake. It would take about 600 years for that small watershed to produce enough runoff to fill the lake to its present level. Hundreds of years would pass before runoff from the drainage basin could flush pollutants out of the lake.

Even thoroughly treated sewage effluent is a problem because it contains dissolved nutrients such as phosphate and nitrate that fertilize algae. Algae are already much more abundant than they were a few decades ago. The only solution seems to be to export sewage effluent from the drainage basin, to regulate septic tanks, and to take extreme care in land management. The alternative is to watch Lake Tahoe slowly turn into a repulsive, green algal soup.

East Shore of Lake Tahoe

U.S. 50 and Nevada 28 between South Lake Tahoe and Tahoe City closely follow the east shore of the lake in Nevada. The views west across the lake to the alpine valleys and glacially carved crags of the high Sierra Nevada are magnificent.

Bedrock along the east shore of the lake between South Lake Tahoe and Crystal Bay is almost entirely Sierran granite. The only exceptions are a small patch of young volcanic rocks at Glenbrook Bay and an eroded volcanic neck about 7 miles north of Stateline. Cave Rock is part of that volcanic neck; waves eroded caves into it during the last ice age, when the lake level was much higher than now.

Weathered granite 3 miles north of U.S. 50, on the east side of Lake Tahoe.

All the bedrock between Crystal Bay and Tahoe City is volcanic rocks erupted within the past 10 million or so years, presumably as part of the Basin and Range pattern of activity. This is the natural dam that impounds Lake Tahoe. Much younger lake sediments and sand dunes bury the volcanic bedrock along the north sides of Crystal and Agate Bays, making flat landscapes.

West Shore of Lake Tahoe

California 89 hugs the western shore of Lake Tahoe, where the lake laps onto the steep eastern slope of the Sierra Nevada. East, across the lake, is the Carson Range, composed of granite detached from the Sierra Nevada along the West Tahoe fault, which trends north.

Bedrock along the southern half of the west shore is mostly granite, largely buried under glacial moraines composed mainly of granite boulders. Granite in the vicinity of D. L. Bliss State Park escaped glaciation. Watch for the deeply weathered rock and thick sandy soils. Ice age glaciers scoured that cover off the country they scraped.

The Sierra Nevada snatches so much snow out of the clouds that little remains for the Carson Range. That rain shadow developed before the ice ages because the Carson Range shows no sign of glaciation. Its peaks are full and rounded, not craggy and jagged like those of the heavily glaciated high Sierra.

Glaciation

Imagine huge glaciers glittering in the summer sun all along the west side of the lake and rising as a nearly continuous wall of white to the high crest of the Sierra Nevada. Glacial moraines faithfully record the outlines of the glaciers that descended from the high snowfields of the Sierra Nevada into the much larger, ice age version of Lake Tahoe. Icebergs cracked off the floating lower ends of those glaciers and set sail across the lake.

A short but steep valley glacier descending from the high granite ridge to the west scoured out the basin of Emerald Bay. It dumped its load of debris in the big glacial moraines that embrace the bay. Cascade Lake, just south of Emerald Bay, and Fallen Leaf Lake, just west of South Lake Tahoe, have similar origins, except that their moraines completely isolate them from Lake Tahoe.

California 140

Merced—Yosemite National Park

78 miles

The highway crosses an eastern slice of the San Joaquin Valley between Merced and the western edge of the Sierra Nevada, where the land becomes hilly, increasingly wooded, and less cultivated. Sediments deposited during Tertiary and Pleistocene time cover the valley floor. Rocks in the Sierra Nevada are older, more varied, and much more experienced than those in the San Joaquin Valley.

The road in the Sierra Nevada west of Mariposa crosses slightly recrystallized metamorphic rocks, mostly slates of the Western Jurassic terrane, and a mass of granite. The road east of Mariposa crosses a broad expanse of slates in the Calaveras complex and farther east a narrow belt of more drastically recrystallized rocks of the Shoo Fly complex. Granite of the main mass of the Sierra Nevada batholith appears along the road just west of the park boundary.

Slate

In the westernmost 6 miles of the Sierra Nevada, the highway crosses slate and phyllite, part of the Western Jurassic terrane. The original mudstones were deposited on the ocean floor during Jurassic time, then heated and recrystallized into metamorphic rocks as they were jammed into the Sierran trench.

Slate is only slightly metamorphosed but much harder than the original mudstone. It cleaves easily along surfaces parallel to the imaginary surfaces that would bisect the folds into mirror images. That cleavage direction is not in any way related to the original sedimentary layering.

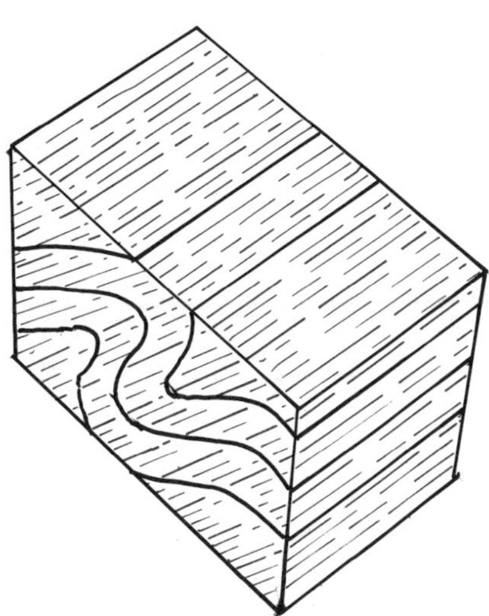

The relationship between the cleavage direction in slate and the symmetry of folds.

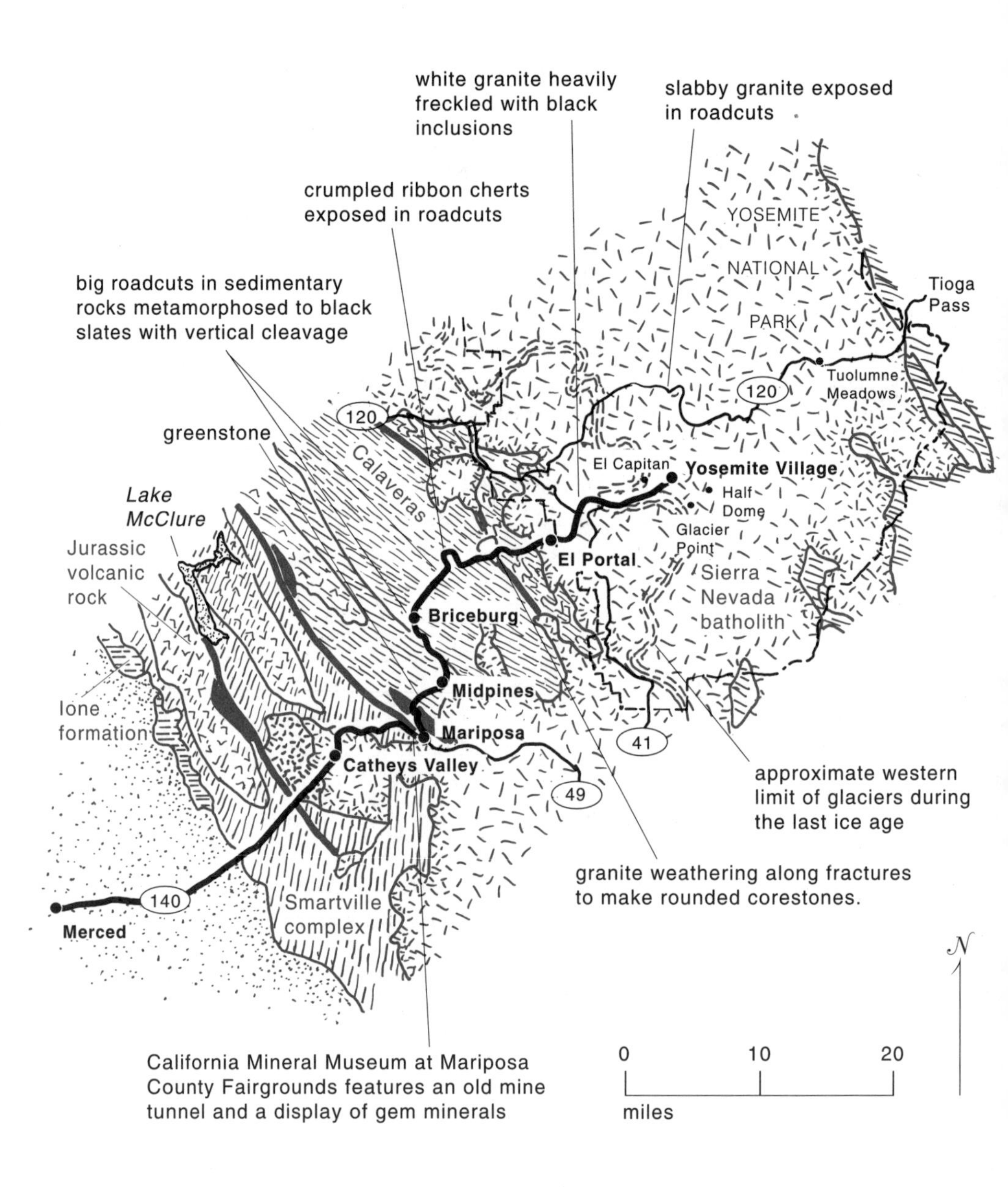

Rocks along California 140 between Merced and Yosemite National Park.

Phyllite differs from slate in containing larger grains of mica. If the rock looks almost like slate but has a silky sheen, call it phyllite. If its mica crystals are large enough to see without a magnifier, the rock is schist.

Gabbro, Diorite, and Granite

Between the area about 3 miles west of Catheys Valley and that 1 mile east, the highway crosses two kinds of dark igneous rocks of Jurassic time. One kind is gabbro, a dark rock that would be basalt if its mineral grains were much smaller. The other kind is diorite, which is also dark though it contains a lot of white plagioclase feldspar.

Farther east, the road crosses several miles of pale granite that intrudes the dark gabbro and diorite. Age dates place it at 140 million years, an early Cretaceous date.

Melones Fault Zone

Mariposa is in the Melones fault zone, the boundary between the Western Jurassic terrane to the west and the Calaveras terrane to the east. Metamorphosed sedimentary rocks west of the fault belong to the Mariposa formation, a deposit of Jurassic time. Green serpentinite within 1 mile southwest and northeast of Mariposa is in the Melones fault zone.

The rusty greenstone exposed in Mariposa and northeast to Midpines is metamorphosed basalt, part of the Jurassic ocean floor. The black slates with vertical cleavage exposed in Bear Creek Canyon began as sediments deposited on top of the oceanic basalts during Jurassic time.

Calaveras Complex

From about 1 mile east of where California 140 reaches Briceburg to El Portal, the highway follows the Merced River through the valley it eroded in late Paleozoic rocks of the Calaveras complex. They are black slates that weather to rusty colors, along with lesser amounts of sandstone and ribbon cherts.

Watch for the ribbon cherts about 5 miles northeast of Briceburg. They are strikingly layered white rocks with thin black interlayers. The original sediments accumulated on the deep ocean floor as a deposit of siliceous ooze, the silica shells of microscopic animals. That happens only on remote reaches of the ocean floor, beyond the long reach of terrestrial sediments. These ribbon cherts were folded twice, first into nearly horizontal folds, then those were folded vertically.

Granite

About 2 miles west of El Portal, California 140 passes roadcuts in massive granite in the valley of the Merced River. The ridges above are in

Ribbon chert exposed along California 140.

the late Paleozoic metamorphic rocks of the Calaveras complex. Many such examples show granite in the valleys and metamorphic rocks on the ridges exist; the Sierra Nevada batholith underlies large parts of the Calaveras complex, much of it at rather shallow depth.

Yosemite National Park

Almost all the rocks in Yosemite National Park are variations on the theme of granite. All are part of the great, Sierra Nevada batholith, a veritable sea of granite. The Yosemite Valley is America's archetypical glaciated landscape, an object lesson in what enormous valley glaciers can do with an enormous mass of granite.

The first geologist to draw a really good geologic map of the park was F. C. Calkins, who worked there during the First World War and was still mapping rocks forty years later. His map shows several large masses of granite, each crystallized from a separate intrusion of molten magma, and each full of internal complexities, which include large variations in the rocks.

Ordinary granite consists mostly of white or pink feldspar and quartz with a light peppering of black minerals. It is pale. The extremely pale rocks in the park are tonalite, which contains considerably more quartz than granite. The darkest rocks are diorite, which is mostly black minerals and

Pegmatite and thin seams of aplite in granite.

white feldspar with little or no quartz. Other rocks include pegmatite, a variant of granite that consists of grossly oversize crystals, and aplite, another variant of granite that consists of extremely small crystals.

It is generally, but not universally, true of masses of granite that their oldest parts are the darkest. That appears to be true of the individual masses of granite in Yosemite National Park.

The pegmatites and aplites are the youngest rocks within any mass of granite. Both make seams of pale gray or pink rock that formed as their magma injected fractures in the newly crystallized granite. Aplites resist weathering better than ordinary granite, so they tend to stand up in low relief as narrow ridges a few inches wide.

Although they look different, pegmatites and aplites actually crystallize from the same magma, in some cases within the same narrow seam. They crystallize from an extremely watery residual magma, the last remaining melt after all the rest has crystallized into granite. If that last fluid residue crystallizes with its water content intact, it becomes pegmatite. If it loses much of its water, the sudden drop in pressure causes it to crystallize as aplite.

Granites of the Yosemite area, like others in the Sierra Nevada, commonly contain inclusions of dark rocks. It is easy to imagine that granite magma might break older rocks into chunks, then engulf them. That is probably how dark inclusions containing traces of sedimentary layers

Exfoliating granite. —H. W. Turner photo, U.S. Geological Survey

formed. More commonly, the dark inclusions are fine grained and show no layers. They probably began as blobs of basalt magma that were trapped in the granite magma.

Domes

Domes are common in large expanses of granite. They exist in many parts of the world, in all sorts of climates. Stone Mountain, Georgia, is a granite dome, as is Sugarloaf Mountain, which so famously overlooks the harbor at Rio de Janeiro. So the granite domes in the Sierra Nevada are neither unusual nor surprising, but that does not make them easy to understand.

Their origin may have something to do with the great curving shells that so conspicuously spall off granite, a process called exfoliation. Most geologists believe the fractures that define the exfoliating slabs opened as erosion removed the rock that once covered the granite. They contend that the granite expanded as the weight of the overburden was removed, opening fractures parallel to the ground surface.

Landscape

The Yosemite landscape is not so much typical of glaciated mountain valleys as it is supertypical—almost a cartoon version of the usual glaciated valleyscape. The valleys are much straighter than in most glaciated

A glacier plucked off one side of Half Dome, block by block.

mountains, their walls are much higher and steeper, and their waterfalls are more dramatic. The difference is in the fractures.

All rocks of any kind contain fractures, which typically come in several parallel sets. The granite in Yosemite Park contains strongly developed sets of vertical fractures. Glaciers erode bedrock mostly by freezing fast to it, then plucking out the blocks between the fractures as the ice moves downslope. That explains why so much of the Yosemite valleyscape follows vertical fracture sets in the granite. The straight valleys follow them, as do the high cliffs along their walls, such as the towering face of El Capitan.

Waterfalls

Deeply glaciated mountain landscapes typically contain waterfalls, but those in Yosemite National Park are higher and more spectacular than most. Tributary streams cascade into the main valley because the smaller ice age glaciers that filled their valleys did not carve them as deeply as the much larger glacier that filled the main valley. When the ice melted, the floors of the tributary valleys were left well above the floor of the main valley. Geologists call them hanging valleys. Now the tributary streams pour down waterfalls at the mouths of their higher valley floors.

Bridal Veil Falls pours from the mouth of a hanging valley.

Huge glacial erratics along Tioga Pass Road.
—G. K. Gilbert photo, U.S. Geological Survey

California 180

Fresno—Kings Canyon National Park

88 miles

East of Fresno, California 180 crosses the eastern part of the San Joaquin Valley, where no rocks are visible. The road farther east traverses about three-quarters of the Sierra Nevada, with excellent glimpses of the bedrock.

Most of that bedrock is granite, the rest metamorphic rocks. The granite is pale and massive. The metamorphic rocks probably belong to the Calaveras complex. Most are much darker than the granite, and they are slabby and streaky. They occur in roof pendants—rocks caught between granite plutons—where they were strongly heated, then recrystallized almost beyond recognition.

Outcrops in the western foothills tend to be weathered and bouldery, in some areas covered with the red soil of the old landscape that existed before the Sierra Nevada rose. Grassy hills cloaked in oak trees eventually give way to the ponderosa pine, sugar pine, and sequoia forests of the national park. Roadcuts in granite are in some places fresh, in others deeply weathered to loose debris. The spectacular alpine landscapes around the upper end of Kings Canyon tell of glaciers carving hard bedrock.

Dark Rocks in the Western Sierra Nevada Batholith

At the eastern edge of the San Joaquin Valley, the highway passes between Jesse Morrow Mountain on the north and Campbell Mountain on the south. These western outposts of the Sierra Nevada rise above the sediments that fill the valley floor. Their rocks are gabbro and variations on the theme of gabbro—dark rocks that consist mainly of pale plagioclase feldspar, black augite, and black hornblende. They crystallized during early Cretaceous time, about 115 million years ago.

The low hills between those mountains of gabbro and the community of Squaw Valley contain more dark rocks, mostly gabbro and quartz diorite. They are part of a belt of dark, early Cretaceous igneous rocks that appears along the western base of the Sierra Nevada for a distance of about 75 miles. Drill cores suggest that the belt actually continues for more than 300 miles, mainly buried under sediments along the eastern edge of the Great Valley.

About 3 miles east of Squaw Valley, the road passes just south of Bald Mountain, which contains the southern part of the Kings River ophiolite. Watch along the highway for a few exposures of dark pillow basalt—all you are likely to see without a hike. Other components of the oceanic crust—serpentinite, layered gabbro, and basalt dikes—are exposed north of the highway. This slab of oceanic crust dates from Jurassic time, sometime around 200 million years ago.

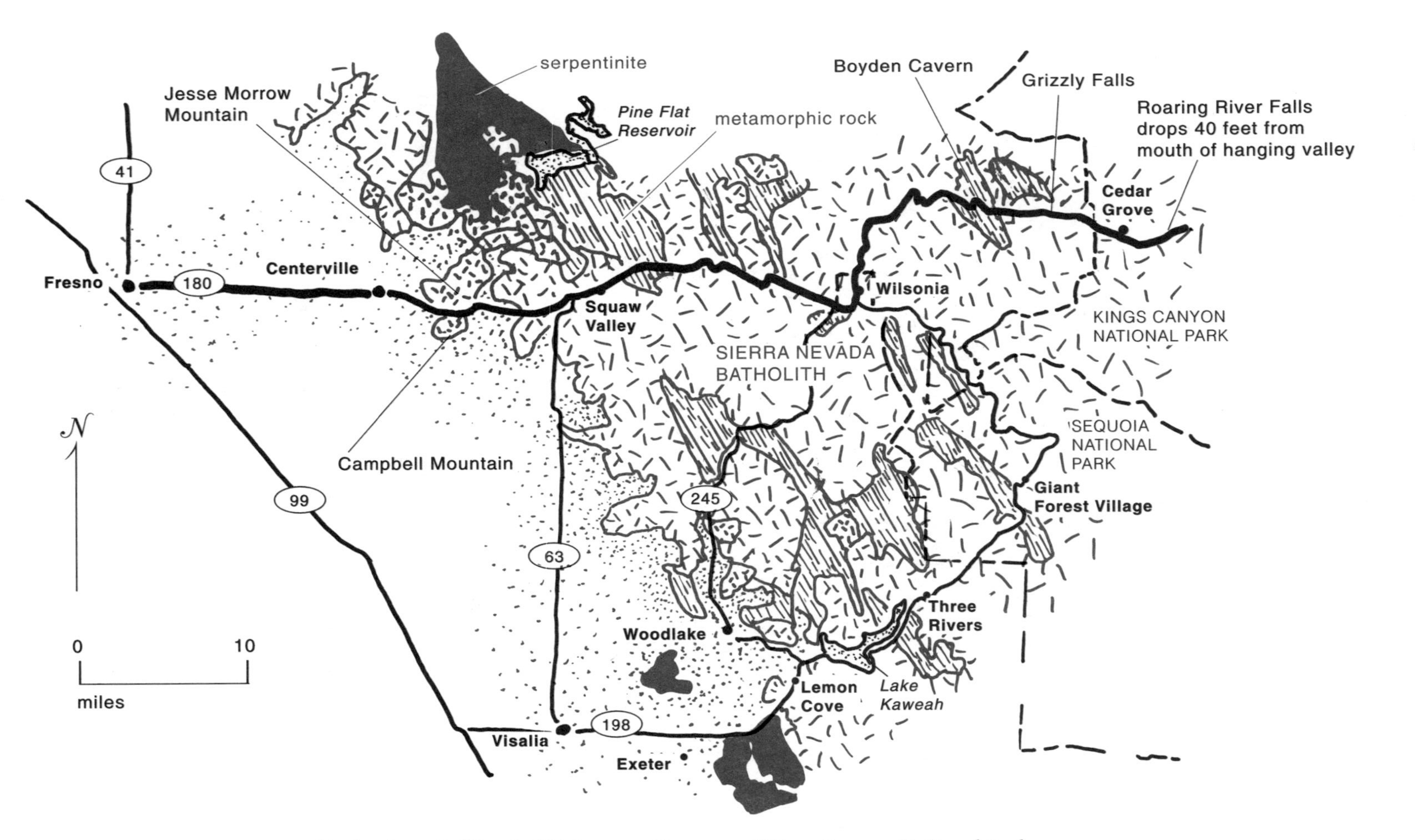

Rocks along California 180 between Fresno and Kings Canyon National Park.

Granite weathering into grus in Kings Canyon National Park.

Boyden Cavern Roof Pendant

The road enters the Boyden Cavern roof pendant from the east, a few miles before it drops deep into Kings Canyon. The roof pendant is a relic of the older rocks that existed before the magmas of the Sierra Nevada batholith arrived. The road follows the South Fork of the Kings River through it, between the area a mile or two west of Horseshoe Bend and a mile or two east of Boyden Cavern. Watch for big roadcuts in layered gneisses about 1 mile east of the bridge across the Kings River at Boyden Cavern. The rocks recrystallized almost beyond recognition in the heat of the granite magma.

A chaotic mess of various kinds of rocks, set in a nondescript matrix of finely crystalline rock, separates the eastern and western parts of the roof pendant. Rocks in the western part are quartzite, marble, and schist; those in the eastern part are an assortment of sandstones, slate, and metamorphosed volcanic rocks. The marble makes a prominent light bluish gray band that crosses the canyon at the bridge across the Kings River.

Boyden Cavern is in the band of marble in the western part of the roof pendant. Marble is metamorphosed limestone, also made of calcite and soluble in slightly acidic water. Groundwater erodes caves into marble just as it does into ordinary limestone.

The road between the area about 20 miles east of Squaw Valley and Horseshoe Bend is generally west of the quartz diorite line. That line, actually a zone of transition, separates the medium gray quartz diorite to the west from the much lighter granite to the east. Quartz diorite consists mostly of black pyroxene set in a matrix of white plagioclase feldspar, with small amounts of gray quartz. The granite farther east is about half pink orthoclase feldspar along with white plagioclase, glassy quartz, and a light peppering of black minerals.

Granites east of the Boyden Cavern roof pendant are all fairly pale and contain generous amounts of quartz and pink orthoclase feldspar, but they are not all the same. Geologists have found that they belong to several masses of granite, each intruded at a different time, each distinctive enough to recognize and plot on a map.

3

Klamath Mountains

People call them the Klamaths and Siskiyous, these rugged mountains of northwestern California and southwestern Oregon. To keep things simple, we will simply call them the Klamath block. They really are a single, if extremely complex, block of the earth's crust.

The mountains of the Klamath block guard their geologic secrets well beneath a deep mantle of soil and a cover of lush greenery. Good bedrock outcrops are hard to find. The occasional exposures offer tantalizing

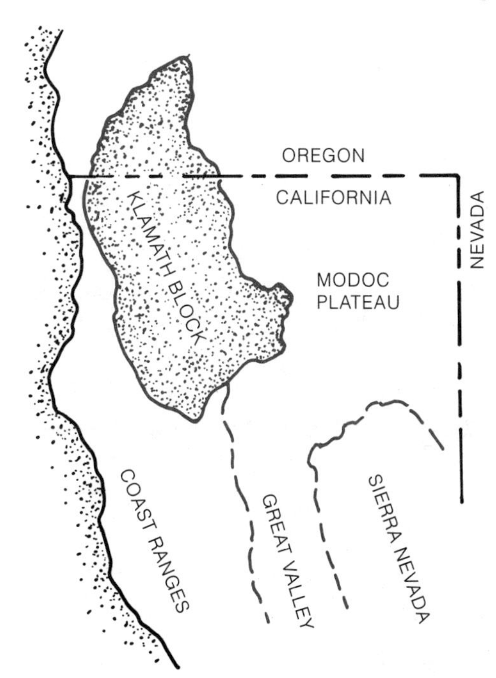

The Klamath block.

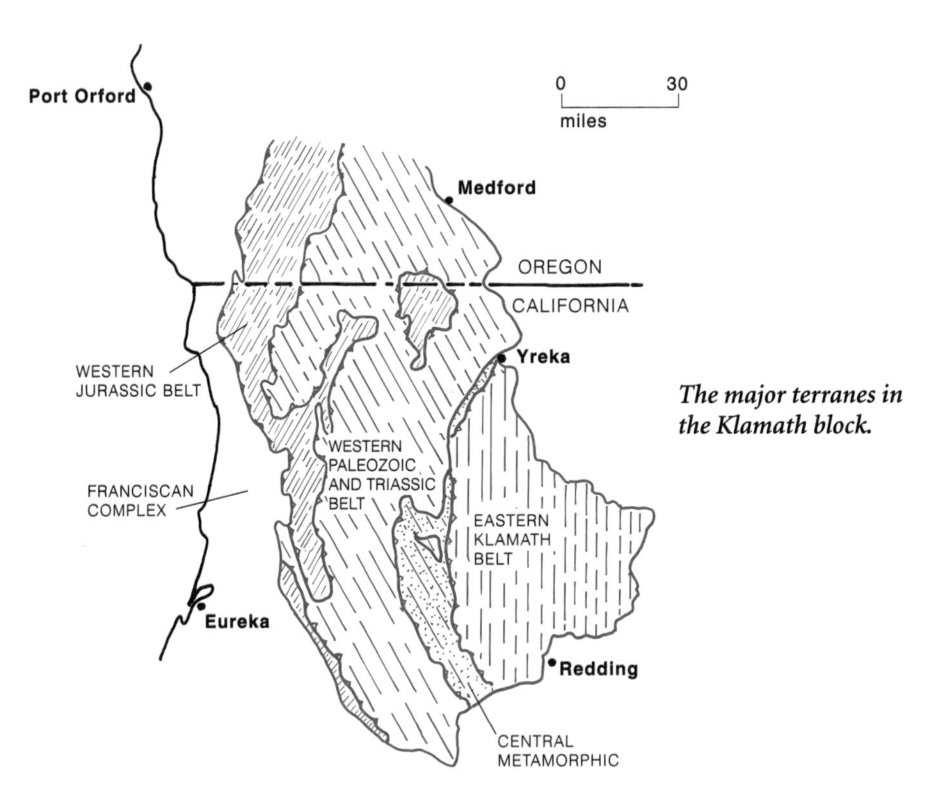

The major terranes in the Klamath block.

The Klamaths west of Castle Crags.

glimpses of fascinating rocks that hint of geologic stories still untold. But despite the difficulties, the basic architecture of the Klamath block seems reasonably clear. The details will occupy generations of geologists still unborn.

TERRANES

Geologists broadly lump the complex rocks of the Klamath block into four belts, terranes basically similar to those of the northern Sierra Nevada. Most California geologists refer to the Klamath terranes as belts. Each belt contains within itself a complex, bewildering assortment of rocks, all of which appear to share related origins and a common history.

The number of terranes or belts that geologists recognize depends largely upon the level of detail they bring to their study of the rocks. Geologists who like to think broadly tend to minimize the number of terranes they consider; those who are more interested in the details break larger terranes into smaller units, or subterranes. We will limit this discussion to the four major belts.

As in the northern Sierra Nevada, each of the broad belts of rocks in the Klamath block has the general form of a great slab of rock tilted gently down to the east. Each lies on a fault on which it overlaps the next slab to the west, like shingles on a roof. The oldest belt is in the eastern part of the Klamath block, and they become progressively younger westward. That arrangement places the oldest slabs on the top of the pile, the youngest on the bottom, exactly contrary to the normal order of stacking. Belts stacked in that way must have moved along faults to where we now see them. Either the older slabs moved west, riding over the younger ones, or, if you prefer, the younger slabs moved east, dragging under the older ones.

Eastern Klamath Belt

The mangled rocks of the Eastern Klamath belt include a wild assortment of old ocean floor, volcanic rocks, and sedimentary formations deposited as early as Ordovician time, about 450 million years ago. They include sequences of layered cherts that were surely laid down on remote reaches of the deep ocean floor. All these rocks were deposited along, or offshore from, the continental margin that originated about 800 to 700 million years ago. Many resemble the rocks in the Shoo Fly terrane of the northern Sierra Nevada and are the same age.

Their ages clearly mean that the rocks in the Eastern Klamath belt were in the way when North America collided with the ocean floor during the Antler mountain building event of Devonian and early Mississippian time. That, no doubt, was when they were jammed against the western edge of

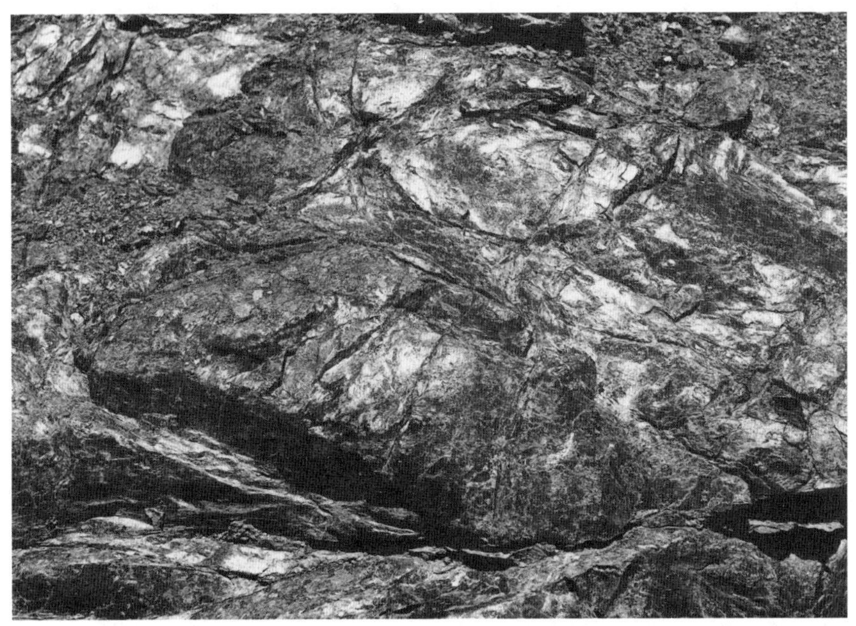

Serpentinite in the Trinity ophiolite, southwest of Yreka. View is 5 feet across.

the continent, tightly folded, broken along faults, and recrystallized into slates, schists, and marble.

The Trinity ophiolite, which separates the Eastern Klamath belt into halves, is a big piece of Ordovician ocean floor. Some geologists have suggested that many of the rocks in the Eastern Klamath belt are the remains of an old volcanic chain and that the Trinity ophiolite is the ocean floor on which those volcanoes stood. If so, we can imagine an oceanic trench developing offshore when North America began to collide with the ocean floor as the Antler event began. A chain of volcanoes, probably volcanic islands, rose along a line parallel to the trench and were finally crushed into the western edge of North America as it moved west.

Central Metamorphic Belt

The Eastern Klamath belt lies on the Trinity thrust fault, which separates it from the Central Metamorphic belt beneath and to the west. The Central Metamorphic belt is another bewildering complex of metamorphic rocks, mostly schist, marble, and dark amphibole gneiss. It seems reasonably clear it slid along the Trinity thrust fault as it was dragged beneath the Eastern Klamath belt.

Many geologists now argue that the rocks of the Central Metamorphic belt are the same as those in the Eastern Klamath belt, just more thoroughly metamorphosed as the sinking ocean floor dragged them into hotter regions beneath the Eastern Klamath belt. If that is indeed what happened, then the age of the metamorphism would reveal the time of movement on the Trinity thrust fault. Age dates on minerals in the metamorphic rocks give figures in the range of 400 to 380 million years, Devonian time. Those ages nicely fit the idea that the Central Metamorphic belt was dragged under the Eastern Klamath belt and metamorphosed during the Antler mountain building event.

Western Paleozoic and Triassic Belt

The Western Paleozoic and Triassic belt is a complex mess of mostly dark, oceanic rocks that correspond to those in the Calaveras complex of the northern Sierra Nevada. It includes great slices of oceanic crust and a variety of sedimentary rocks, mostly of kinds that typically accumulate on the ocean floor.

Rocks of the Western Paleozoic and Triassic belt were squashed into the Sierran trench during early to middle Jurassic time and dragged under the Central Metamorphic belt along the Siskiyou thrust fault. They include blueschists, which contain minerals that could only have crystallized under the pressure of 12 to 20 miles of rock. Evidently, they were rapidly dragged deep under the continental margin.

Western Jurassic Belt

Rocks in the Western Jurassic belt closely resemble those in the Western Jurassic terrane, west of the Melones fault zone in the Sierra Nevada, and probably correspond to them. The Western Jurassic rocks include old

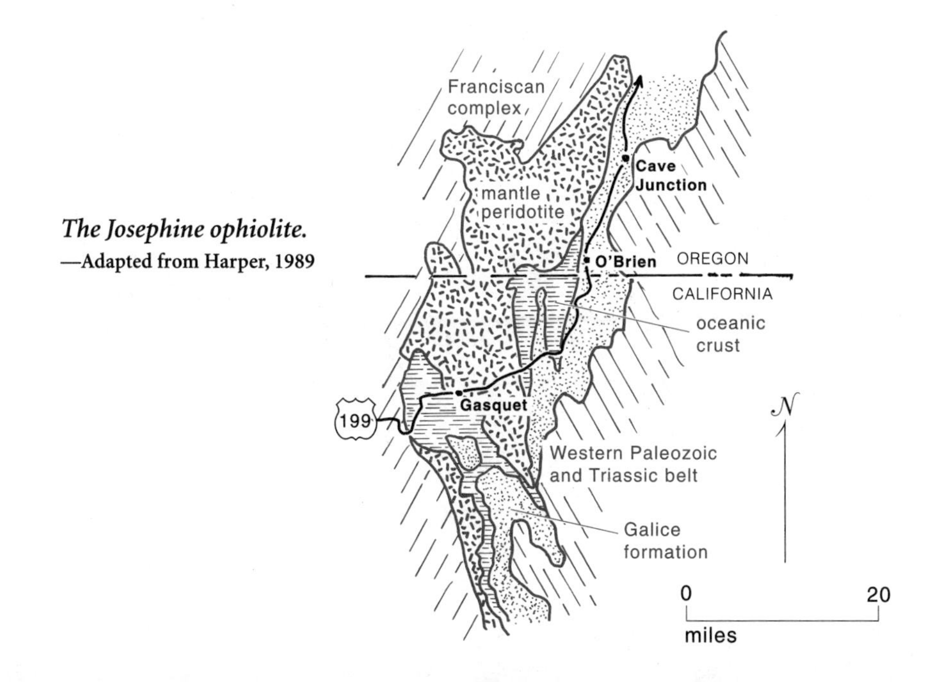

The Josephine ophiolite.
—Adapted from Harper, 1989

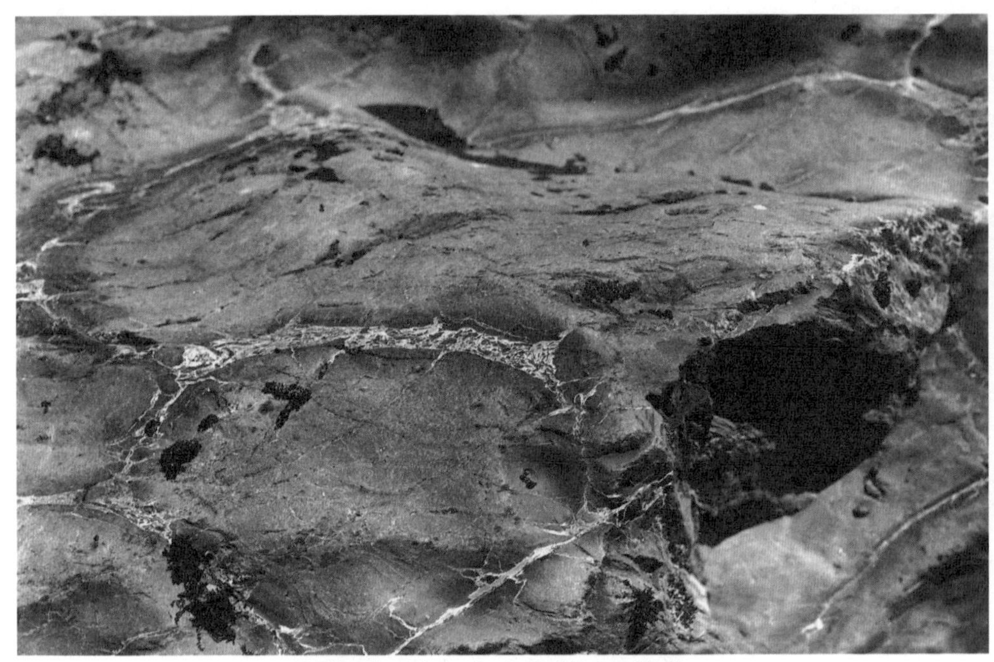

Pillow basalt in the Josephine ophiolite.
Pale yellowish matrix of clay outlines pillows.

Part of a basalt dike in the Josephine ophiolite.

oceanic crust and the sediments laid down on it. Many geologists think these rocks include the crushed remains of a volcanic chain that grew parallel to the oceanic trench during Jurassic time. The whole mess was crumpled into folds as it jammed against the advancing western edge of North America before Jurassic time ended. The rocks are only slightly metamorphosed, so they are still perfectly recognizable.

The centerpiece of the Western Jurassic belt is the Josephine ophiolite, one of the largest and most spectacular slabs of old oceanic crust in North America. It makes up much of the western part of the Klamath block and extends far into Oregon. The Josephine ophiolite contains a complete section through the oceanic crust, all nicely preserved. It is a geologic classic.

Franciscan Complex

The Franciscan rocks of the Coast Range line the extreme western edge of the Klamath block. They make a fringe only about 5 miles wide at Crescent City, its narrowest point. From a geologic point of view, they are the northern continuation of the Coast Range, even though they do not make a separate range of mountains.

The Coast Range ophiolite and the Great Valley sequence abut the south end of the Klamath block. They do not continue north along the eastern flank of this long northern prong of Franciscan rocks. This part of the Franciscan complex was jammed against and dragged under the western edge of the Klamath block along the South Fork Mountain thrust fault. The part of the Franciscan complex south of the Klamath block was jammed against and dragged under the Coast Range ophiolite and the Great Valley sequence.

The Franciscan rocks east of the Klamath block are mainly sandstones and mudstones in dull shades of medium and dark gray and tan. They also include ribbon cherts, ophiolites, and masses of greenish serpentinite.

KLAMATH BLOCK MOVES WEST

The match between the belts in the Klamath block and the terranes of the northern Sierra Nevada is very close—compelling evidence that they formed as one. Few geologists doubt that the Klamath block is the detached northern end of the Sierra Nevada, moved west about 60 miles. They differ only in their ideas about when, why, and how that movement happened.

We find it easiest to imagine the Klamath block moving west at the same time the line of collision between North America and the floor of the Pacific Ocean jumped from the Sierran to the Franciscan trench. We think that probably happened sometime during late Jurassic time, before the Franciscan complex was added to the western fringe of the Klamath block. The separation was complete by early Cretaceous time because Great Valley sediments of that age lap onto the southern flank of the Klamath block. It is not clear why and how the Klamath block broke off the northern end of the Sierra Nevada and moved 60 miles west.

MINING

Prospectors found placer gold in the streams of the Klamath block in 1848, the same year such deposits were found in the northern Sierra Nevada. As in the northern Sierra Nevada, the early prospectors promptly set out in search of the bedrock mother lode but soon found that most of the gold in the modern streams came from fossil placers in ancient streams. They worked the placers in hydraulic mines without causing nearly as much environmental devastation as did those in the northern Sierra Nevada. That was because most of the hydraulic mines in the Klamath block worked along streams that drained to the Pacific Ocean, not to the rich farmland in the Sacramento Valley. The Sawyer Decision that ended hydraulic mining in the northern Sierra Nevada did not apply to the northern mountains.

The early prospectors also found gold in the bedrock, the mother lode, soon after they found it in the streams. Mines were producing from the bedrock by 1852. Altogether the various mines in the Klamath block produced about 20 percent of the four million or more ounces of gold that came out of the mountains of California.

Metals from Peridotite

Any region that contains large areas of peridotite from the upper mantle also contains chromite. The problem is not to find some chromite but to find it in a minable quantity. The Klamath block produced chromite almost entirely under government subsidies, much of it during the two world wars when foreign sources were in jeopardy.

Black placers of heavy minerals in the sand bed of any stream that drains peridotite include grains of chromite. Ordinary black sand owes its color mainly to magnetite. Drag a magnet through it, and it will come out covered with a black fuzz of magnetite. Chromite is not magnetic so it will stay behind in the sand. Waves also concentrate the heavy minerals in beach sands into placers. Many of the beaches along the Klamath block contain placers with large concentrations of chromite. According to some reports, the beach at Crescent City contains as much as 7 percent, or did at one time. The main trouble with mining beach placers is their tendency to relocate or disappear during heavy storm

Peridotite in the Josephine ophiolite.

Chromite tends to linger in the residual soil that forms on weathering peridotite and may reach minable concentration and quantity. The soil on the top of Gasquet Mountain contains as much as 2 percent chromite weathered from the Josephine ophiolite. The same soil also contains lesser amounts of nickel and cobalt.

Platinum and its closely related elements are as inevitable in peridotite as chromite but not nearly as abundant, or as easy to find. Gold washed from stream and beach placers in the Klamath block commonly contains a few tiny flakes of platinum.

Copper and Zinc

The West Shasta Mining District is in the Eastern Klamath belt near Shasta Lake. It produced some 340,000 tons of copper and 40,000 tons of zinc from massive sulfide deposits within the Balaklala rhyolite. The rhyolite erupted during Devonian time, presumably as part of the Antler event.

As the name implies, massive sulfide deposits consist mainly of metallic sulfide minerals. They are the most spectacular of all ore bodies—glittering masses of metallic minerals, exactly the sort of thing children imagine when, with stars in their eyes, they think of valuable ores. In this case, the sulfides are mainly pyrite, an extremely common and nearly worthless iron sulfide mineral. The copper came from chalcopyrite, another sulfide mineral with a rich golden color. The zinc came from sphalerite, a sulfide mineral that does not look at all metallic.

Rhyolite does not ordinarily contain many sulfide minerals, so those glittering masses in the Balaklala rhyolite must have replaced the original rock. The Copley greenstone, a nearby body of altered basalt, is the likely source of the metals. Circulating hot water probably dissolved the sulfide minerals from the greenstone, meanwhile altering it to its greenish color, and deposited them in the rhyolite.

KLAMATH MOUNTAINS ROADGUIDES

Neither the rugged landscape nor the scanty population of the Klamath block inspire much road building. Deep soils and lush greenery make good views of the rocks hard to find, except in the roadcuts, which help immensely. Treasure them.

Interstate 5

Redding—Dunsmuir

52 miles

Interstate 5 follows the Sacramento River in its winding course through the eastern part of the Klamath block between Redding and Dunsmuir. Most of the rocks along this route began their careers as sedimentary formations deposited during late Paleozoic time, or as volcanic rocks erupted during that same interval. They belong to the Western Paleozoic and Triassic belt—if they were in the Sierra Nevada, they would be in the Calaveras complex. Bedrock between Castella and Dunsmuir is mostly black peridotite and serpentinite in shades of dark green—old oceanic crust of the Trinity ophiolite. The bedrock is also part of the Western Paleozoic and Triassic belt.

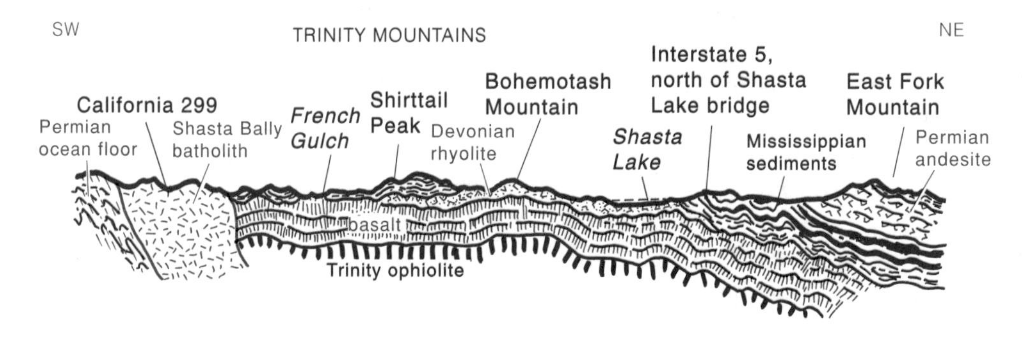

Geologic section across the Klamath Mountains from California 299 near Weaverville to Shasta Lake. —Adapted from Irwin and Dennis, 1979

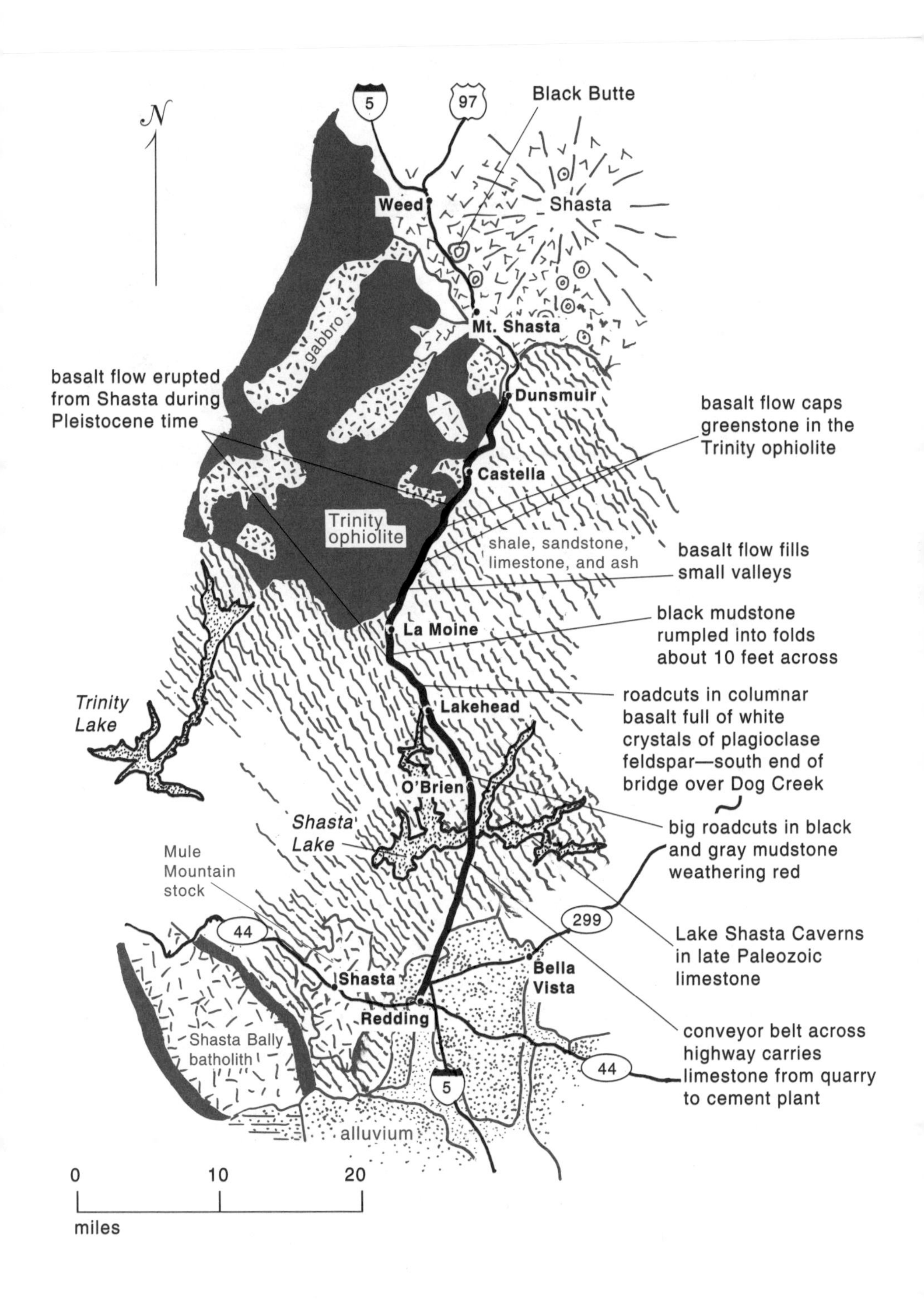

Rocks along Interstate 5 between Redding and Dunsmuir.

Castle Crags looking west from Interstate 5.

Lassen and the Ruins of Tehama

Look directly east from the highway where it passes through the higher parts of Redding for an excellent skyline profile of Lassen Peak and the wreckage of Tehama, a large volcano that nearly destroyed itself. Lassen, which is basically a rhyolite dome, looks a bit like a giant gumdrop. The jagged remains of Tehama almost embrace it from north and south. People in Redding enjoyed wonderful views of Lassen in eruption in 1915.

Castle Crags

Watch for Castle Crags just west of the highway, a few miles south of Dunsmuir. Trees grow among picturesque crags, spires, pillars, and boulders of pale gray granite. It is easy to imagine them as the wreckage of some mythical castle.

Castle Crags is a spectacular example of a granite mountain that lost its soil cover to erosion. The rock weathered most rapidly along fractures, while the solid granite between fractures remained as rounded masses embedded in the soil. Then something, perhaps a fire, destroyed the plant cover, depriving the soil of its protective umbrella of leaves. Rain splash and surface runoff erosion stripped the soil, leaving the fresh rock between fractures as exposed boulders and pillars of granite, core stones. Many granite mountains would look like Castle Crags if they were to lose their cover of soil.

Use a magnifier to look closely in the granite for tiny angular cavities lined with perfect little crystals. They are gas cavities that formed as the rock crystallized, miarolitic cavities if you like technical terms. They indicate that the granite crystallized at low pressure, probably within a few thousand feet of the surface, presumably in the roots of a volcano.

The Castle Crags granite is enclosed within a large mass of serpentinite. It is part of the Trinity ophiolite complex, oceanic crust that formed during Ordovician time, some 450 million years ago. Its age places it in the Eastern Klamath belt.

It seems incongruous to see a mass of granite intruding an ophiolite complex. Ophiolites are old ocean floor and contain nothing with the right composition to melt into granite magma. The Castle Crags granite magma must have melted somewhere below the ophiolite, within a mass of rocks with a generally granitic composition. The schist in the Central Metamorphic belt is the most obvious candidate for the honor. The Castle Crags granite provides another reason for believing that the Central Metamorphic belt was dragged under the Eastern Klamath belt.

Trinity Ophiolite

The Trinity ophiolite is a slab of oceanic crust that was on the ocean floor during Ordovician time. Sedimentary formations deposited on it are now the metamorphic rocks west of the road between Weed and Yreka, well north of Dunsmuir.

Most of the rocks in the part of the Trinity ophiolite near the highway are peridotite, along with generous masses of gabbro. The gabbro is the lowermost part of the old oceanic crust; it is a dark rock that would be basalt if it were not made of such large crystals. The peridotite is a slice of the upper mantle that was below the oceanic crust. Peridotite is normally black, but much of this is altered to serpentinite, in various shades of dark green. Serpentinite would have the same composition as peridotite if it did not contain some water.

Laterite

The conspicuously reddish soil along much of the way between Redding and Dunsmuir is laterite, the typical soil of warm and wet regions. Its main constituents are kaolinite clay and aluminum oxide, which are white, and iron oxide, which stains it all red. Millions of years of warm rain washed everything soluble out of these soils, leaving them with a scanty supply of natural fertilizer nutrients. Many kinds of trees flourish in laterite soils, but most food crops need more fertilizer nutrients to grow well and yield nutritious produce.

U.S. 199
Crescent City—Oregon
41 miles

This short stretch of highway in the northwestern corner of California winds through mountainous terrain, along the Middle Fork of the Smith River. The rocks of the Franciscan complex are exposed in a band about 5 miles wide along the coast. They almost certainly continue east beneath the Western Jurassic belt.

The Josephine ophiolite occupies most of the geologic map. It is part of the Western Jurassic belt, which lies on the South Fork Mountain fault. Its rocks are mainly peridotite, partly altered to serpentinite. The several masses of gabbro crystallized in the base of the oceanic crust. The slates of the Galice formation were deposited on the Josephine ophiolite when it was still part of the ocean floor. They also belong to the Western Jurassic belt.

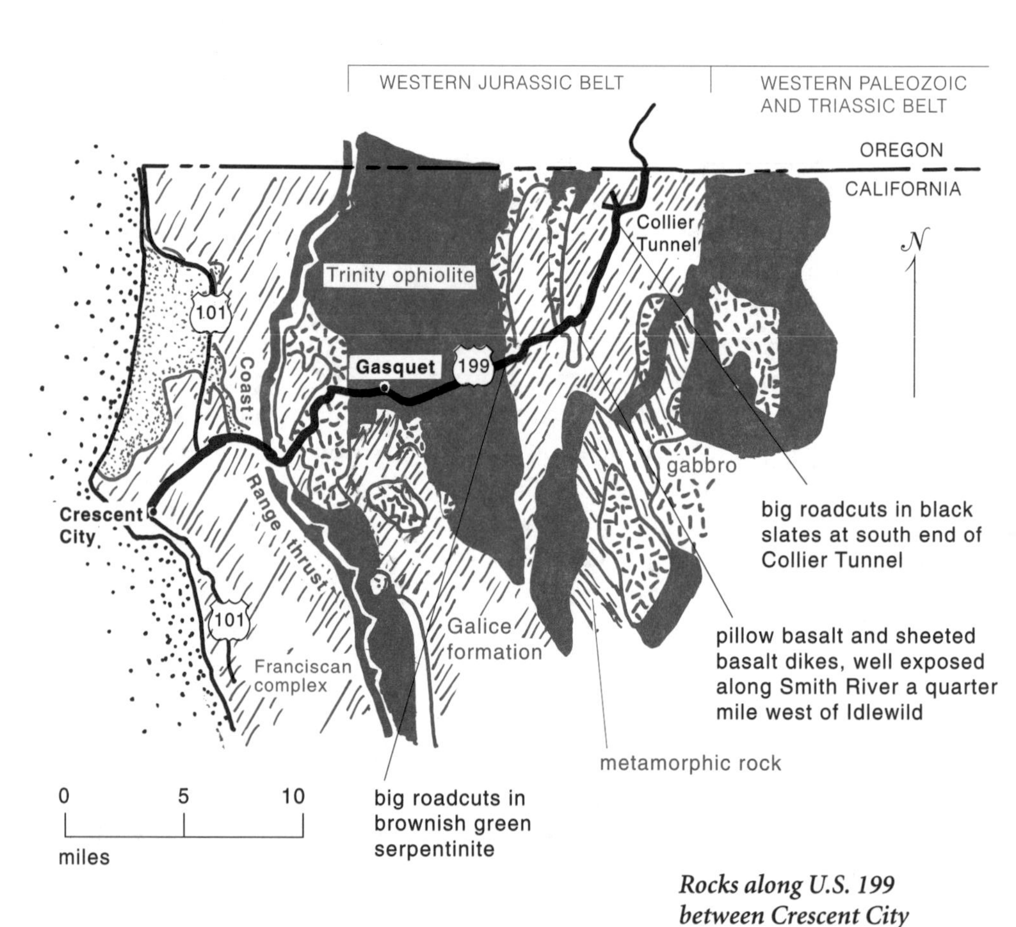

Rocks along U.S. 199 between Crescent City and the Oregon line.

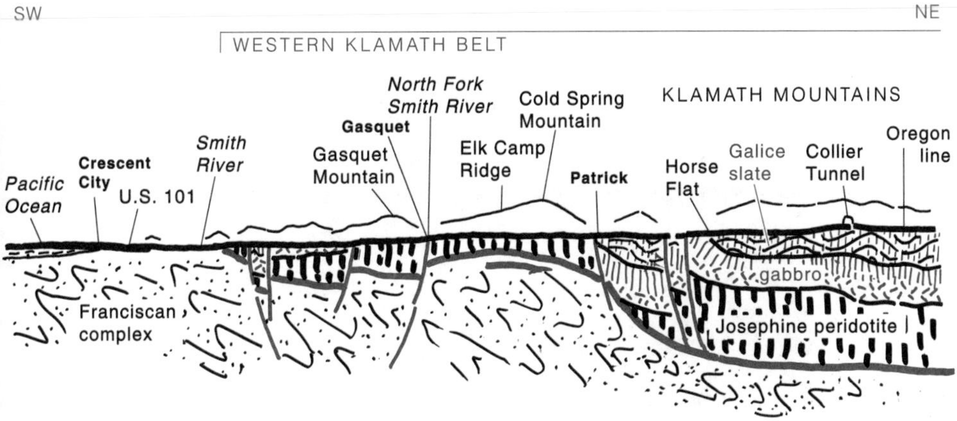

This section, between Crescent City and the state border, shows the Josephine ophiolite lying on the South Fork Mountain thrust fault, with rocks of the Franciscan complex beneath.

Josephine Ophiolite

The Josephine ophiolite is the star geologic attraction of the western part of the Klamath block. It is a big slab of old ocean floor—basalt and gabbro—over peridotite of the upper mantle. This series of rock, the ophiolite, lies on the South Fork Mountain thrust fault, above the younger Franciscan rocks. The highway crosses the old oceanic crust several times in the 20 to 25 miles of its traverse across the folded slab of ophiolite.

The peridotites are a slice of the upper mantle and the lower part of the ophiolite. They are greenish black rocks that weather to brown or orange. Some of the peridotite is altered to serpentinite, which is mainly responsible for its slightly greenish color. But the amount of serpentinite is less than in most ophiolites. The peridotite is rich in magnesium but so poor in potassium, phosphate, and trace element nutrients that the soils weathered on it support only a sparse plant cover, despite the abundant rainfall.

The best exposures of the old oceanic basalt of the Josephine ophiolite are in smooth outcrops along the Smith River, right below the road. Look for dark gray pillows about 2 feet across with fragmental debris between them. Gabbro, the lowest part of the old oceanic crust, is exposed along the road in the 4 miles northeast of the Smith River. Watch for dikes of black basalt cutting through the gabbro and some of the basalt. They were the feeder channels for the basalt flows, the geologic plumbing of the oceanic crust.

Serpentinized peridotite 1.5 miles north of Gasquet.

Gabbro and basalt of the old ocean floor exposed in the gorge of the Middle Fork of the Smith River.

Black slates of the Galice formation at the south end of Collier Tunnel.

The highway crosses mantle peridotite and patches of pillow basalt in the 5 or 6 miles along the Middle Fork of the Smith River northeast of Gasquet.

Rocks along the stretch of 14 miles between the area 1 mile east of Patrick Creek and the Oregon border are black slates and basalt altered to greenstone, the Galice formation. The basalt was the top of the old oceanic crust, and the slates began their careers as sediments deposited on the basalt. Both are especially well exposed at the south end of Collier Tunnel.

California 299

Arcata—Redding

144 miles

The route of California 299 between Arcata and Redding winds through the northern Coast Range and the rugged mountains of the Klamath block. This is a wild and thinly peopled part of California. Its rugged terrain, dense forest, and deep soils greatly impede geologic fieldwork. The lovely roadcuts in the eastern part of this route provide a rare treat in their good exposures of the bedrock.

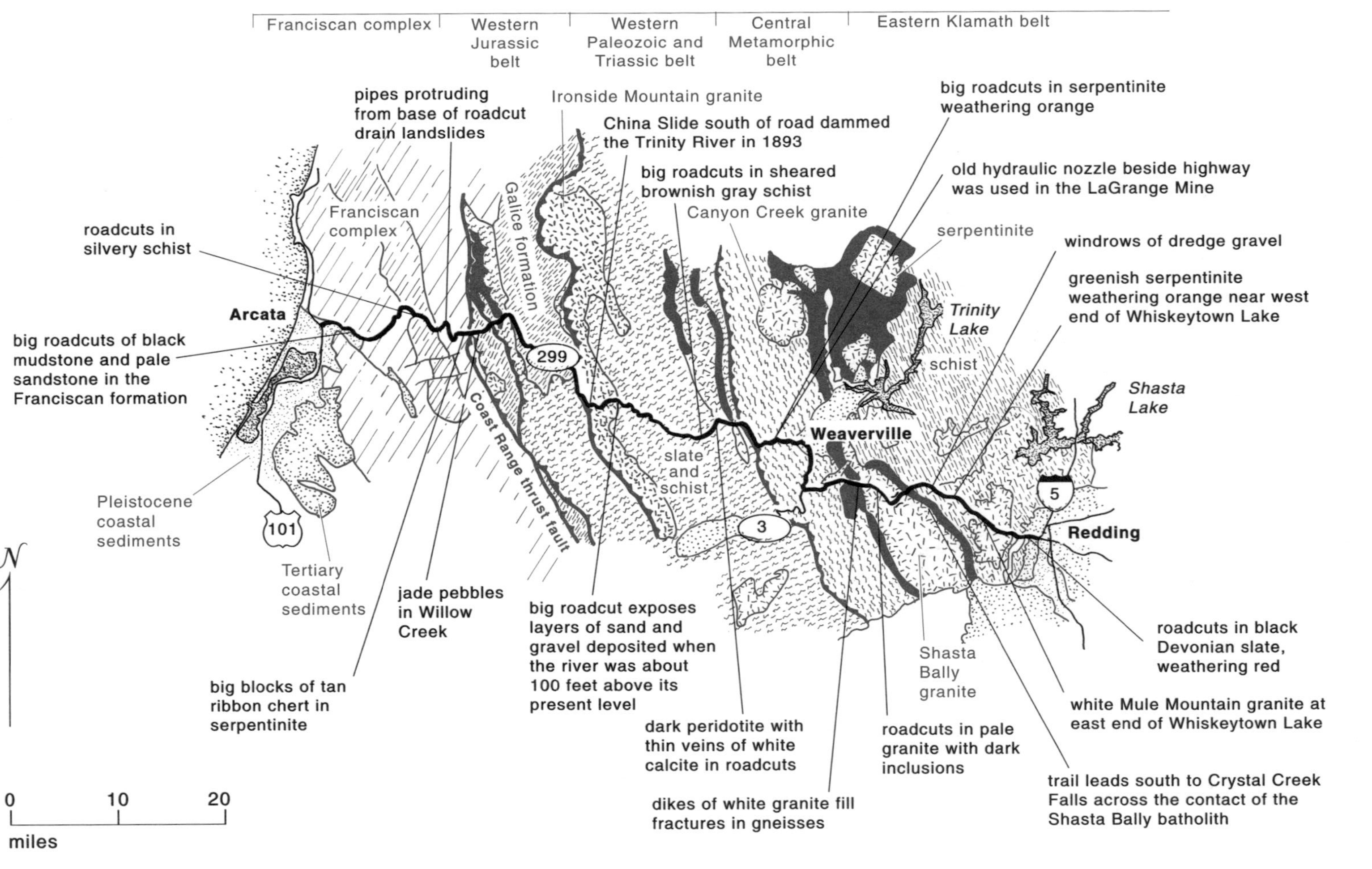

Rocks along California 299 between Arcata and Redding.

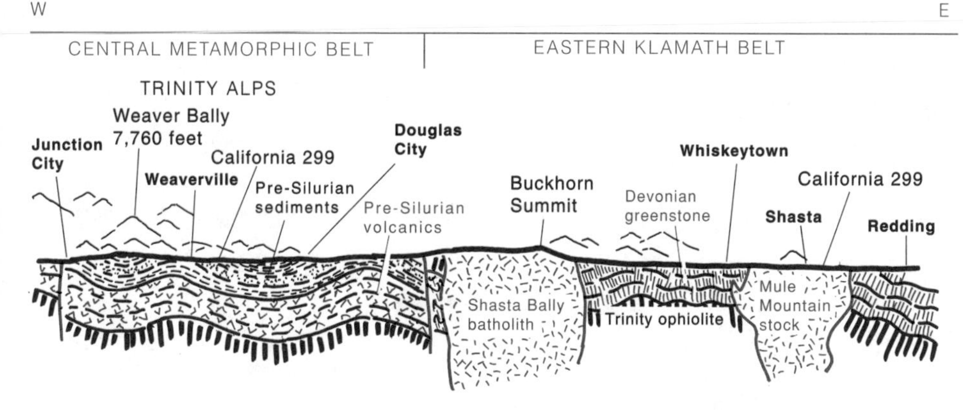

Geologic section approximately along the line of California 299 between Junction City and Redding.

Franciscan Rocks

The highway crosses the South Fork Mountain thrust fault on the slope west of Berry Summit. All the rocks west of it belong to the Franciscan complex, those to the east to the Western Jurassic belt. Between the pass and Arcata, the highway winds through hills eroded in Franciscan rocks, monotonous mudstones and muddy sandstones like those in the Coast Range. Some of them show the effects of slight metamorphism in their slabby fracture pattern and broken surfaces that glisten with minute flakes of mica.

Western Jurassic Belt

The highway passes through two small belts of mantle peridotite along the slope between Berry Summit and Willow Creek. They are part of the Josephine ophiolite. Watch for the formidable outcrops of thoroughly sheared peridotite of the upper mantle partially altered to serpentinite. It was part of the bedrock floor of the Pacific Ocean until it somehow got scrambled into the Klamath block while the ocean floor was sliding into the Sierran trench, about 200 to 150 million years ago. Brownish gray muddy sandstones in the same area are the sediments dumped on the same ocean floor.

California 299 follows the canyon of the Trinity River between Willow Creek and Junction City. Mudstones and muddy sandstones exposed in this area were deposited earlier, and scraped off the seafloor sooner, than those farther west, but they look much the same.

The enormous roadcuts between the area 1 mile west of Burnt Ranch expose massive greenstone, metamorphosed oceanic basalt. A rusty wash

A huge cut in serpentinite mélange, with blocks of tan chert from the old ocean floor. About 30 miles east of Arcata.

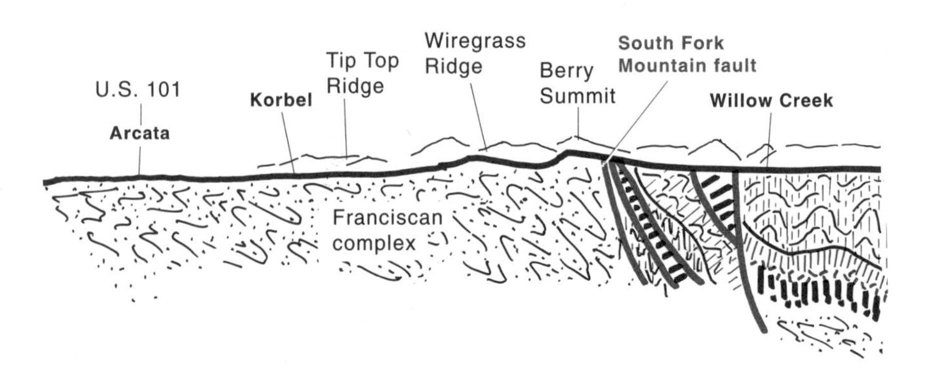

of iron oxide stains some of the surfaces. It formed as the black pyroxene in the basalt weathered.

The road between Burnt Ranch and Del Loma crosses the Ironside Mountain batholith. It is a large body of granite that shouldered its way into the sedimentary rocks as a mass of molten magma about 165 million years ago, in Jurassic time.

Stream gravel on hill slope debris just east of Big Bar. The Trinity River laid the gravel down before it eroded its bed down to the present level.

E
WESTERN PALEOZOIC AND TRIASSIC BELT
CENTRAL METAMORPHIC BELT
LAMATH MOUNTAINS
Ironside Mountain 5,255 feet
TRINITY ALPS
Jurassic Galice slate
Del Loma
Big Bar
Junction City
Ironside Mountain batholith
Mesozoic-Paleozoic schist

Geologic section approximately along the line of California 299, across four terranes.

Big slabs of dark gray schist 1 mile east of Big Flat.

Heaps of dredge spoils just west of Junction City date from the 1890s. It is hard to say much in favor of dredge spoils except that they provide a good source of road metal and construction aggregate, neither of which is in great demand in these thinly populated mountains.

East of its junction with California 3, California 299 climbs out of the canyon to cross hills eroded into schist and gneiss. They began their careers as muddy sedimentary rocks, were heated, and then recrystallized into metamorphic rocks that hardly resemble the original sediments. Look for spangly crystals of black or white mica that sparkle in the sun and for little needles of glossy black hornblende. In places, the gneisses got hot enough to melt into granite magma that injected fractures to make dikes. The large area of granite about 1 mile farther east may well have crystallized from a large volume of the same magma.

Mud, mudstone, schist, and granite all have the same general chemical composition and consist of essentially the same minerals, but in varying proportion. It is much easier to tell them apart by looking at the rocks than by analyzing them chemically. The relationship between mudstone and granite is like that between a cake mix in the box and a cake on the table. A bit of cooking does wonders.

Granite

The eastern part of the route crosses rocks of the Western Paleozoic and Triassic belt, the Central Metamorphic belt, and into the Eastern

Klamath belt. The granite that appears along several miles on either side of Buckhorn Summit is in the Shasta Bally batholith. The ghost town of Shasta is on granite of the Mule Mountain pluton, which extends about 2 miles on either side of the town. Both of these granites are weathered almost beyond recognition in roadcuts along the higher parts of the route. Fresh rock appears only deep in the valley west of Buckhorn Summit. Evidently, the weathering happened before the streams eroded their deep valleys into fresher rock beneath.

Three or four miles east of Buckhorn Summit, the highway crosses the zone of streaky metamorphic rocks that surrounds the Shasta Bally batholith. They were deposited as sediments in Devonian or Mississippian time, between 400 and 350 million years ago, so they belong to the Western Paleozoic and Triassic belt. If they were in the Sierra Nevada, they would belong in the Calaveras complex. They recrystallized into metamorphic rocks during Mesozoic time, baked in the heat of the molten granite magma of the Shasta Bally batholith. Metamorphic rocks around the Mule Mountain granite have a similar story.

Gold

The French Gulch and Backbone mining districts are in the hills about 10 miles west of Redding, around the town of French Gulch. Prospectors found placer gold in 1849, immediately started looking for the mother lode, and found it in gold quartz veins the next year. Most of the veins are in slate, a few in metamorphosed volcanic rocks. As so often happens, granite lurks nearby, in this case the Shasta Bally batholith.

Bedrock mining in quartz veins started about 1850, continued at a brisk pace during the decades between 1880 and 1914, revived during the Great Depression of the 1930s, then terminally declined. Estimates place total production at something more than $30 million, little enough to show for a substantial investment and all those decades of hard work. As usual, the estimates of total production do not reveal whether the district ever showed much of a profit.

Redding is in the northern end of the Sacramento Valley, just south of the torn southern boundary of the Klamath block. Rocks in the northernmost Sierra Nevada, about 60 miles east of Redding, probably match those in the mountains directly north of town.

Miners washed quite a lot of placer gold out of the stream gravels during the gold rush and more during the Great Depression, when dredges reworked some of the gravel. A number of small underground mines have produced from gold quartz veins in a belt that extends southwest from Redding to Centerville.

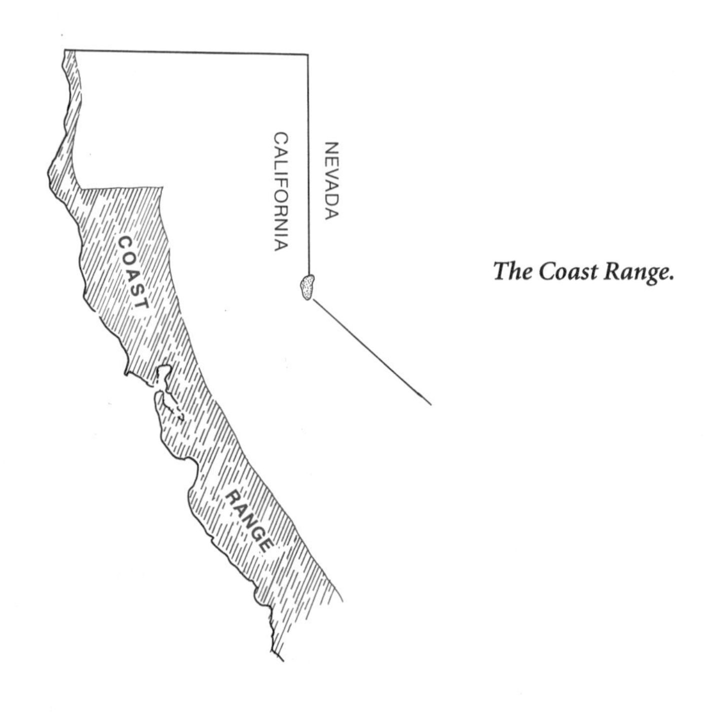

The Coast Range.

Crumpled ribbon chert southwest of New Almaden.

4

Coast Range

The Coast Range, sometimes called the Coast Ranges, consists of many small mountain ranges and ridges that run parallel with the coast. The ridges are steep and wooded; the valleys broad and flat. The larger valleys hold towns and farms, the hills are thinly populated. Roads are few, especially those that cross the northwest grain of the landscape, and many of them carry few cars. The Coast Range offers zones of amazing tranquillity within a busy state.

In broadest outline, the Coast Range consists of two major groups of rocks: the Franciscan complex and the Great Valley sequence. A few big patches of coastal and bay sediment cover parts of the Franciscan rocks, and lesser deposits of sediment fill the valley floors. Some of those deposits were laid down in seawater, some on dry land. The Sonoma and Clear Lake volcanic fields add variety to the geologic mix.

Most of the long ridges and valleys of the Coast Range are fault slices that moved horizontally in the San Andreas system of faults, the side to the west invariably moving north. Many of the fault slices are still moving north, to the rumbling accompaniment of an occasional earthquake.

FRANCISCAN COMPLEX

The whole point of geology is to make sense of the rocks. Geologists normally begin by making geologic maps, carefully plotting the outcrops to see what patterns emerge. In most areas, a coherent pattern does emerge, and geologists can understand the main events that brought the rocks to their present location and condition. That approach does not work well in the Coast Range, where too many of the rocks do not make sense in the traditional way. If all rocks resembled the Franciscan complex, no science of bedrock geology could have emerged. Most of the early geologists who worked with Franciscan rocks concluded that their creator could have used some advice in the art of assembling a proper package of rocks.

The Franciscan complex is one of the world's grand messes. It is a wild assortment of sedimentary rocks, deposited in seawater at many depths and in widely separated parts of the ocean, along with generous slices of the basalt ocean floor. These are rocks at their most frustrating—complicated and poorly exposed. The rocks in one outcrop too often seem unrelated to those in the next.

During the late 1960s, geologists finally accepted that large parts of the Franciscan complex really are almost hopelessly scrambled. They agreed to call those chaotic jumbles mélanges. Most mélanges are chunks of solid rock embedded in a matrix of soft clay, which acquires a scaly look. Others have a matrix of serpentinite, which breaks into a sheared rubble of chunks with polished surfaces.

Recognition of mélanges was, in a way, an admission of defeat, but it was also a long step toward understanding what happens in an oceanic trench. Now that they are high, dry, and deeply eroded, we can see exposed in the Franciscan rocks the deep interior of an oceanic trench and imagine what happens there.

Oceanic sediments scrape off the basalt of the bedrock ocean floor as it sinks out from under them. They join the chaotic mix already stuffed into the trench, making the accumulation grow steadily wider from back to front—from east to west in California trenches. Meanwhile, most of the

Fossils of oceanic animals in Franciscan muddy sandstone on Mount Diablo.

bedrock ocean floor continues down into the earth's mantle, but occasional slabs shear off into the accumulated sediments in the trench.

The most abundant Franciscan rock type is muddy sandstone—graywacke, to use an old European miner's term. These are typically dark rocks, somber and nondescript. A closer look in a good exposure reveals layers that range from a few inches to a few feet thick—if the rock is not so pervasively sheared that no trace of the original bedding survives. Many of the layers contain coarse sand and pebbles near the base and grade up into clay at the top. Those graded beds are typical of muds deposited on the ocean floor within a few hundred miles of continental margins.

Many geologists call those oceanic sedimentary rocks with graded beds turbidites because turbidity flows deposit them. Turbidity flows are great clouds of muddy water that occasionally pour down the continental slope and billow over the ocean floor, dropping sediment like winter snow. The heavier objects, small pebbles and coarse sand, settle to the bottom first, then the finer silt and clay follow, hours or even days later.

Oozes are the main sediment that rains on the remote parts of the ocean floor, beyond the long reach of the clouds of muddy water that lay down the turbidites. They consist of the shelly remains of small animals and of microscopic algae that drift in the surface waters. Most of those shells consist of calcite, some of silica.

At moderate depths, the calcite shells dominate to become calcareous ooze, white or pale gray mud that looks and feels like cheap toothpaste. It hardens into limestone, which is rare among Franciscan rocks. Too bad. Limestone is an important industrial commodity, and California could use more, especially in the Bay Area.

At truly abyssal depths, seawater is cold and under great pressure. Those conditions enable it to absorb enough carbon dioxide to become slightly acidic, just corrosive enough to dissolve the minute calcite shells. Subtract the calcite sediment and all that remains is silica, the dainty shells of various microscopic algae and animals. That delicate stuff accumulates ever so slowly to form deposits of siliceous ooze, which eventually solidifies into colorful ribbon cherts. Chert is an extremely fine-grained sedimentary rock, a felted mass of microscopic quartz crystals that forms in many ways and in many situations. It is most familiar to many people as the common material of Indian arrowheads.

Occasional outcrops of brightly colored ribbon chert greatly relieve the Franciscan monotony of dark sandstones. Chert comes in piles of layers 1 or 2 inches thick, in various shades of red, yellow, green, and white, generally in that order of abundance. Ribbon cherts commonly buckle into folds with sharply angled crests and troughs. Such folds are typical of

Sharp folds in ribbon chert near the north end of the Golden Gate Bridge.

formations in which thin layers of rigid rock alternate with even thinner layers of weak rock. Look closely at a clean sea cliff or roadcut in ribbon chert to see the thin layers of weak clay between the stiff layers of hard chert.

Ribbon chert is the most convincing single item of evidence that many of the Franciscan rocks began their careers as sediments laid down on the remote reaches of the deep ocean floor. Those outcrops of ribbon chert tell of enormous expanses of ocean floor, perhaps thousands of miles, that sank through the oceanic trench into the earth's mantle.

A small proportion, generally just a few percent, of the oceanic crust shears off the descending slab and into the mass of sediments accumulating in the trench to become ophiolites. The Coast Range is full of them. Watch for their darkly greenish or black pillow basalts and occasional outcrops of dark peridotite, more or less altered to greenish serpentinites.

Watch as you drive through the Coast Range for those strange patches of ground where the soil is orange, little grass grows, and digger pines filter the light through their long needles as though they were puffs of pale blue smoke. Those poor soils are on serpentinite, the California state rock and one of earth's most peculiar rocks. Large roadcuts reveal a dark green rock that tends to break into blocks ranging in size from citrus fruit to melons. Their slickly polished surfaces come in dull shades of green and yellowish green.

Occasional masses of serpentinite evade their naturally ordained fate in the earth's mantle by squeezing up through the scramble of oceanic sediments accumulating in the trench, slithering along any easy passageways they can find, most commonly along faults. The small chunks within the intruding masses of slippery serpentinite slip past each other as though they were so many wet watermelon seeds in a bag. That is how they acquire their polished surfaces.

Blueschist and Eclogite

Watch for dark blocks of blueschist or eclogite scattered across the ground above the serpentinite. Most range in size from walnuts to large watermelons, but some are larger than a bus. They stand out in the landscape because they better resist erosion than the serpentinite that contains them.

Blueschist blocks are almost black but with a bluish cast imparted by needles of an intensely blue amphibole mineral called glaucophane, which is rich in sodium. Eclogite blocks contain scattered crystals of red garnet set in a matrix of greenish pyroxene called jadeite, which is also rich in sodium. Both rocks feel notably dense in the hand. And it is surprisingly hard to break them with a hammer, the quality geologists call toughness. Eclogite is among the toughest of all rocks.

Mineralogists found long ago that they could grow the peculiar minerals of blueschist and eclogite synthetically, if they cooked the ingredients under conditions of very high pressure and relatively low temperature. Such high pressures prevail deep beneath the surface, but the low temperatures exist at much shallower depth. At the time, that combination seemed

A roadcut in serpentinite. The shiny surfaces are typical.

Tilted Great Valley sequence at the eastern edge of the Coast Range.

impossibly contradictory because both pressure and temperature normally increase with depth. How could a rock be deep enough to recrystallize into blueschist or eclogite without also being too hot?

Like a great many of the other old mysteries of geology, that one unfolded with the coming of plate tectonic theory during the late 1960s. When geologists finally realized that the ocean floor sinks through trenches at an average rate of several inches per year, they could easily imagine that the sinking plate might descend to depths where blueschist and eclogite can form before it could absorb enough heat to prevent their forming. Then, rising masses of slippery serpentinite regurgitating from the depths picked up chunks of blueschist and eclogite and carried them into the upper part of the trench filling.

GREAT VALLEY SEQUENCE

The Coast Range ophiolite, a great slab of ocean floor tilted steeply down to the east, lies on the Franciscan complex all along the eastern side of the Coast Range. Serpentinite generally lies at its base. The contact between the Coast Range ophiolite and the Franciscan complex is nowhere simple; invariably it is sheared.

Like almost any slab of ocean floor, old or new, the Coast Range ophiolite has a cover of oceanic sediments—in this case, the Great Valley sequence.

Pillow basalt from the old ocean floor, now in the Franciscan complex in the eastern foothills of the Coast Range west of Willows.

These sedimentary rocks are quite similar to those in the Franciscan complex and are the same age, late Jurassic to Cretaceous. They are tilted but otherwise intact, with little deformation. Most of the Great Valley sequence consists of layers of tan sandstone sandwiched between layers of black shale. Graded layers and minor structures in the sandstones suggest that they were deposited from turbidity currents.

How did that almost undeformed slab of ocean floor come to lie above similar, but severely deformed, oceanic rocks of the same age? How could the Great Valley sequence remain intact, while the Franciscan rocks beneath the Coast Range ophiolite were so thoroughly scrambled?

It is now clear that the Franciscan complex consists of oceanic sediments and slices of bedrock ocean floor that were stuffed into the Franciscan trench. The Coast Range ophiolite is the oceanic crust that was on the east side of the trench, not sinking. Some of the Franciscan rocks were jammed against it, some dragged under it. The base of the serpentinite that separates the Coast Range ophiolite from the Franciscan complex is the surface along which the sinking ocean floor sank, the former plate boundary.

Most of the rocks between the Coast Range ophiolite and the coast belong to the Franciscan complex. Its eastern part is mostly mudstone and dark sandstone with some basalt, along with some ribbon chert. Some of those rocks are recrystallized to schists and laced with veins of white quartz.

Its central part is an incoherent mess, a mélange of sheared shale with blocks of dark sandstone, greenstone that was once basalt, chert, limestone, and some serpentinite. The western part is mainly sandstone and mudstone, with some pebble conglomerate. It contains the youngest rocks, the least deformed because they were the last stuffed into the trench.

PLATE BOUNDARIES

Forty million or so years ago, an oceanic ridge was adding new oceanic crust to the western edge of the Farallon plate, which was moving southeast. Meanwhile, the North American plate was moving west, riding over the Farallon plate, which was sinking through the Franciscan trench. That was an inherently unstable situation.

The western edge of North America finally reached the oceanic ridge sometime between 26 and 20 million years ago. Then the trench and ridge both disappeared as the trench swallowed the last of the Farallon plate. As the trench and ridge met, the North American and Pacific plates met along a new transform plate boundary, the San Andreas fault. Meanwhile, the Coast Range rose.

The San Andreas fault gradually lengthened northward and the Coast Range rose progressively to the north. The oceanic ridge met the trench piecemeal, segment by segment. Their first encounter was along the coast of southern California, probably during Oligocene time, about 35 million years ago. Then the new transform plate boundary, the San Andreas fault, lengthened as more and more segments of oceanic ridge met the trench. That happened in what seems to have been an orderly way until the middle of Miocene time, about 16 million years ago. Then the process suddenly accelerated, and the San Andreas fault quickly reached nearly its present length.

The ocean floor sinking through a trench is all that holds the light rocks that were swept into it at such great depth. When the ocean floor stopped sinking through the Franciscan trench, the rocks that had been swept into it floated to the elevation their density dictated. They became the Coast Range. Thus, the Coast Range rose as the oceanic trench met the oceanic ridge and the San Andreas fault appeared. The sediments that had been stuffed under the western edge of the Coast Range ophiolite and the Great Valley sequence tilted both rock groups steeply down to the east as they rose. The Coast Range ophiolite and the Great Valley sequence almost certainly continue beneath the floor of the Great Valley to the Sierra Nevada.

Three plate boundaries and three plates meet at a single point at the Mendocino triple junction, off Cape Mendocino in northern California. The San Andreas fault meets the modern trench, the last remnant of the

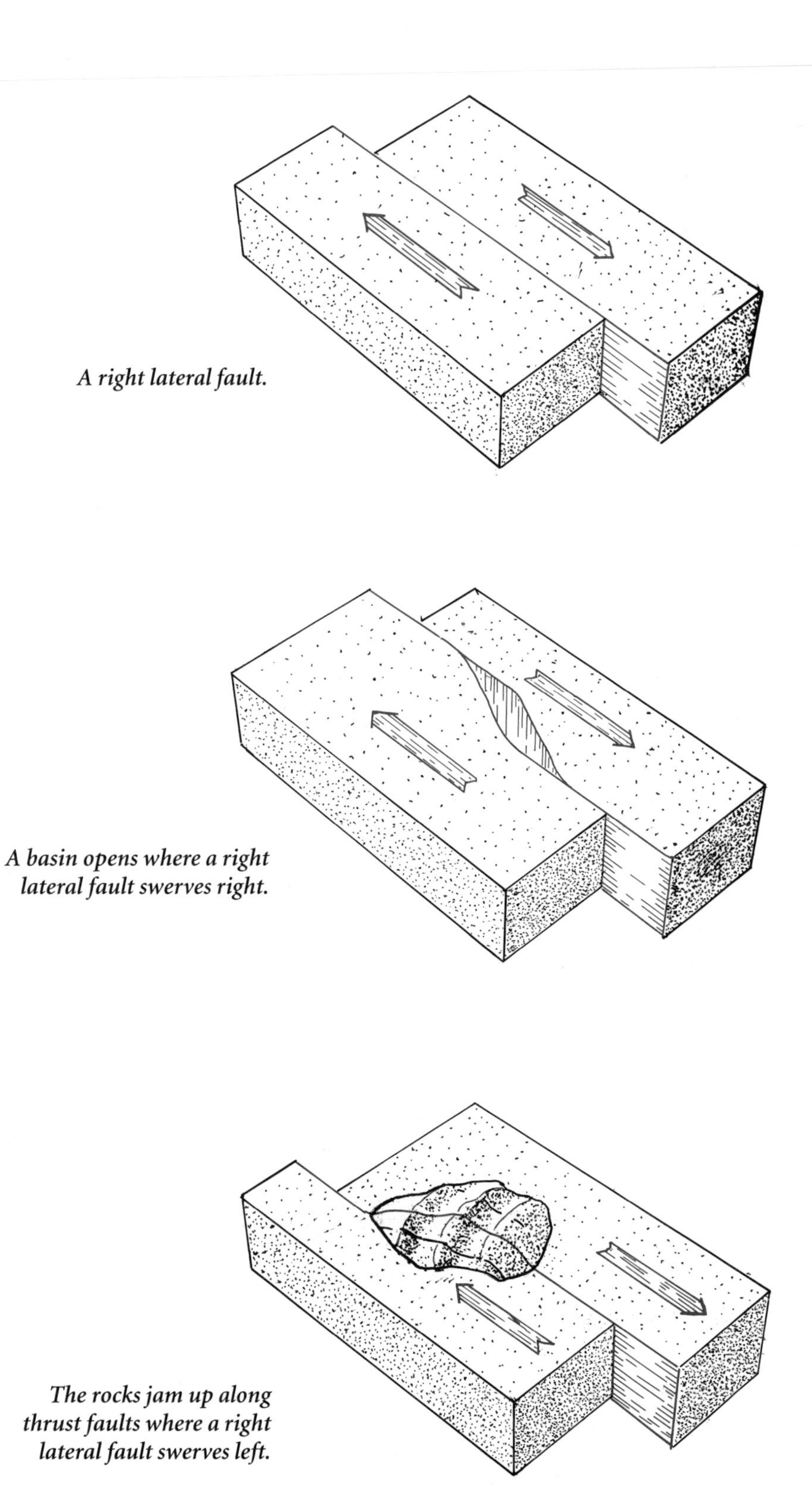

A right lateral fault.

A basin opens where a right lateral fault swerves right.

The rocks jam up along thrust faults where a right lateral fault swerves left.

Franciscan trench. The fault turns there to head straight west across the ocean floor as the Mendocino scarp. The Pacific plate meets the last remnant of the Farallon plate, which also meets the North American plate. Some believe the Mendocino triple junction has had a major influence on events that reach far inland from the West Coast. Others think not.

If present trends continue, the San Andreas fault will lengthen as the trench continues to swallow the last remnants of the Farallon plate, now off the coast of the Pacific Northwest, and the Mendocino triple junction migrates north. As the last scrap of the Farallon plate disappears, the San Andreas fault will meet the Queen Charlotte fault off the coast of British Columbia. Then the transform plate boundary will reach all the way to the Aleutian trench, south of Alaska. We can all eagerly await those events, which will probably occur about 15 million years from now. Meanwhile, the chain of volcanoes in the High Cascades will snuff out from south to north.

Although most geologists identify the San Andreas fault as the transform boundary between the North American and Pacific plates, that is not quite true. The boundary is actually a swarm of parallel faults that trend northwest. Some move as much as several inches per year, some much less rapidly, and some have stopped. Collectively, they move the Pacific plate northward about 2 to 3 inches per year. New faults appear, move for a few million years, then stop as others appear. The San Andreas fault is simply the one now most active. Geologists think of the swarm as a soft and somewhat shifting plate boundary.

Invariably, the side west of the faults in the San Andreas swarm moves north because the Pacific plate is moving north past the North American plate. If you stand on any of those faults and look either way along it, the side on your right will move toward you. Or, if you look across one of those faults, the far side will move to your right. Geologists call that a right lateral sense of movement.

Basins and Thrust Faults

Faults that move horizontally, like those in the San Andreas system, trace a remarkably straight line across the countryside. But they do curve in places, and those curves cause interesting complications. If a right lateral fault swerves to the right, continued movement will open a gap in the fault trace, a basin. Most, if not all, of the broad valleys in the northern California Coast Range are probably basins of that sort. Their floors fill with sediment as they open.

The opposite happens if a right lateral fault swerves left. Continued movement jams the rocks on opposite sides of the swerve into each other. The jammed rocks fold, and the original fault becomes a thrust fault as

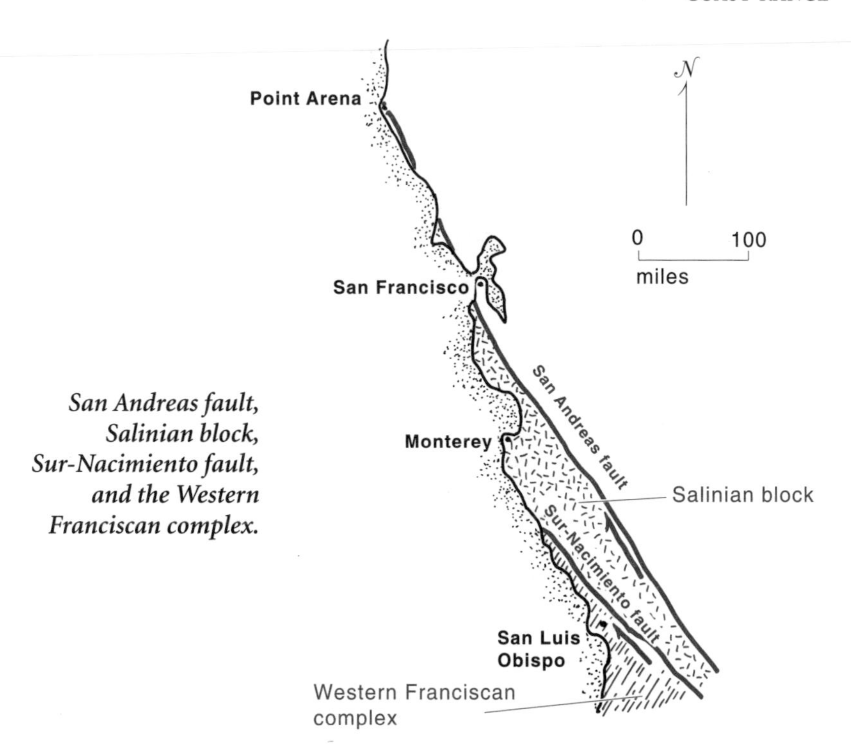

San Andreas fault, Salinian block, Sur-Nacimiento fault, and the Western Franciscan complex.

the rocks on one side ride up over those on the other. Thrust faults that developed along left swerves in faults in the San Andreas system are common in southern California, less so in central and northern California.

Sur-Nacimiento Fault

Geologists call the sliver of California that is moving northwest along the west side of the San Andreas fault the Salinian block. Along most of the coast, it consists mainly of granite, probably Sierran granite that started its career somewhere in southern California or perhaps Mexico. But the Sur-Nacimiento fault adds an unhappy complication to a situation that was already quite complicated enough.

The Sur-Nacimiento fault follows a northwest trend through the southern Coast Range. It defines the western boundary of the Salinian block. West of the Sur-Nacimiento fault lies the Western Franciscan complex. Rocks in the Western Franciscan complex look very much like the Franciscan rocks east of the San Andreas fault. The only thing that clearly distinguishes them is their position west of the Salinian block.

Most geologists agree that the Sur-Nacimiento fault moved in response to the same plate movements that now drive the San Andreas fault. Many think it may be an earlier version of the San Andreas fault. If so, it must have moved north, just as the Salinian block west of the San Andreas fault

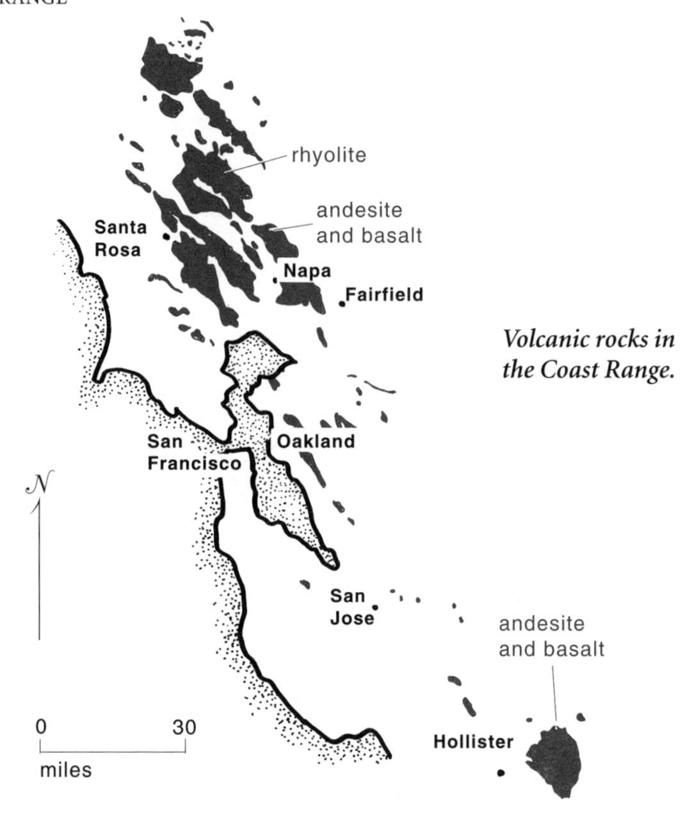

Volcanic rocks in the Coast Range.

is now moving north. Most geologists also agree that the Sur-Nacimiento fault has not moved for such a long time that it no longer poses any danger of earthquakes.

YOUNGER ROCKS

Long after the ocean floor stopped sinking through the trench, and Franciscan rocks that filled the trench rose to become the main part of the Coast Range, younger sediments accumulated in low areas. Some of those deposits were laid down in shallow seawater, in bays that flooded parts of the Coast Range. Others were laid down on dry land, in the floors of broad valleys, some of which had been bays. Most, perhaps all, of those low areas are basins that opened as faults sliced up the Coast Range and made the mess even messier.

Volcanic rocks cover the Franciscan rocks in the Sonoma and Clear Lake volcanic areas in the northern Coast Range and in the Quien Sabe volcanic area east of Hollister. These volcanic rocks consist mainly of pale andesite or rhyolite, along with much smaller amounts of basalt. Their location and age strongly suggest that the volcanoes are associated in some way with the San Andreas system of faults. Exactly how they are associated is unclear.

EARTHQUAKES

All of the Coast Range is earthquake country, from one end to the other, mostly because of the San Andreas system of faults. The most fascinating geologic question in the San Francisco Bay Area concerns the Big One. Will it happen? If so, when? And where? How hard will it strike? Yes, it will happen. No one knows when. Yes, it will be devastating.

Andrew C. Lawson was one of the commanding figures of North American geology in general, and of California geology in particular. After the great San Francisco earthquake of 1906, he led the newly formed California Earthquake Commission that had been appointed to discover why the ground shook.

He and the other members of the commission soon discovered the San Andreas fault, which had not been known. They plotted its trace on maps and found that the ground west of it had moved north during the earthquake in a sudden jump of as much as 22 feet. That pointed to the fault as the culprit. They concluded that sudden movements on faults cause earthquakes. Subsequent experience has consistently confirmed that theory.

An easy way to visualize the theory is to imagine two thick slabs of soft foam rubber sliding past each other with their cut edges touching. The cut edges are the fault. Imagine the slabs hitching along in sharp jerks as their edges catch and stick together until the rubber stretches enough to snap them free. Many faults behave in much the same way. Rocks on their opposite sides catch, bend as they accumulate strain, then snap free and move past each other as they straighten. The sudden movement offsets the rocks and releases the accumulated strain energy in an earthquake.

If the opposite sides of a fault are not stuck, they move more or less freely, releasing harmlessly small earthquakes. Streets, sidewalks, and walls built across the fault move as the offset of opposite sides continues. That kind of movement poses no danger of an earthquake. It is time to worry when the movement stops because the opposite sides of the fault have stuck. That is when the rocks begin to accumulate strain, which they will eventually release as an earthquake.

Given a few years, it is possible to watch the rocks near a stuck section of fault bend. Survey lines, roads, and fences slowly curve. Their curvature shows how far the rocks have bent and how fast they are bending. When the fault finally snaps loose, the curves suddenly become straight lines, offset along the line of the fault. The growing curvature of formerly straight roads or rows of trees that cross the fault show how far the fault will move, when it moves, and approximately how big the earthquake will be, when it happens.

Some buildings in the Mission district of San Francisco sank to about half the height of their doorways as the soft mud beneath them compacted during the 1906 earthquake.

The San Andreas fault has been stuck in the Bay Area since the late 1930s, and the rocks on either side of it have been slowly bending. Their stored energy is already enough to cause a strong earthquake, and the potential grows with every year. The long absence of earthquakes on the San Andreas fault in the Bay Area is more alarming than reassuring. Someday, it will snap loose, releasing the next great San Francisco earthquake, the Big One that everyone dreads. It will almost certainly happen, but no one can say when. The longer we wait, the bigger it will be.

Forecasting and Managing Earthquakes

The future will bring more big earthquakes to California just as surely as next winter will bring blizzards to Montana. Those are good predictions, but they are not a precise forecast of date and place. No one can now forecast when or where the next earthquake will strike, any more than anyone can forecast in July the dates of the blizzards of the following winter.

Various studies suggest that the San Andreas fault is likely to produce a devastating earthquake in the Bay Area at intervals somewhere between every 100 and every 1,000 years. The Hayward fault, an important feature of the East Bay, seems likely to release somewhat less devastating shocks considerably more frequently, perhaps every 10 to 100 years. Geologists base such estimates on what they can discover about the fairly recent history of the fault and the assumption that the same pattern of activity will continue. Their estimates are likely to mislead.

Consider the peculiar case of the missing Parkfield earthquake. A series of remarkably similar earthquakes, all with a magnitude of about 6.0, happened there in 1857, 1888, 1901, 1922, 1934, and 1966. That is a remarkably consistent recurrence interval of about 22 years. Seismologists were eagerly poised with millions of dollars worth of equipment when the next earthquake came due in 1988. No earthquake. The U.S. Geological Survey thought it had detected the symptoms of imminent movement in 1992 and issued its first official earthquake forecast. No earthquake. Another Parkfield earthquake will surely happen sometime, but neither the regular schedule of the past nor the latest forecasting technology revealed when that time may come.

Some experts dread the day when someone finally learns to recognize the symptoms of an impending earthquake. They foresee thousands of panicky people fleeing with no place to go, leaving their homes and businesses at the mercy of looters before the earthquake actually strikes. They fear that an earthquake forecast might cause almost as much economic and social chaos as the earthquake itself.

At first thought, the idea of managing earthquakes seems at the outermost fringe of human presumption. We can hardly hope to direct the stirrings in the bowels of the earth that break its crust along faults, but we may someday manage to release stuck faults. In fact, that has been done on an experimental basis by drilling wells into stuck faults and injecting water to lubricate them.

Many geologists and engineers think it should be possible to release the San Andreas fault slowly, one little bit at a time, but no one is sure that releasing a short segment would not trigger the Big One. Better to wait until after that event, then consider drilling wells to lubricate the fault to prevent it from sticking again. Of course, we can hardly dream of lubricating faults that are stuck at depths beyond the reach of drilling tools. Nor will it be possible to deal with unknown faults, which abound. Most of those will not make themselves known until they cause an earthquake.

Land Use Planning and Building Codes

Good land use planning and strict building codes provide the stoutest armor against earthquakes. In much of California, it is too late to argue that people should avoid building on or near an active fault. San Francisco was a big city long before anyone knew the San Andreas fault existed. Even

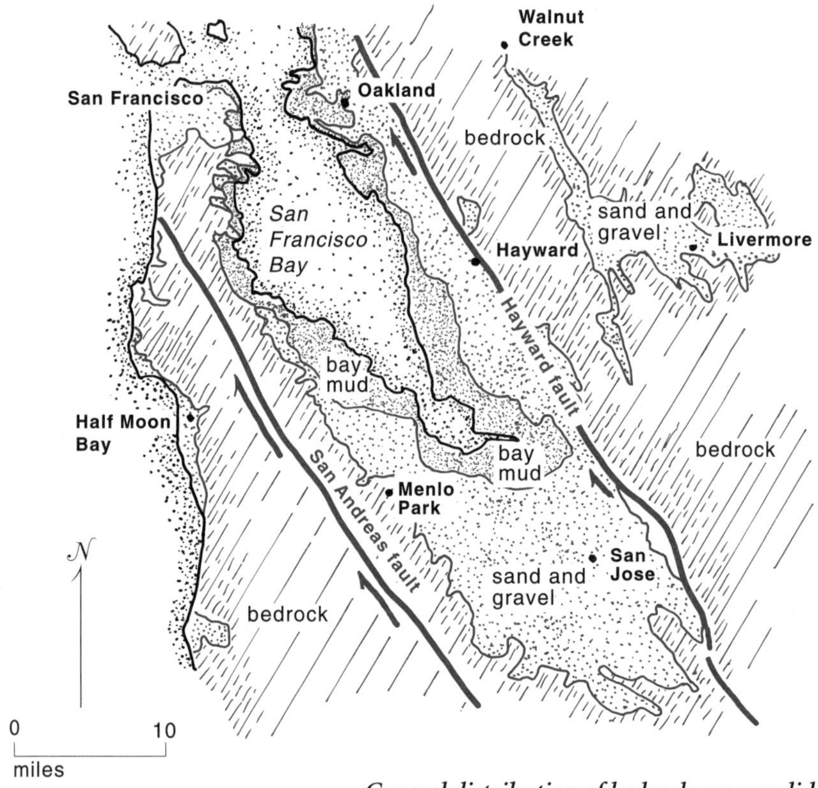

General distribution of bedrock, unconsolidated sand and gravel, and sloppy bay muds. —Adapted from Borcherdt and others, 1975

so, it is still possible to zone some of the most dangerous areas against the most vulnerable kinds of development.

Areas on and near a major fault are especially dangerous. So are areas of loose ground, even miles from the fault. Unconsolidated sediments, such as mud around the bay, shake more violently during an earthquake than solid bedrock, for the same reason that gelatin quivers more than hard ice cream. The soft muds under the Marina district of San Francisco, for example, shook far more violently than nearby hills eroded in solid bedrock, both during the earthquake of 1906 and again during the much smaller Loma Prieta earthquake of 1989.

Buildings that depend on masonry walls for their support are likely to fall during an earthquake. Properly reinforced concrete and buildings made with wood or steel frames will probably shed some plaster and broken glass, but they rarely tumble down. California began to enact building codes that forbid simple masonry construction after the San Francisco earth-

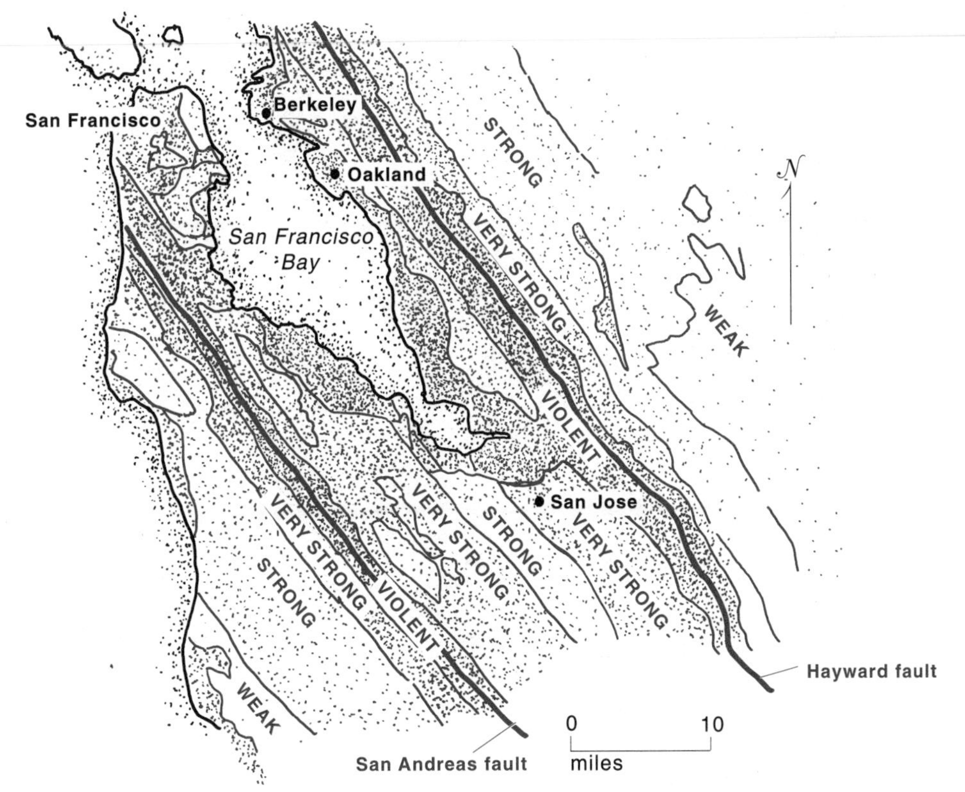

Violence of ground motion that accompanies an earthquake on the San Andreas or Hayward faults. Areas along the fault or on unconsolidated sediment would suffer the greatest damage, those in the bedrock hills the least. —Adapted from Borcherdt and others, 1975

quake of 1906. Seismic bracing was not required until after 1933, primarily as a reaction to the great Tokyo earthquake of 1923.

LANDSLIDES

The Coast Range is landslide country, from one end to the other. Watch its grassy slopes for wrinkled and rumpled surfaces and its roads for those curving patches on the downhill side. Most landslides are modest little affairs, easy to ignore as long as they move someone else's house. But the thousands of landslides that destroy roads and buildings every year probably inflict more aggregate property damage than the occasional devastating earthquake.

Deep soils cover most Coast Range slopes, and the Franciscan rocks, especially the serpentinites, are so closely fractured that they are almost as weak as the soils. Some of the younger formations that locally cover the Franciscan complex, especially in the East Bay Area, are just as weak

Small landslide on California 20, east of Clear Lake.

Water dripping from these little pipes may save the slope above this roadcut from sliding.

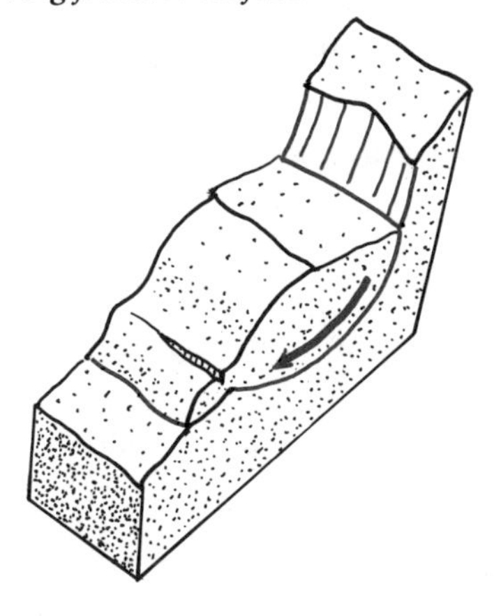

Most landslides move on a curving fracture surface.

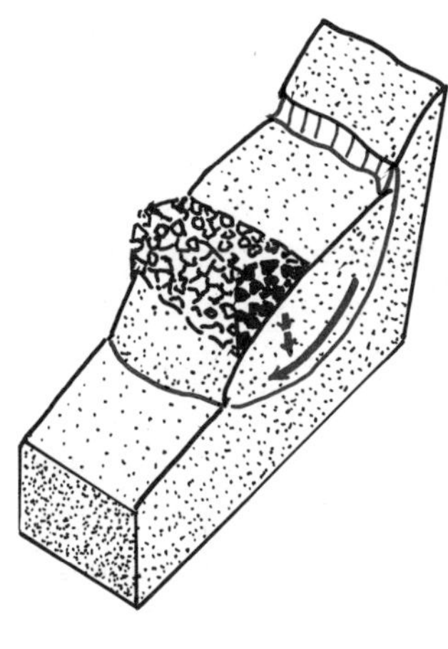

A load of rocks on the lower end of a slide increases friction on the slide surface.

as the Franciscan rocks. Add water, shake the ground a bit, and those broken rocks and soil are likely to come down the slope.

Most landslides start with a curving fracture in the ground, shaped about like the bowl of a spoon. As the mass of rocks and soil above that fracture begins to move, cracks open in the ground at its upper end, while the slope bulges at the lower end.

The slide mass above the fracture may move slowly, in occasional small hitches, or not at all. Most slides hang precariously on the slope until something triggers them into catastrophic movement. That may be a series of heavy rains, an earthquake, a sonic boom, or some unfortunate combination. When it finally goes, the slide mass may move as a coherent piece of ground, or it may stir itself into a sloppy porridge and come down the slope as a mudflow. The spectrum of possibilities is wide.

It is possible to stabilize a landslide if it is detected early. One good way to do that is to drain the water from the slope. Rows of little pipes installed in countless roadcuts in the Coast Range drain the slopes well enough to prevent sliding. Another common sight is a big pile of rocks dumped on the slope above or below the highway. Their weight increases friction on the slip surface beneath, which prevents the slide from moving.

Whether they move slowly or catastrophically, landslides typically leave a curving scar, the fracture surface, on the hillside where they started. And they land as a hummocky dump of debris somewhere below that scar. Hills with those curving scars on them and with hummocky dumps on their lower slopes are as unsuitable for development as they are picturesque.

SAN FRANCISCO BAY AREA

It would be hard to write a description of the rocks and landscapes for the roads in the Bay Area, even harder to read one. So we will consider them as an area, instead of as individual roads.

The geologic map shows large areas of Franciscan rocks east of the San Andreas fault on the San Francisco and Marin Peninsulas. Some of the rocks west of the fault are granite of the Salinian block, exposed near Montara Point and San Pedro. Others are sedimentary formations deposited on that granite since the end of Cretaceous time. Those sedimentary rocks are complexly folded and broken along faults, although not nearly as pervasively as the Franciscan rocks. The Great Valley sequence in the East Bay east of the Hayward fault is also steeply tilted in places and also likely to slide. Elsewhere, it is buried under Tertiary sedimentary rocks, which are also complexly folded and broken along faults.

The most significant rocks from the human point of view are the soft muds and sands around the bay, which will shake violently during any large earthquake. Their distribution shows where earthquake damage will be most severe. Large areas of these shaky sediments are densely developed for both residential and commercial use.

San Francisco Bay floods a broad structural depression in the Coast Range where the Sacramento and San Joaquin Rivers meet to flow into the ocean through the Golden Gate. Some of the record exists in the Merced formation, a relic of Pleistocene time. All the sand grains in its lower part, which was deposited in seawater, came from locally exposed rocks. The upper part, a portion of which was deposited on land, contains sand grains eroded from the northern Sierra Nevada. Evidently, the Bay Area was still below sea level when the lower part was laid down, then filled to sea level as Pleistocene time continued. The transition from sand originating locally to sand originating from the Sierra Nevada apparently records the establishment of the modern Sacramento River drainage system, about 0.5 million years ago.

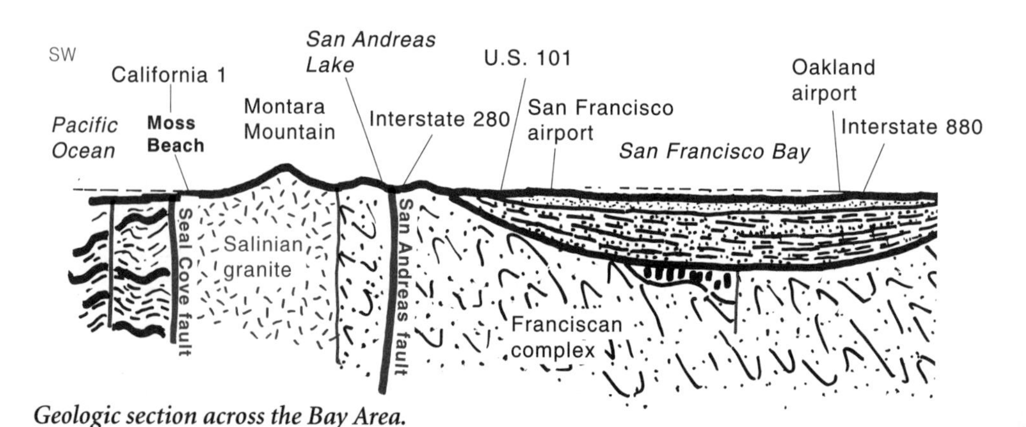

Geologic section across the Bay Area.

What is now bay was dry land during the ice ages, when sea level dropped more than 300 feet as water accumulated in the great continental glaciers. The Sacramento River picked up several tributaries as it flowed through that coastal lowland. Then it flowed through the last mountain ridge in a deep canyon that is now the Golden Gate.

The coastal lowland filled with water as sea level rose, when the great continental glaciers melted at the end of the last ice age. San Francisco Bay assumed its present form about 10,000 years ago, when sea level returned to its present stand. The bay is now about 55 miles long from north to south and from 3 to 12 miles wide, an area of about 435 square miles. The deepest parts are about 350 feet deep, but more than 80 percent is less than 12 feet deep. The water in most of the bay is brackish, with about 2.8 percent dissolved salts, 15 percent less than normal seawater.

Water flows in and out of San Francisco Bay with the tides. On average, every rising and ebbing tide moves enough water through the Golden Gate to flood about 1.25 million acres to a depth of 1 foot. The incoming currents reach speeds as great as 4 miles per hour; the outgoing flow is much slower. Those currents keep the Golden Gate swept free of sediment to a depth of about 350 feet. Seafarers who navigated under sail considered them a major problem.

Ever since the rising sea level flooded the Golden Gate and converted the lower part of the river valley into a bay, the Sacramento and San Joaquin Rivers have been filling it with sediment. The filling speeded up after early placer and hydraulic mining operations in the Sierra Nevada started more than 2 billion cubic yards of sediment down the rivers. More than a billion cubic yards of that sediment have now reached the bay, greatly enlarging the Sacramento Delta and filling large areas of shallow water. About 8 million cubic yards of sediment wash into the bay every year. Soft sediment laid down within the past few thousand years lies beneath most of the flat land around the bay. Thousands of years from now, the rivers will have filled

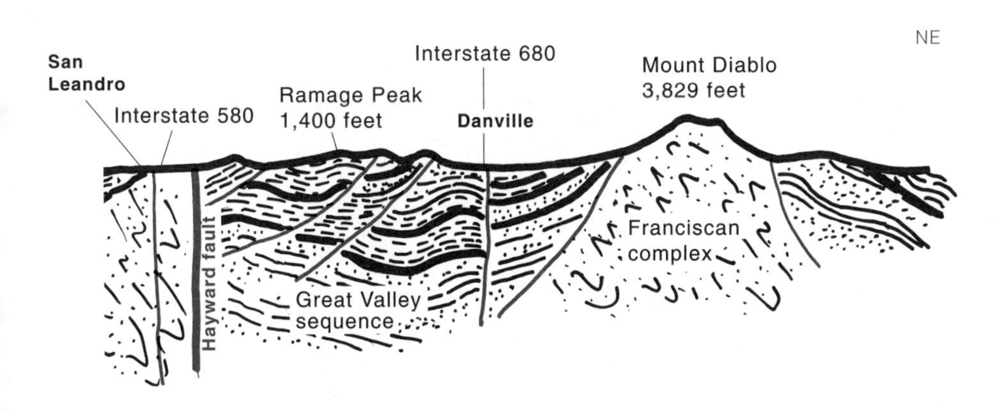

most of San Francisco Bay with sediment and will again flow through the Golden Gate. Grass will grow where ships now pass, and the chilly fog will drift over broad reaches of marshland.

People who built the cities around the bay covered large areas of soupy sediment with rubble and garbage to make a more sharply delineated shoreline. Every earthquake shakes those deposits of soft sediments and unconsolidated fill like the proverbial bowl full of jelly. The buildings that now cover them are at great risk.

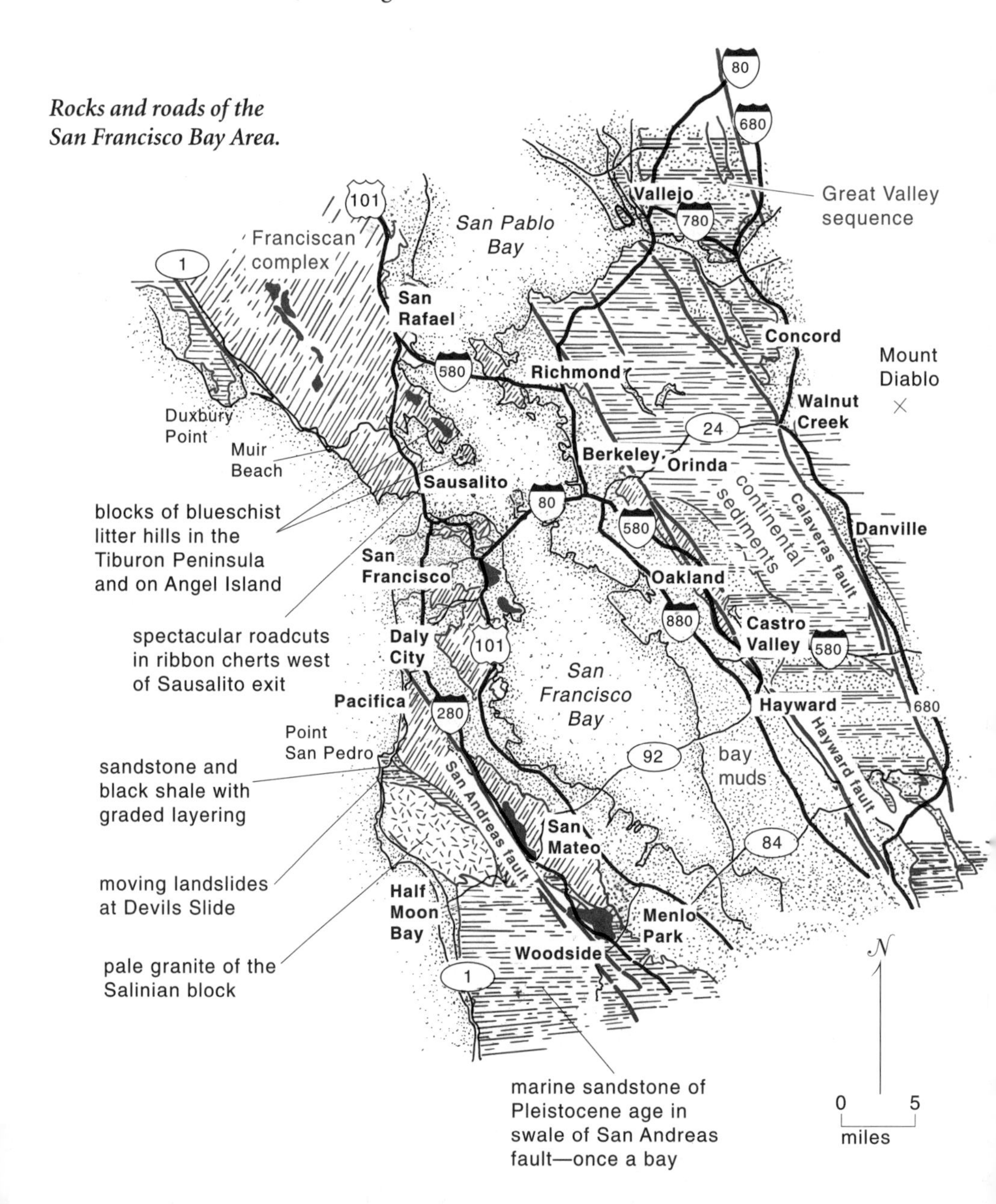

Rocks and roads of the San Francisco Bay Area.

Creep on the Hayward fault offset this curb in downtown Hayward.

San Andreas System of Faults

The most active faults in the Bay Area, those most likely to cause earthquakes, all trend northwest, as do most of the older faults that no longer move. All are part of the somewhat diffuse boundary between the North American and Pacific plates.

Skyline Boulevard nearly follows the San Andreas fault along the border between Daly City and Pacifica and in northwestern San Bruno. The swale along the fault just west of Skyline Boulevard was devoid of houses until the 1970s. Through some failure of planning, it is now covered with them. It should have become a park. The Big One will bring tragic consequences when it finally strikes.

The San Andreas fault is most obvious from the 9 miles of Interstate 280 between San Bruno Avenue and the interchange to the San Mateo Bridge. Look down the wooded slope to the west to see San Andreas Lake near the north end of that stretch and the Crystal Springs Reservoir farther south. San Andreas Lake was impounded behind an earthen fill dam in 1870 and was later raised. Crystal Springs Dam was one of the largest in the world when it was finished in 1890. It is 145 feet high, made of concrete blocks, and stands on fractured Franciscan sandstones. The Hetch Hetchy aqueduct carries water from the northern parts of Yosemite National Park to Crystal Springs Reservoir.

Although both dams are in the San Andreas fault zone, neither was damaged in the 1906 earthquake. The older dam is just west of the line of rupture, and the Crystal Springs dam is about 1,000 feet east of it. An offset fence and line of cypress trees about a half mile north of the Crystal Springs Reservoir showed that the fault moved 9 feet during the earthquake of 1906.

View south along the San Andreas fault, 1965. Housing subdivisions in Daly City and Pacifica stand on the fault. —R. E. Wallace photo, U.S. Geological Survey

The Hayward fault runs through the East Bay on a course parallel to the San Andreas fault. You can see it on any road map because it defines the sharp inland border of dense urban development at the steep western face of the Berkeley Hills. The Hayward fault is not stuck in Hayward, where it creeps at an average rate of about 0.2 inches per year. It is stuck elsewhere, most famously in Berkeley where it runs directly beneath the University of California football stadium. Geologists estimate that the Hayward fault has about a 25 percent probability of releasing an earthquake of magnitude 7 sometime before the year 2020. The Hayward fault may cause more damage and loss of life than the San Andreas fault because it draws a longer trace through heavily populated areas on unconsolidated fill around the bay. And it is as close to downtown San Francisco as the San Andreas fault.

Santa Cruz Mountains

The hills of the San Francisco and Marin Peninsulas consist mostly of Franciscan rocks, except in the southern part, where they also include some granite. The San Andreas fault cuts through the Santa Cruz Mountains from Pacifica southeast. The side west of the fault is part of the Salinian

Serpentinite of the south abutment of the Golden Gate Bridge. View is about 3 feet across.

block, moving north with the Pacific plate; east of the fault is the North American plate. So the Santa Cruz Mountains are not a coherent geologic entity, just a chance topographic juxtaposition of one block of rocks to another. The San Andreas fault will slice the range into two slowly separating ridges, each moving north, but at different rates.

Rock Foundations of the Golden Gate Bridge

The San Andreas fault passes the Golden Gate about 2 miles offshore. That was too close for the complete comfort of the engineers who designed and built the Golden Gate Bridge. At the time, people argued that any earthquake capable of wrecking the bridge would also level much of San Francisco. Why fret?

The north pier of the bridge stands on a broad base of reasonably solid Franciscan rocks—no cause for concern. The south pier stands on serpentinite, a notoriously weak and slippery rock—no cause for confidence. Bailey Willis, a distinguished geology professor at Stanford, expressed doubts about the stability of the south pier while the bridge was under construction. The builders turned to the even more distinguished Andrew C. Lawson of Berkeley, who was not known to admire Willis. Professor Lawson descended into the deep bedrock excavation behind the coffer dam, rapped

Franciscan sandstone at Baker Beach, just west of the south end of the Golden Gate Bridge.

Folded ribbon chert on Twin Peaks. View is about 6 feet across.

on the serpentinite with his hammer, and declared it perfectly sound, an excellent foundation for a bridge pier. The truth will emerge when the Big One strikes.

Twin Peaks and the Hills of San Francisco

Twin Peaks probably offers the best panoramic view of the city and its bay. Spectacular exposures of reddish ribbon cherts exist on both peaks, especially on their southwest sides. The chert beds are inclined at steep angles and locally wrinkled into tight folds. The thin layers of darker rock that separate the chert layers are shale, probably laid down from the farthest edges of the clouds of muddy sediment that deposited graded beds closer to shore.

The great block of the San Francisco Mint, on Market Street northeast of Twin Peaks, rests squarely on a big mass of green serpentinite. Watch for it along the sidewalk. More serpentinite is exposed on Potrero Hill and along the axis of Hunters Point. Most of the other hills in the city are Franciscan sandstone.

Sand Dunes

Early settlers in San Francisco found a spectacular tract of shifting sand dunes in much of what is now the coastal part of the city, inland almost to Twin Peaks. Most of the dunes are now covered with buildings, except for some along the beach and a few lonely survivors near the zoo. Watch

Sand still blows across the Great Highway at Ocean Beach near the San Francisco Zoo.

for the vague outlines of old sand dunes here and there in Golden Gate Park. Otherwise, all that remains are sandy backyard gardens and the occasional sandy hump.

The Sacramento River brought all that sand through the Golden Gate during the ice ages, when sea level was low and the great sand trap of San Francisco Bay did not exist. Waves coming from the north swept the sand south, down the beach. Then the strong sea breeze blew it off the upper beach when the tide was out and into the dunes. Now that sea level is high, the river dumps its sand into the inland reaches of the bay, mostly into the advancing edge of the Sacramento Delta.

Fort Baker Area

When traffic is stopped, watch for the splendid exposures of Franciscan rocks in roadcuts immediately north of the Golden Gate Bridge, between the Marin abutment and the Sausalito interchange. The roadcuts include basalt flows, originally black but now streaked with shades of dull green, and numerous layers of muddy sandstone. Layers and occasional globs of red are chert. All of the layers tilt steeply down to the east and show obvious signs of having been torn, sheared, and crumpled. That happened as they were scraped off the descending seafloor and stuffed into the Franciscan trench, about 100 million years ago.

Large masses of serpentinite that include beautiful chunks of dark blue blueschist and dark green eclogite make the backbone of the Tiburon Peninsula, where houses now cover most of the best rocks. The same serpentinites exist on Angel Island, where good specimens of blueschist and eclogite are still easy to find in the state park, where they are protected.

A winding road leads west from the north end of the Golden Gate Bridge to the grounds of old Fort Baker and to the lighthouse at Bonita Point. Roadcuts along the way provide one marvelous view after another of Franciscan ribbon cherts. Most are weathered just enough to cover the original colors with a rich stain of reddish brown iron oxide, the typical color of Coast Range soils. Thin beds of chert are intricately rumpled into sharply angled folds. These cherts precipitated far out on the deep ocean floor and were later jammed into the continental margin during the collision between the continent and ocean floor.

Rodeo Beach dams the lower end of a valley to impound Rodeo Lagoon. The dark and pebbly sand consists largely of grains of red and green Franciscan chert, like that so widely exposed in the surrounding hills. Rodeo Lagoon now traps any sand coming down the creek long before it can reach the beach. But the lagoon did not exist during the ice ages, when sea level dropped more than 300 feet, shifting the coast out to somewhere near the

Farallon Islands. Then the stream could have carried sand down the valley to the present site of Rodeo Beach and far beyond.

East Bay Area

The straight line of the Hayward fault sharply separates the busy flatlands near the bay from the abrupt rise of the Berkeley Hills to the east. Few faults are as obvious in the landscape as the Hayward fault, or exert such a strong effect on the pattern of development.

The Diablo Range east of Walnut Creek extends about 130 miles southeast from Carquinez Strait almost to Coalinga. It consists basically of a series of broad arches in the Great Valley sequence. Streams have eroded through some of the arches to expose Franciscan rocks in their cores. The Calaveras fault and the San Ramon Valley separate the Berkeley Hills from the Diablo Range. A variety of younger formations deposited during the 65 million years of Tertiary time cover large areas of the older rocks.

Mount Diablo makes a curiously jumbled patch on the geologic map, an especially perplexing puzzle. Many geologists consider it a mélange of mangled Franciscan rocks embedded in serpentinite that rose as a great plug through a cover of much younger rocks. Some think it may still be rising. Others doubt the whole story.

Ribbon chert exposed on upper Claremont Avenue in the Berkeley Hills.

Mount Diablo is prominent from the south.

Rocks east of Mount Diablo include the Domengine formation, which was deposited during middle Eocene time, about 50 million years ago. It contains a variety of rock types, including sandstone clean enough to use as glass sand, good ceramic clays, and very poor brown coal. But any coal was good coal in the days before oil, especially in California, which has so little. The Domengine formation west of Antioch provided most of the coal mined in California between 1860 and 1920. The mines went out of business as cheap oil and natural gas entered the market.

The broad Livermore Valley is a low area between the Berkeley Hills and the Diablo Range. It drains into San Francisco Bay through Alameda Creek, which flows through the southern part of the Berkeley Hills by way of Niles Canyon, a deep gorge that cuts right across the topographic grain of the landscape. It seems clear that Alameda Creek was established in its valley before folding raised the Berkeley Hills. Surveys show that the hills are now rising about 0.06 inches per year, 6 inches per century, an almost unbelievably high rate. So far, Alameda Creek has met the challenge.

Bayshore in China Camp State Park

The delightful drive through China Camp State Park winds across and around low bedrock hills made of Franciscan rocks and passes close to large expanses of salt marsh along the edge of San Pablo Bay. If you mentally subtract the salt marsh from the scene, you reconstruct the shore as it was when sea level rose to its present stand at the end of the last ice age, when the waves lapped directly against the bedrock. During the years since then, the salt marsh has grown, trapping sediments and taking the first steps in the long process of filling the bay.

COAST RANGE ROADGUIDES

Most of the roads in the Coast Range follow the topographic grain, which trends parallel to the San Andreas system of faults, more or less parallel to the coast. Few wind the more tortuous ways east and west across the mountains.

Coast Range landscapes consist largely of broad valleys between long ridges that trend northwest. The San Andreas system of faults cuts the range into long slices, all of which are moving north. They move like traffic on a highway, those nearest the coast moving fastest and passing those in the lanes farther east. The broad valleys are basins that open where right lateral faults swerve right.

Most of the ridges in the Coast Range are more ragged and irregular than ridges normally are in an erosional landscape. They owe their raggedness partly to landslides but mainly to the drastically different ways the various Franciscan rocks weather and erode.

Except for a few patches of volcanic rocks and younger bay sediments, nearly all the rocks exposed east of the San Andreas fault belong to the Franciscan complex, the Great Valley sequence, and the Coast Range ophiolite. Those exposed west of the San Andreas fault as far north as Fort Ross are granite of the Salinian block and sedimentary rocks deposited on it. Sedimentary rocks exposed north of Fort Ross probably lie on oceanic crust, which is also moving north, but may or may not be part of the Salinian block.

U.S. 101

San Francisco—Leggett

188 miles

The highway winds through the forested hills of the Coast Range. Almost all the bedrock is Franciscan rocks, mainly dark sandstones but also serpentinites and ribbon cherts. Sonoma volcanic rocks that erupted between about 2.1 million and 10,000 years ago make up the hills between Novato and Healdsburg. Much younger deposits of sand and gravel fill the floors of the Sonoma and Alexander Valleys. Deposits of much older bay muds appear in the terraces around the Sonoma Valley.

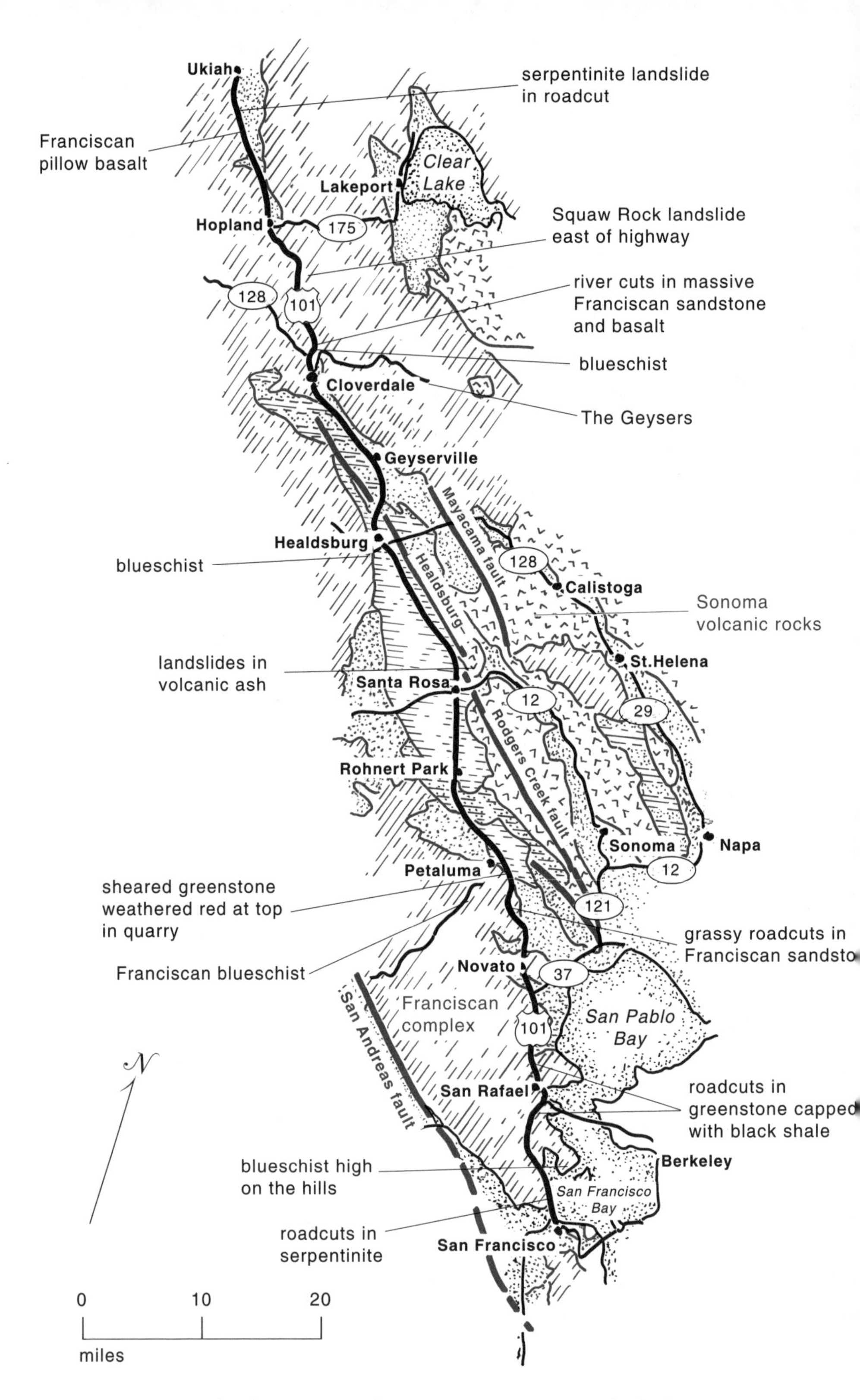

Rocks along U.S. 101 between San Francisco and Ukiah.

Cotati Valley

U.S. 101 crosses Franciscan rocks, locally exposed in roadcuts, between San Francisco and the Petaluma interchanges. That is at the south end of the broad Cotati Valley, which the road crosses between Petaluma and Healdsburg.

The Cotati Valley is long north to south, parallel to the trend of the San Andreas fault system. Its shape suggests that movement along those faults opened it, probably several million years ago. Younger sediments now cover the evidence that might clarify its origin.

Low hills along the western rim of the Cotati Valley, west of Santa Rosa Creek, are eroded into beds of soft sedimentary rock, layers of mud, sand, and gravel. They cover the intensely deformed Franciscan rocks over an area of several hundred square miles that reaches almost to the coast. Abundant fossils of animals that lived in shallow seawater several million years ago, during Pliocene time, show that the Cotati Valley was then a shallow bay.

The Sonoma Mountains along the eastern rim of the Cotati Valley are the eroded remains of the Sonoma volcanic pile. It consists mostly of pale volcanic ash that erupted while the valley was a bay. Rhyolite ash flows overwhelmed and buried redwood forests, now preserved as petrified logs and stumps. Together, the rocks and fossils conjure a picture of a broad saltwater bay enclosed on its north and south sides within hills eroded in Franciscan rocks, and nestled beneath high volcanoes to the east. A dark cloak of forest covered the land.

After the bay drained and the volcanoes died, streams eroded broad valleys deep into the rocks that recorded them. Then, broad aprons of gravel, washed in from the surrounding hills, partially filled those valleys, probably during a time when the climate was much drier than now. More vigorous streams flowing in the wetter, modern climate are now cutting new valleys deep into those gravel deposits. Remnants of the gravel are the basis for the well-drained soils that support the flourishing vineyards.

Tolay Volcanic Rocks

The highway crosses the Tolay fault at the north edge of Petaluma and just grazes the Tolay volcanic pile near Rohnert Park. The volcanic rocks are about 10 million years old, which makes them seem distinctly out of place this far north in the Coast Range. But that is also the age of the Quien Sabe volcanic pile east of Hollister, which contains remarkably similar rocks. Many geologists believe that a branch of the Hayward fault sliced the western edge off the Quien Sabe volcanic pile and moved it north to Cotati. If so, the fault moved 120 miles during the past 10 million years. That would require an average speed of about 0.3 inches per year, quite reasonable for one of the faults in the San Andreas swarm.

Santa Rosa Earthquake

In October 1969, two earthquakes of magnitudes 5.6 and 5.7 struck Santa Rosa within less than two hours. They severely damaged about one hundred buildings, mostly older houses that did not conform to the modern building codes. Casualties were light.

The epicenters of the two main earthquakes, and of the aftershocks, align nicely along a northwest trend through Santa Rosa that certainly defines the trace of the responsible fault. Detailed studies of the earthquake records show that the ground west of the fault moved north, the usual picture in the Coast Range. The fault movement was too deep to break the ground surface.

Alexander Valley and the Russian River

The highway between Healdsburg and Cloverdale follows the Russian River through the long Alexander Valley, its floor nearly covered with vineyards. The Alexander Valley looks on the map like a narrow prong reaching north from the Cotati Valley, between the Mendocino Mountains to the west and the Mayacamas Mountains to the east. It appears to be another fault basin but much smaller than the Cotati Valley, and it never held a bay snuggled at the base of high volcanoes. Its surrounding hills are eroded Franciscan rocks.

The Russian River flows south through the Alexander Valley to near Healdsburg, where it abruptly turns west to flow through a deep gorge in the Mendocino Mountains, then enters the ocean near Jenner. California 116 follows the river through the gorge.

Why did the Russian River turn to erode a narrow valley through hard bedrock, right across the topographic grain of the landscape? Why not

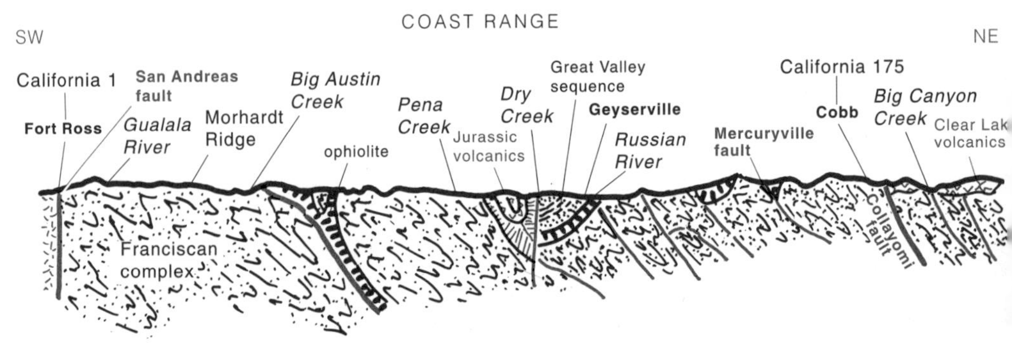

Geologic section across the Coast Range near Geyserville. —Central part adapted from Suppe, 1979

continue south through the much lower landscape and softer rocks in the Cotati Valley? Some geologists think the river was in its valley before the mountains rose across its path; the river eroded its channel as fast as they rose. Others contend that the river was flowing in its valley before the Cotati Valley sank. Either way, it is a strange turn of the river.

The Geysers

The name misleads. No erupting geysers like those in Yellowstone Park exist at The Geysers. The area did contain hot springs and fumaroles, which are natural steam springs. A hunter discovered them in the hills about 10 miles east of Cloverdale in 1847. They were the centerpiece of a popular resort until the early 1900s.

The steam comes from basalt and Franciscan sandstone about 8 miles northeast of Geyserville, along the north side of Big Sulphur Creek. It probably owes its existence to the large volcanic field in the Clear Lake area to the east and northeast, which was erupting as recently as a few thousand years ago. The volcanic field still has plenty of hot rock beneath the surface, perhaps even molten rock.

Natural steam produced from wells began to drive turbines that generate electricity in the 1920s. The big geothermal power plants now running began to operate in 1955 and for decades generated nearly half of San Francisco's electricity. The steam reserves were expected to last for several centuries, but production has already begun to decline, and it now seems unlikely that they will last even one century.

Between Cloverdale and Leggett

The long road between Cloverdale and Leggett winds through wooded hills along the rugged spine of the Coast Range. All of the rocks are Franciscan muddy and sandy sediments deposited in the Franciscan trench and the ocean floor beyond, then jammed under the Coast Range ophiolite and the Great Valley sequence. Rocks near the northern end of the route are less deformed than most Franciscan sediments, probably because they were late into the crush of the trench. The gray and brown sandstone and the black shale are Franciscan sediments. The blocky greenish rocks are slices of the ocean floor on which the Franciscan sediments were deposited.

Watch between Cloverdale and the area a few miles north of Hopland for outcrops of dark rock, mostly dark green serpentinites broken into blocks with slick surfaces. Serpentinite contains a great deal of magnesium but virtually none of the other mineral nutrients that plants need. That is why the orange or red soils that typically develop on serpentinite are so infertile. They support a sparse and straggly cover of digger pines and stunted shrubs that cloak the soil in patches and leave it exposed elsewhere.

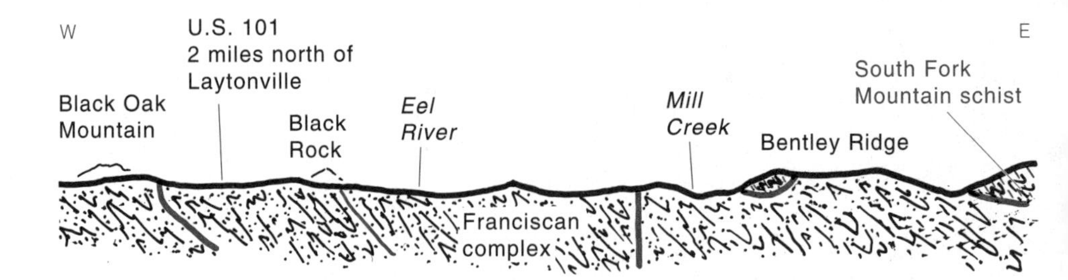

Geologic section drawn east across the Coast Range near Laytonville. —Adapted from Etter and others, 1981

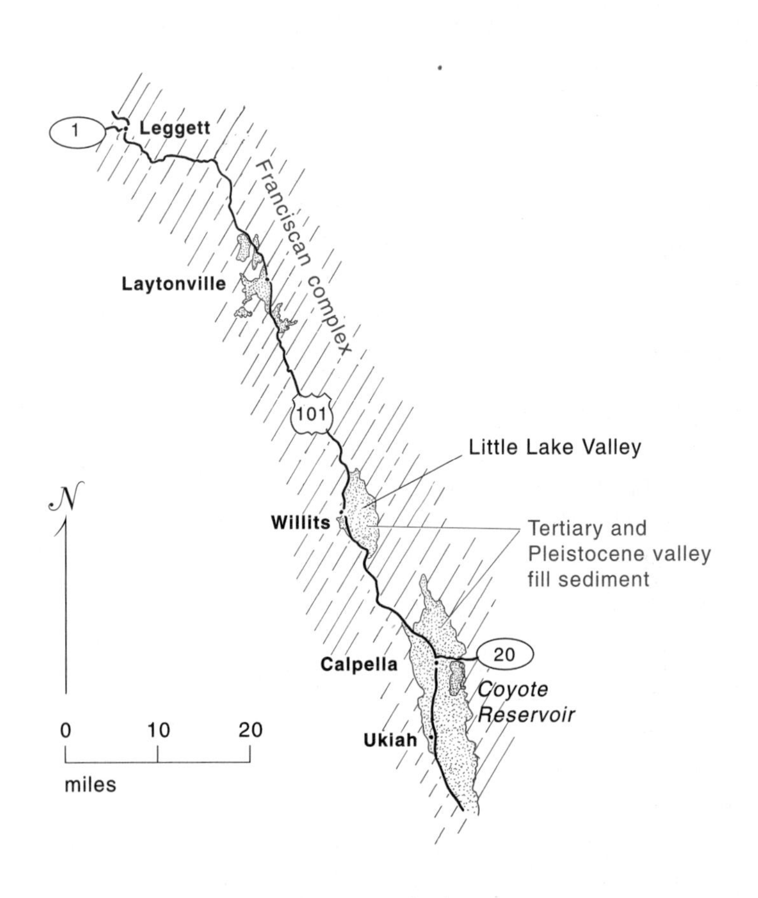

Rocks along U.S. 101 between Ukiah and Leggett.

Roadcut in Franciscan shale and sandstone, 6 miles south of Leggett.

Pillows in Franciscan basalt in a small quarry just west of U.S. 101, about 3 miles north of Calpella.

The road follows the Russian River between Hopland, Ukiah, and Calpella, past more vineyards. The valley between Hopland and Ukiah traces one of the faults that slices the Coast Range parallel to the San Andreas fault. The road leaves the valley of the Russian River just north of Ukiah, crossing northwest through hills eroded into Franciscan rocks to reach the Little Lake Valley at Willits. The valley is an oval about 7 miles long from north to south and about 4 miles wide, probably a basin dropped between the offset ends of two faults.

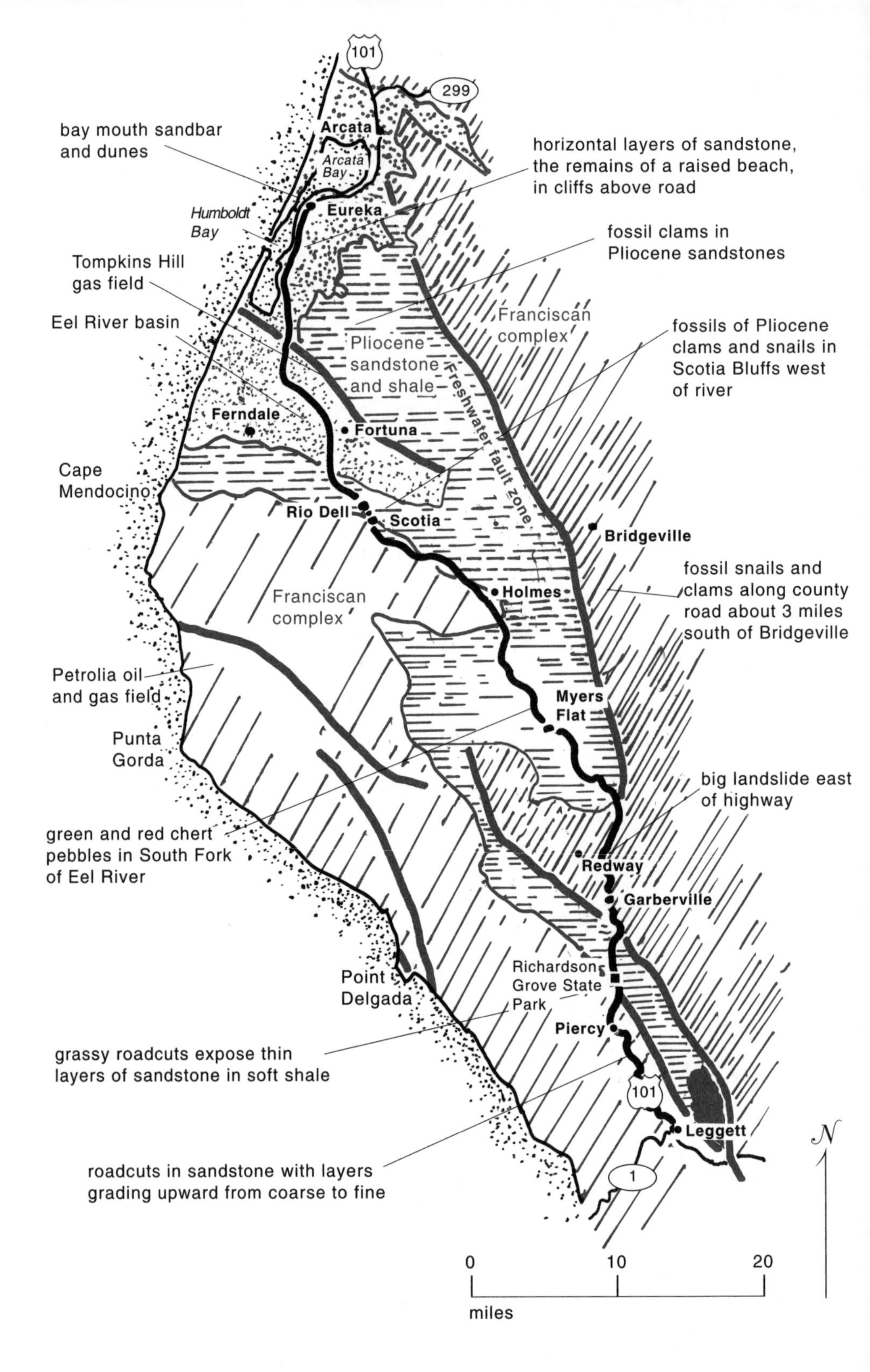

Rocks along U.S. 101 between Leggett and Eureka.

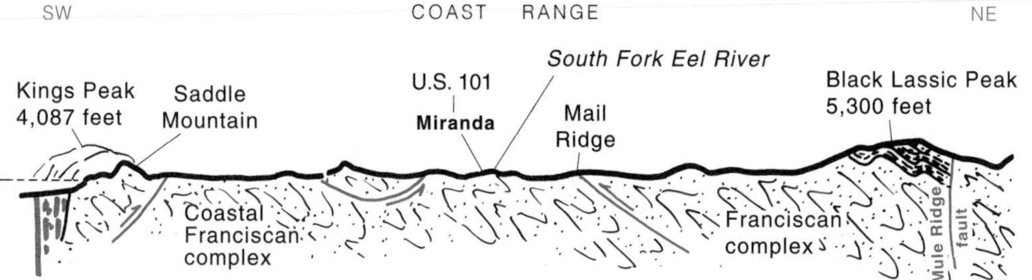

Geologic section across the Coast Range near Miranda.
—Adapted from Blake and others, 1985

Dark sandstone and shale near Richardson Grove.

U.S. 101
Leggett—Eureka
93 miles

The highway follows the deep and winding canyon of the South Fork of the Eel River along most of the route between Leggett and Rio Dell. Large outcrops make steep bluffs. Occasional small outcrops, high on the hillsides, mark old landslide scars, where the bedrock was exposed as the soil slid off the slope.

Franciscan rocks lie beneath most of the area. Rocks in the Eel River basin are soft sedimentary formations deposited during Tertiary time, the 65 million years that have elapsed since Cretaceous time ended. The more

Clam shells in soft buff sandstone of Scotia Bluffs. View is 6 inches across.

open pattern in the western part of the geologic map shows the distribution of coastal Franciscan rocks, which are younger than those farther inland and less thoroughly mangled. They were the last jammed into the trench. Conspicuous outcrops between Leggett and Garberville expose these layers of sedimentary rocks. Roadcuts reveal sedimentary layers that are gently tilted, not tightly crumpled and broken. The rare fossils are younger than those in the Franciscan rocks farther east.

Eel River Basin

The highway between Rio Dell and Holmes follows the broad valley the Eel River eroded along the Russ fault. Rocks north of this fault, including some near the road, are mostly soft muds deposited about the time the Coast Range rose, perhaps in a shallow bay. Their softness explains the width of the valley.

Most of the route between Rio Dell and Eureka crosses the nearly flat surface of the Eel River basin. It contains soft muds and sands deposited in fairly shallow water during the past 4 or 5 million years, long after the Coast Range rose. Watch for these muds and sands in the big river cuts at Rio Dell. Fossil seashells abound in any good outcrop.

The Eel River basin is a trough enclosed between faults that trend nearly east, nearly parallel to the great Mendocino fracture on the floor of the Pacific Ocean just a few miles to the south. The bounding faults bend around to the south at the east end of the basin to become parallel to the San Andreas fault.

Any thick accumulation of muddy sediments deposited in seawater is a likely source of crude oil. The Eel River basin looks especially tempting because apparently similar basins in the southern California Coast Range have produced enormous quantities of oil. Drilling in the Eel River basin has revealed precious little petroleum in any form, and hope for a big discovery fades with every new dry hole.

A strong earthquake with several damaging aftershocks struck the Eel River basin in April 1992. It shook many of the lovely old Victorian houses in Ferndale off their foundations and inflicted heavy damage on Scotia and Rio Dell. The scarcity of serious casualties testifies to the importance of building codes. Wood-frame houses may shake off their foundations, but they rarely collapse. A Ferndale built with brick structures would have fallen into heaps of rubble, burying any number of people.

The Eel River at Scotia carries an almost incredible 4,330 tons of sediment every year from every square mile of its drainage basin. On average, 4 to 8 inches of soil is washing off the slopes every hundred years. That is the highest regional rate of erosion ever measured in the United States, more than thirteen times the national average. Most of that phenomenal load goes down the river in about 6 days of the largest discharges during the winter floods. The Eel River drains about 3,100 square miles of wet countryside, on weak rocks poised on steep, unstable slopes. Landslides certainly contribute a large part of the river's sediment load.

California 20

Calpella—Williams

79 miles

California 20 crosses Franciscan rocks along most of the route. The main exceptions are the Great Valley sequence, along the east flank of the Coast Range, and valley fill sediments in the northern Redwood Valley, near Clear Lake, and in the Sacramento Valley. The road passes some lovely volcanic rocks in the vicinity of Clear Lake and passes Mount Konocti, a shapely volcano still very little eroded.

The junction with U.S. 101, at Calpella, is in the north end of the Russian River valley. The valley probably owes its existence to movement along faults parallel to the San Andreas fault. Valley fill sediments were deposited in its floor within the past several million years, then erosion carved them into a gentle landscape of softly rolling hills, now nearly covered with vineyards. The steeply rugged hills that surround the Redwood Valley are eroded into much harder rocks, mainly Franciscan sandstone. Franciscan

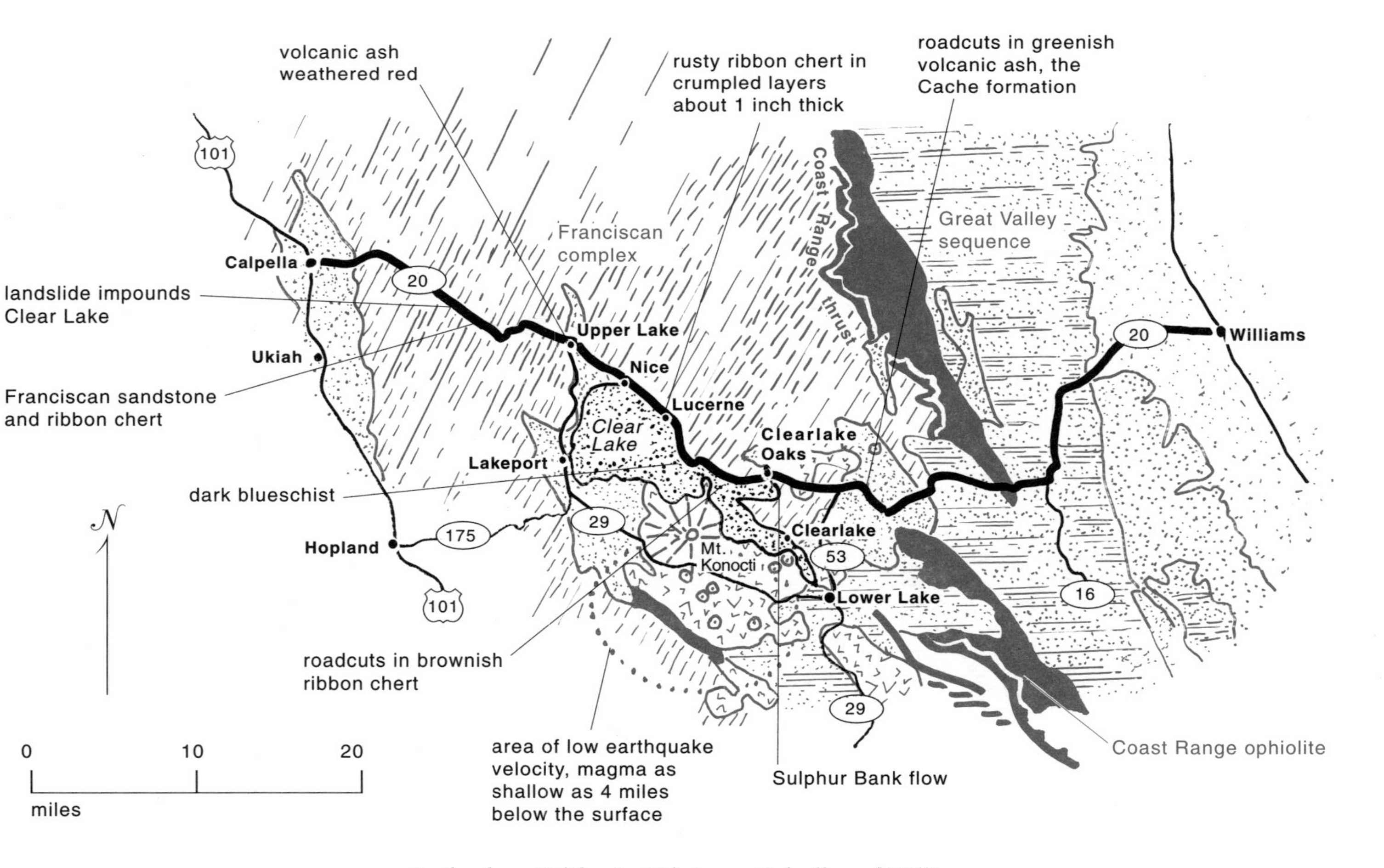

Rocks along California 20 between Calpella and Williams.

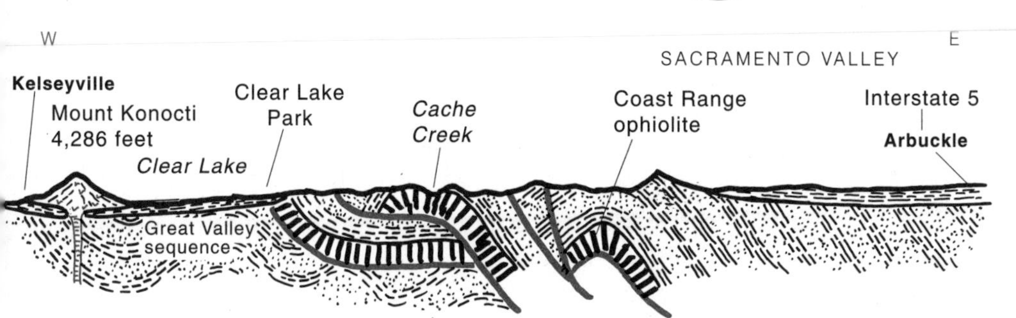

Geologic section across the Coast Range between Kelseyville and the Sacramento Valley.

rocks make up all the mountains between U.S. 101 and Clearlake Oaks. Lake Mendocino, just southeast of Calpella, is a reservoir impounded at its south end behind the Coyote Dam. California 20, east of Lake Mendocino, follows the eastern base of the Cow Mountains through the valley of Cold Creek almost to Clear Lake.

Clear Lake

Clear Lake floods a broad valley that originally drained west along the line of California 20, through Cold Creek and into the Russian River. Then a large landslide plugged the drainage, sometime within the past few thousand years. Water impounded behind the landslide filled the valley until it spilled through a new outlet into Cache Creek, which drains east

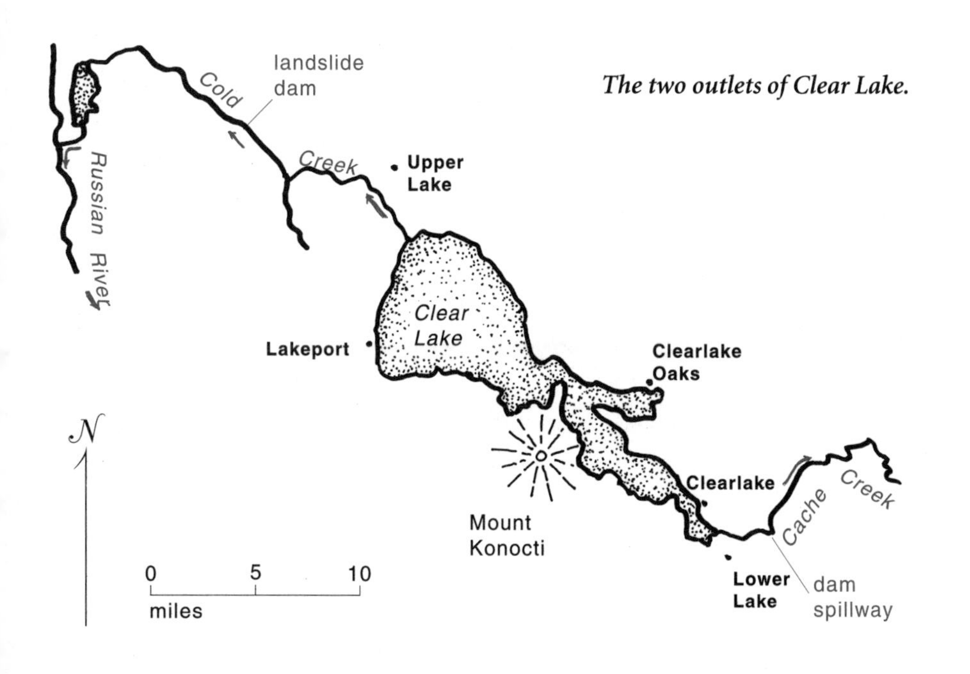

The two outlets of Clear Lake.

from the southeastern tip of the lake into the Sacramento River. Some geologists think that was actually the original drainage before a lava flow dammed it, impounding an earlier version of Clear Lake with an outlet to the Russian River.

A dam built across Cache Creek in 1915 to increase water storage capacity raised the lake level about 11 feet, just enough to start a trickle of water over the landslide dam and down the old outlet through Cold Creek into the Russian River. So Clear Lake now has two outlets.

Some geologists argue that the Clear Lake basin is basically a caldera that collapsed as the eruptions that built the Clear Lake volcanic field emptied a magma chamber beneath it. If so, then earlier versions of the lake certainly existed long before the landslide dammed the modern version. Studies of sediments deposited in the lake reveal that some are as much as 150,000 years old, which suggests that it did indeed begin as a caldera.

Whatever its history, Clear Lake now survives because its main outlet stream happens to flow over the hard basalt. It will be a long while before that stream can cut its outlet channel deep enough to drain the lake. Had the spillway started to flow across the broken and more easily erodible rock of the landslide dam, it would have drained the lake long ago.

Between Nice and Clearlake Oaks, California 20 hugs the north shore of Clear Lake at the western base of High Valley Ridge, with wonderful views across the water. Numerous roadcuts north of the highway expose Franciscan rocks, most of them deeply weathered to the warm shades of reddish brown typical of the Coast Range soils.

Volcanoes

Eruptions began in the Clear Lake volcanic field about 1.5 million years ago and nearly ended sometime before 0.5 million years ago. They created most of the landscape south and west of Clear Lake within the past million years, some of it within the past few thousand years as the late eruptions sputtered on.

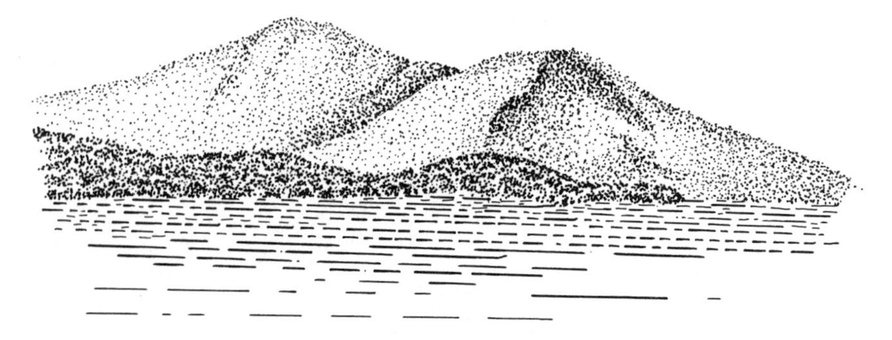

Mount Konocti overlooks Clear Lake.

The dominating bulk of Mount Konocti, 4,286 feet high, hovers above the west side of the lake, a large volcano with several peaks. It retains most of its original volcanic form, despite a big landslide scar on one slope and some less dramatic erosional insults since its last eruption. Age dates show that its rocks erupted between about 430,000 and 350,000 years ago. One of the almost magical activities of geophysicists involves making incredibly precise measurements of the earth's gravitational field, to reveal almost infinitesimal variations in its strength. Those measurements reveal a roughly circular area of slightly weaker gravity beneath and south of Mount Konocti. It may be the signature of a mass of molten magma within a few thousand feet of the surface. If so, Mount Konocti may not be quite as extinct as it looks and as the age of its rocks suggests.

About 2 miles east of Clearlake Oaks, California 20 passes between a matched pair of cinder cones, each about 1 mile in diameter and 400 feet high. Watch for the quarry on the north side of the road, an enormous red scar in the side of the cinder cone. It produces basalt cinders for road metal, construction aggregate, and garden rock. Faint layers slope away from the vent.

The basalt cinders were black when they erupted, but now they are bright red. Steam rising through them oxidized some of the iron in the black mineral pyroxene to red iron oxide, the excellent pigment used in barn paint. It stains all the surfaces of the cinders. Like most basalt cinder

A red basalt bomb 4 feet across, blown out of the northern of the two cinder cones.

Volcanic ash and gravel eroded into hoodoos,
12 miles east of Clearlake Oaks.

cones, these look much younger than they are. They resist erosion because the cinders are big enough that splashing raindrops can not move them. And the spaces between them absorb water so greedily that none can run off the surface—no surface runoff, no gullies.

The eastern end of High Valley, about 2 miles north of the paired cinder cones, is another area of late volcanic activity. A nifty little cinder cone, about 400 feet high, squats in the midst of a jagged expanse of basalt lava that erupted from it. The lava is so fresh that it still lacks soil, but a straggling cover of adventurous plants suggests that it is older than it looks.

The big Sulphur Bank lava flow is about 1 mile south of the paired cinder cones. It also seems young but is not much to look at. Hot springs and gas vents are still active in the southern part of the flow, presumably the area directly above the vent. The hot water and steam converted much of the lava flow to white opal, leaving deposits of sulfur and mercury that have been mined intermittently for the last century. Sulfur and mercury are the commonest kinds of mineralization in volcanic rocks so young and so near the surface. The water and fish in Clear Lake carry alarming

Pleistocene gravel with lumpy concretions, on California 20, 7 miles east of Clearlake Oaks.

concentrations of mercury. It appears to be a case of natural pollution, which may well come from the Sulphur Bank flow.

Masses of rhyolite magma such as the Sulphur Bank flow erupted quietly only because they contained almost no steam. A good charge of steam would have blown them across the landscape as scalding sheets of white rhyolite ash. Franciscan rocks are notoriously poor sources of groundwater, which probably explains how the rhyolite magma managed to reach the surface without picking up much steam. But a rising mass of magma that happened to intersect a fault zone full of water might erupt with shattering violence.

The existence of so much young volcanic rock at the surface makes it virtually certain that a great deal of hot rock exists at shallow depth. Hot springs actively depositing ore minerals, such as those in the Sulphur Bank flow, show that deep circulation of hot water already exists. They encourage hope for natural steam power like that at The Geysers.

East of Clear Lake, California 20 winds for several miles through low hills underlain by loose volcanic ash, sand, and gravel eroded from the nearby hills and deposited in the floor of a broad valley about 2 million years ago. Their layers slope gently west.

Great Valley sequence on California 20, west of Williams.

Great Valley Sequence

Watch near the line between Colusa and Lake Counties, 15 miles east of the junction with California 53, for a number of good roadcuts in pale green serpentinite. Nearby hills eroded in this rock have a distinctly greenish look in places, and their soil supports a sparse cover of shrubs but no grass. This is the thick zone of serpentinite that separates the Great Valley sequence from the Franciscan rocks beneath.

In the eastern edge of the Coast Range, east of the junction with California 16, the highway passes through the Great Valley sequence, which makes several parallel ridges that trend north. A number of large roadcuts offer good views of the brownish gray, muddy sandstones and shales. Although their layers tilt steeply down to the east, they are not severely contorted like those in the Franciscan rocks. Their outcrops make long, nearly straight ledges across the slopes high above the road.

California 20 crosses the western edge of the Sacramento Valley between the easternmost ridge of the Great Valley sequence and Williams. On a clear day, the view east reveals the jagged low hills of the old Sutter Buttes volcano standing before the distant backdrop of the northern Sierra Nevada.

California 29
Vallejo—Lakeport
100 miles

The broad expanse of flat land that the road crosses between Vallejo and Napa was the northern part of San Pablo Bay, until streams filled it with sediment. The road between Napa and Calistoga follows the length of the Napa Valley, one of the many that slice from northwest to southeast through this part of the Coast Range. The road north of Calistoga twists and winds through the mountains to Clear Lake.

Napa Valley

The Napa Valley is another of those fault basins in the northern California Coast Range. Deep deposits of loose sand, gravel, and volcanic debris washed in from the surrounding mountains fill its floor. The considerable porosity and permeability of these deposits make them a good substrate for the open, well-drained soils that grapevines so deeply appreciate, and they hold enough groundwater to irrigate large expanses of vines.

Grapevines that grow on the valley floor sink their roots into sand and gravel eroded from various mixtures of Franciscan rocks and the younger volcanic rocks that partially cover them. Grapevines that grow on the slopes overlooking the valley floor depend on residual soil weathered from specific kinds of bedrock, which seem to affect the flavors of wines.

Red wines made from grapes grown on soils weathered on the dark rocks of the oceanic crust sometimes have sharp flavors that require some aging to mellow. White wines are not so affected. But white wines made from grapes that grew on soils weathered from serpentinite in some of the side valleys have a sharply acid taste. Soils derived mostly from the abundant younger volcanic rocks, or from granites elsewhere, do not impart such distinctive flavors.

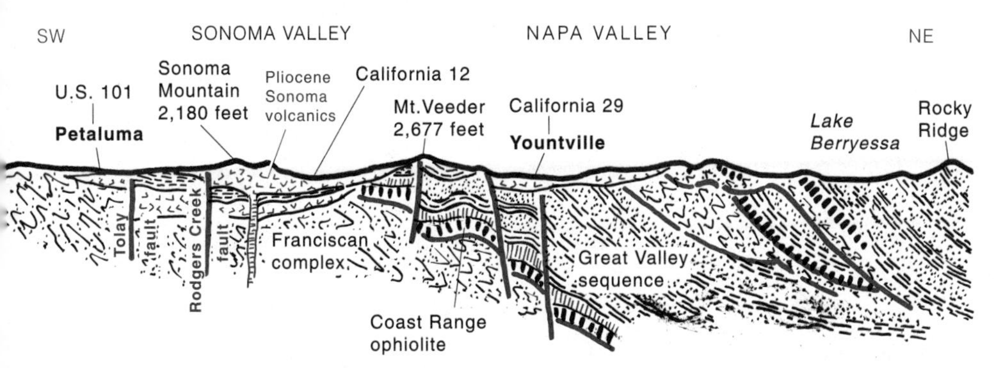

Geologic section northeast from Petaluma, approximately along the line of California 29. The Tolay volcanic rocks are about 10 million years old and many miles from their probable home near Hollister.

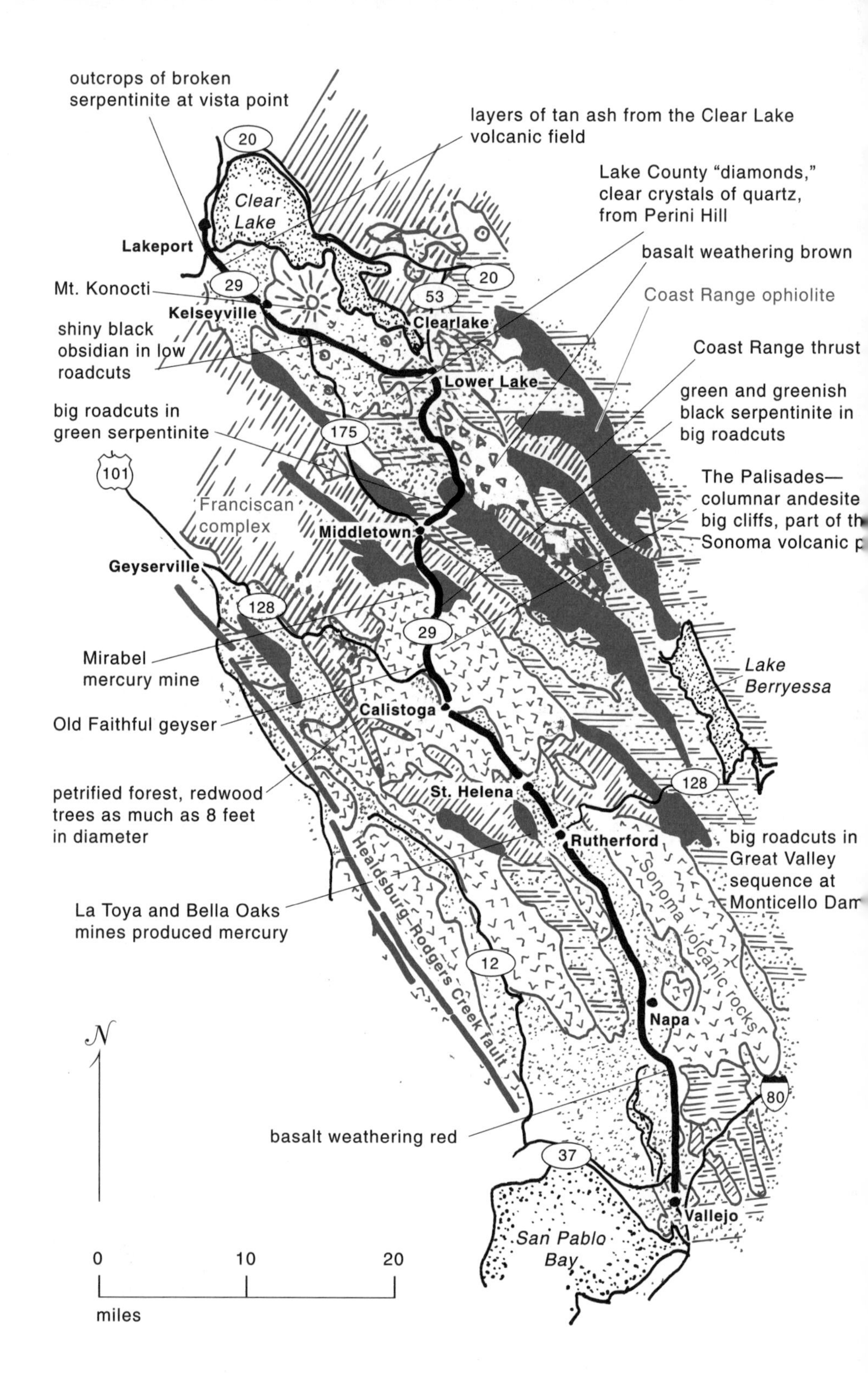

Rocks along California 29 between Vallejo and Lakeport.

Most of the hills around the Napa Valley are eroded into volcanic rocks. They are part of the Sonoma volcanic field, which erupted during latest Miocene and Pliocene time, between about 6 and 3 million years ago. That makes them much older than the rocks in the Clear Lake volcanic field.

Most of the volcanic rocks are rhyolite ash and andesite, in various pale shades of gray, pink, and yellow—the progeny of numerous violent eruptions. Many small roadcuts along the winding highway across the east flank of Mount St. Helena, north of Calistoga, provide the easiest close views of those rocks. The Palisades cliffs, northeast of Calistoga, are part of a huge ash flow that erupted in Pliocene time, between about 5 and 2 million years ago.

From a distance, the great heap of Mount St. Helena looks like a single big volcano, but the wildly varying attitudes of the layers in the rhyolite ash suggest that it is really a complex volcanic center. Mount St. Helena has not erupted for several million years, so it is safely dead.

Few of the original volcanoes in the Sonoma volcanic field survive as recognizable elements of the modern landscape. Movements along faults have dismembered them, and streams have carved most of the pieces into an erosional landscape. The only way to recognize the area as a volcanic field is to look at the rocks.

Pale volcanic rocks commonly contain deposits of silver, so it is no surprise to find old silver mines and prospects in the hills around the Napa Valley. They are souvenirs of the years when the mines produced impressive tonnages of good ore. Robert Louis Stevenson wrote of his honeymoon sojourn at the old Silverado Mine during the summer of 1880. Considerable reserves remain in the ground, patiently awaiting renewed demand. But silver is no longer a monetary metal, and digital recording systems are decreasing its main industrial use in photography.

Calistoga Area

Rhyolite ash buried forests, and then the wood petrified as dissolved silica filled it with colorful chalcedony—agate, if you prefer. Petrified logs, even whole forests of petrified logs and stumps, exist in a number of places in the hills near Calistoga. Pretty pebbles of petrified wood abound in many of the streambeds.

The Petrified Forest is on Petrified Forest Road, several miles west of Calistoga. Rhyolite ash contains a number of petrified logs, complete with petrified bark. Their tops all point southwest. It appears that a rhyolite ash flow leveled, buried, and finally petrified the forest some 6 million years ago. Numerous fossil leaves show that the trees differ only slightly from the modern coastal redwoods.

Petrified trees buried in an ash flow.

Old Faithful geyser at Calistoga.

Old Faithful geyser, 1 mile north of Calistoga at the west side of Tubbs Lane, erupts to a height of about 60 feet at intervals of about 15 minutes in wet seasons, over an hour in dry ones. The water is at a temperature of about 350 degrees Fahrenheit, presumably because it circulates through hot volcanic rocks deep underground.

Although Calistoga's geyser is an accident of well drilling, not a natural geyser, its eruptions are perfectly natural. Circulating water fills cavities in the hot rocks below the surface. As it heats up and expands, water spills out of the top of the well. That relieves pressure on the water at depth, permitting it to flash into steam and erupt. Then the cavities must again fill with water before the next eruption can happen. A percolator coffeepot operates in exactly the same way.

California 29 crosses a broad belt of serpentinite just north of Mount St. Helena. Watch for the darkly greenish rocks in roadcuts, with shiny surfaces that glitter in the sun. As everywhere, the serpentinite weathers to impoverished reddish soils so lacking in fertilizer nutrients that they offer grudging support for a scanty growth of scrubby trees and shrubs, with little grass. But the digger pines flourish, pale blue clouds of trees. More serpentinites appear along the route between Mount St. Helena and Lower Lake, along with some dark Franciscan sandstones that weather to more nurturing soils.

Serpentinite about 12 miles south of Lower Lake.

Clear Lake Volcanic Field

Between Lower Lake and Lakeport, California 29 passes through the southern and western parts of the Clear Lake volcanic field. Some of the volcanoes erupted so recently that it seems reasonable to conclude that the field is still active.

Like volcanic rocks elsewhere in the Coast Range, most of these are pale andesite and rhyolite, mostly volcanic ash. They also include large quantities of black obsidian, rhyolite glass. Obsidian abounds along the road between Lower Lake and Kelseyville, especially between 6 and 10 miles northwest of Lower Lake.

It is surprising that shiny black obsidian has the same chemical composition as ordinary white rhyolite, except in containing almost no water. It owes its color to its iron content, which stains it black in the same way that a few grains of dissolved food coloring stain water. The same amount of iron in ordinary crystalline rhyolite lightly speckles the rock with a few small grains of black magnetite. Given some millions of years, obsidian eventually crystallizes into pale rhyolite.

Clear Lake

The road between Kelseyville and Lakeport crosses the floor of Big Valley, past one vineyard after another. This is another of those dropped fault slices in the Coast Range. Lakeport is on the western edge of Clear Lake, with a view east across its widest expanse of open water, the flooded north end of Big Valley. The only visible rocks are small patches of broken serpentinite at the viewpoint between Kelseyville and Lakeport.

U.S. 101

San Francisco—Salinas

111 miles

U.S. 101 passes through an urban countryside between San Francisco and San Jose. No visible rocks, little original landscape, and only occasional glimpses of the hills to east and west. The more open country south of San Jose offers an occasional view of the landscape but few glimpses of the rocks beneath it. Most of the rocks are too weak to make good outcrops.

The geologic map shows the road following the broad Santa Clara Valley between the San Andreas fault on the west and the Hayward fault on the east. It is one of those long slices between members of the San Andreas system of faults, as are the mountains on either side. The Santa Cruz Mountains consist basically of Salinian granites with a thick cover of

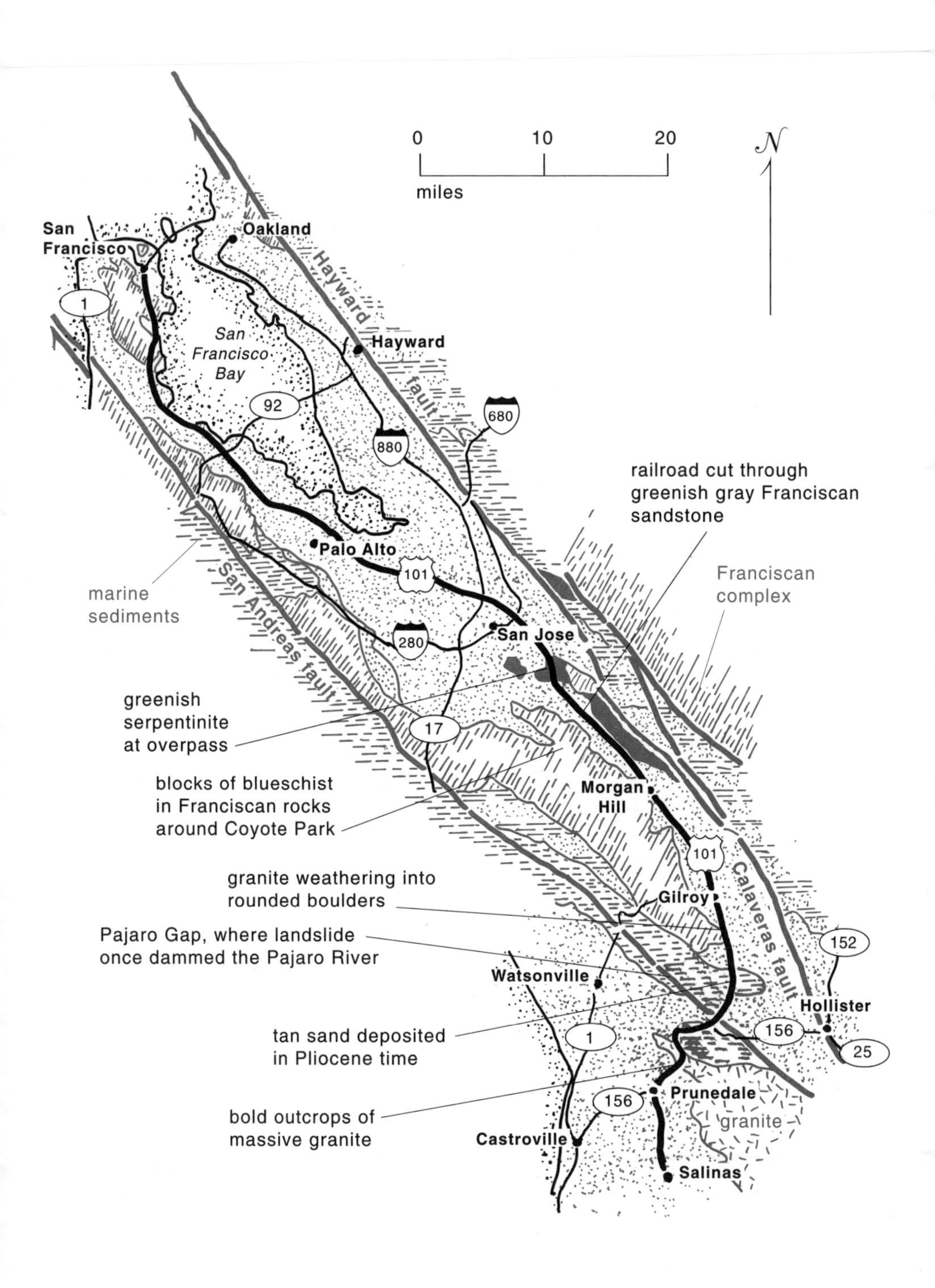

Rocks along U.S. 101 between San Francisco and Salinas.

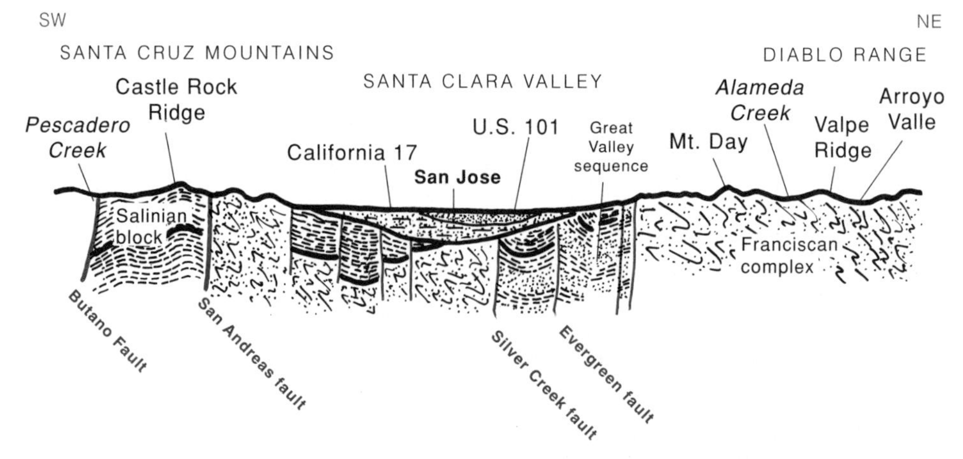

Geologic section across the Coast Range through San Jose.
—Adapted from Wagner and others, 1991

sedimentary rocks, deposited in seawater during Tertiary time. The Diablo Range, to the east, is mostly Franciscan rock with some patches of younger sedimentary formations.

Nearly all of the rocks in the valley floor are poorly consolidated sediments laid down in shallow seawater during Tertiary time. Most contain fossils of animals that lived then. Totally unconsolidated sediments lie, sloppy wet, around the fringes of the bay. All of that weak stuff moves violently during earthquakes. This is perilous ground for major urban development.

About 6 miles south of Gilroy, the highway leaves the Santa Clara Valley to cross the northern tip of the Gabilan Range. The rocks there are deeply weathered granite and sediments deposited in the past 5 million years, both poorly exposed. The highway crosses the San Andreas fault onto the Pacific plate about 1 mile south of the junction with California 129.

The flat ground at the southern end of the route is what remains of an old bay at the mouths of the Salinas and Pajaro Rivers. It would probably have provided a good anchorage for ships when the glaciers melted at the end of the last ice age and the rising sea flooded the lower ends of all rivers. But now it is filled with sediments and is better suited for artichokes and broccoli.

Creeping Calaveras Fault

U.S. 101 follows a route parallel to, and a few miles west of, the Calaveras fault between San Jose and Hollister. This fault appears to have stopped moving in the East Bay Area but not farther south. Unlike the San Andreas fault, which moves in sudden jerks that cause earthquakes, the Calaveras

Slow movement on the Calaveras fault broke the concrete and bent the bolt in Cochran Bridge across Coyote Creek Reservoir near Morgan Hill.

fault south of San Francisco moves more or less constantly, without causing noticeable earthquakes. Coyote Creek Reservoir, east of U.S. 101 at Morgan Hill, floods the swale of the Calaveras fault. The continuous creeping of the fault progressively bends the bridge across the reservoir and cracks its concrete foundations.

Lake Plain

W. C. Brewer noticed in 1864 that the terraces on either side of the Pajaro River are level and concluded that they are remnants of an old lake plain. That was an incredible feat of observation and deduction in a time before topographic maps. It seems clear that a landslide about 4 miles west of the junction with California 129 dammed the river at Pajaro Gap, sometime within the past few hundred thousand years. The flat floor of the Santa Clara Valley a few miles on either side of Gilroy is the bed of the lake, now filled with sediment. Lush vegetable crops, most notoriously garlic, now cover it.

New Almaden

The first mining operations in California were at the south end of Almaden Road, about 10 miles south of San Jose. Watch for the excellent exposures of ribbon chert and greenish serpentinite along the road to Almaden Reservoir.

The New Almaden mines began to produce mercury in 1824 and continued until 1845. They shipped more than one million flasks of mercury, which would have been worth $260 each in 1996 dollars. An old chimney on the grassy hillside west of town is the most obvious relic.

Serpentinite mélange about 1 mile southwest of New Almaden. View is about 4 feet across.

Arrastra that crushed mercury ore at New Almaden.

A small museum in town exhibits some lesser relics of the mining operation, including an arrastra, a rare sight. Arrastras were a primitive type of mill that had the great advantage of requiring almost no capital investment and of being simple to operate. They consisted essentially of a flat pavement of large rocks. Ore was dumped on it, and then oxen or horses dragged a large rock around a pivot in the center of the pavement to crush the ore. Few arrastras survive because the usual last act in the mining operation was to dismantle them to recover the ore that had filtered into the cracks between the stones.

The main ore of mercury is cinnabar, mercury sulfide. It is bloody red in a fresh rock but turns gray upon exposure to sunlight. Cinnabar is so unstable that simple roasting breaks it down, liberating shiny little globules of liquid mercury. More intense roasting drives the metal off as a vapor that can be caught in a condenser. Most early mercury mines had their own recovery plants, many built right on the premises to completely original and novel designs.

Design is important in dealing with something that destroys the central nervous system as efficiently as mercury vapor. Although plenty of people knew that a hundred years ago, few seemed concerned. The old mining literature is full of articles about designs for mercury condensers, which seem to have been considered safe as long as the distinctive odor of the "mercurial vapors" was not too strong. The idea that they might poison the labor force seems to have eluded nearly everyone.

Cement

Limestone rarely forms in deep waters like those that have lain off California during the past hundred million years, so limestone to make cement is in short supply. Even though it is an essential raw material in making cement for concrete, the low price of limestone makes it uneconomic to haul it long distances. A large cement plant on the west side of Cupertino uses limestone quarried from the Franciscan complex. The huge, open cuts are distantly visible from the winding road west of the Stevens Creek Reservoir. Some roadcuts expose similar pale gray limestone.

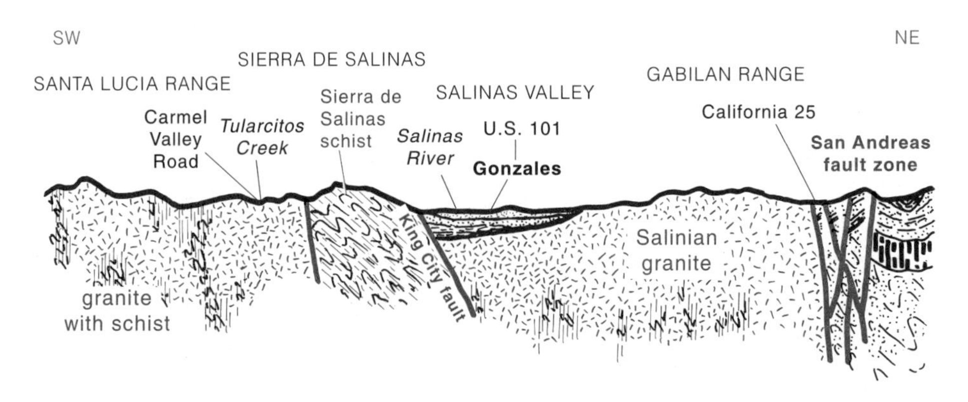

Section drawn south of Atascadero across the Santa Lucia Range, Sierra de Salinas, and Gabilan Range, east to the San Andreas fault.
—Adapted from Ross and McCulloch, 1979

U.S. 101
Salinas—San Luis Obispo
128 miles

The highway follows the Salinas River through the strikingly linear trough of the Salinas Valley, all the long way between Salinas and Atascadero. The rolling hills east of the Salinas Valley are the Gabilan Range. The more rugged peaks to the west are in the northern Santa Lucia Range. The highway crosses the Santa Lucia Range between Santa Margarita and San Luis Obispo.

Gabilan Range

Most of the hard bedrock in the smooth hills of the Gabilan Range, east of the Salinas Valley, is granite, part of the Salinian block. It is distinctive rock, pale and attractively peppered with black flakes of biotite mica and tiny black rods of hornblende. Age dates range from 85 to 72 million years, the same as much of the granite in the Sierra Nevada batholith. Years ago, some geologists suggested that this granite moved north from the southern end of the Sierra Nevada, but it more nearly resembles granite in the Mojave Desert.

The jagged peaks of Pinnacles National Monument punctuate the smoothly rolling granite hills along the eastern skyline near Soledad. They are rhyolite that erupted during Miocene time. The striking contrast between the rounded hills eroded in granite and the jagged peaks in rhyolite shows how differently the rocks erode. That is puzzling because their chemical and mineral compositions are nearly identical. The main difference between them is the larger grain size of the granite.

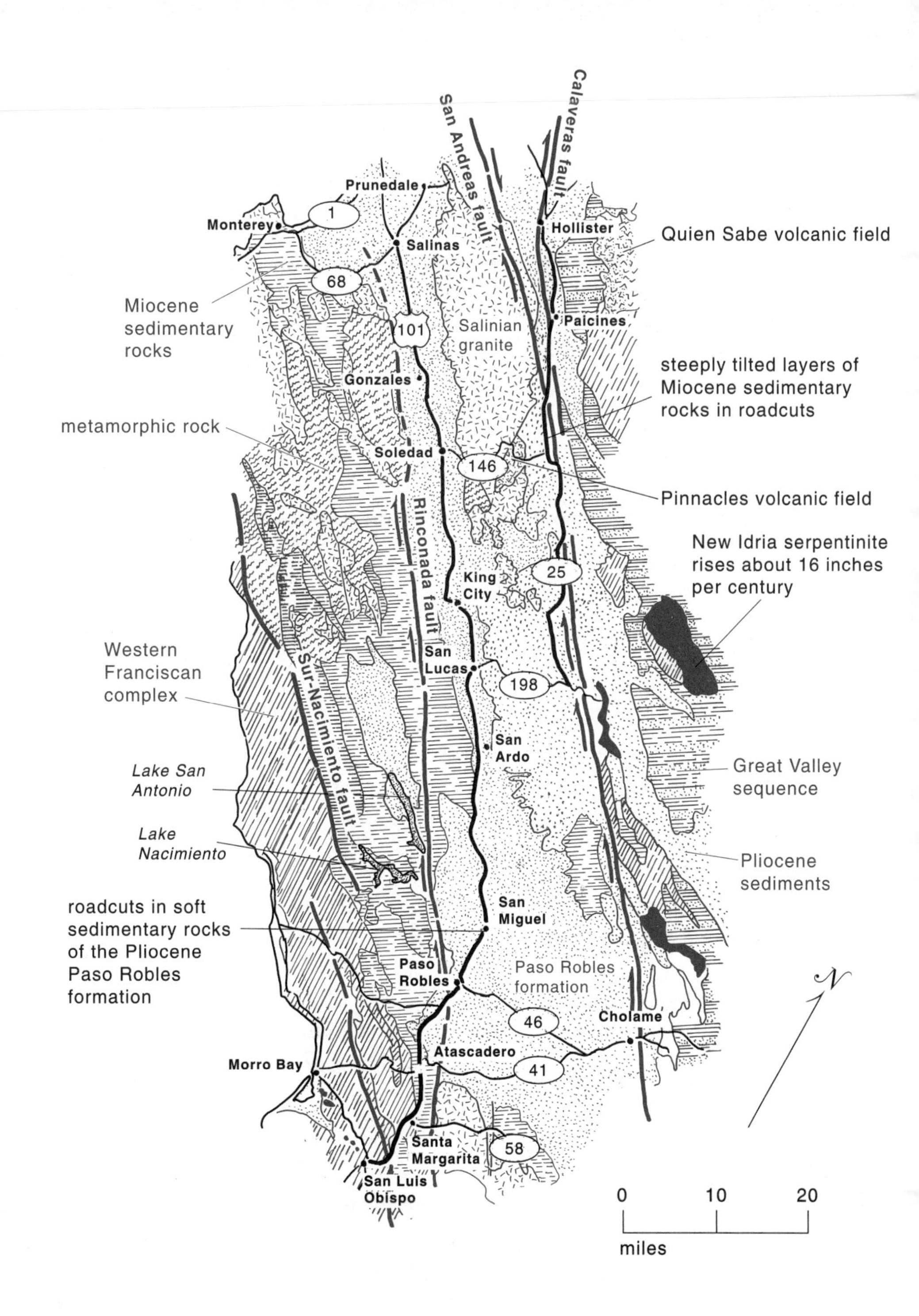

Rocks along U.S. 101 between Salinas and San Luis Obispo.

Sierra de Salinas

The high ridge of the Sierra de Salinas rises nearly 4,500 feet along the west side of the northern Salinas Valley. It consists of the Sierra de Salinas schist, a hard and homogeneous rock composed mostly of quartz and feldspar. It probably began as sandstone rich in feldspar, which was metamorphosed into schist during Mesozoic time, perhaps during the same heating that melted the granite magma of the Salinian block. Then it moved north as part of the Salinian block.

Oil in the Salinas Valley

Salinas is in the northern end of the Salinas Valley. The valley floor was below sea level collecting sediments deposited in seawater until about 2 million years ago. Oil exploration showed that granite bedrock is at least 5,000 feet below the surface, as much as 10,000 feet in some places. It is amazing to find such a deep fill in a valley only 3 to 5 miles wide.

Few geologists thought such a narrow trough was likely to contain petroleum, so most were surprised when oil was discovered near San Ardo in 1947. That is still the only part of the Salinas Valley that produces oil. The oil is trapped where layers of permeable sandstone pass laterally into rocks composed of much smaller mineral grains. Rocks with fine grains have much less permeability. Oil moves through the permeable rock, then stops where it meets the impermeable rock, a stratigraphic trap. Stratigraphic traps are hard to find, so oil fields that produce from them are scarce.

Sur-Nacimiento Fault

The highway crosses the Sur-Nacimiento fault just west of Santa Margarita. It separates rocks of the Western Franciscan complex to the west from those of the Salinian block to the east. Just east of Santa Margarita, the highway crosses the Rinconada fault into low hills eroded in Mesozoic granite, the fundamental basement rock of the Salinian block. That granite crystallized somewhere far to the south, then rode into its present position along the San Andreas and Sur-Nacimiento faults.

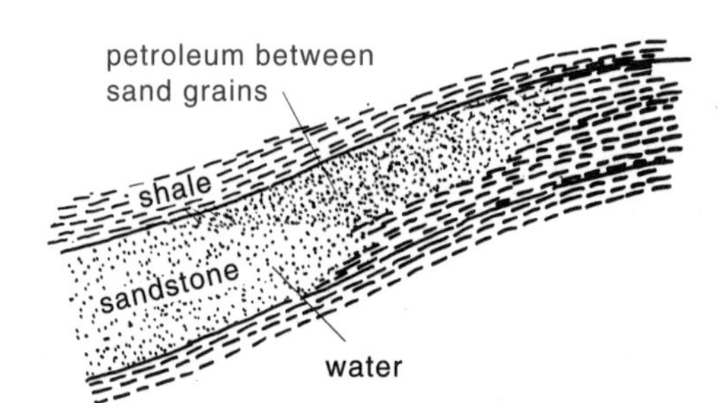

A stratigraphic trap.

Well heads beside U.S. 101 near San Ardo.

Rolling hills north of Paso Robles are eroded in sediments of the Paso Robles formation. They are too soft to make many outcrops.

Santa Lucia Range

The highway climbs through La Cuesta Pass between Santa Margarita and San Luis Obispo, near the southern end of the Santa Lucia Range. The rocks are in the Western Franciscan complex, intensely sheared sediments and several masses of greenish serpentinite. Ridges east of La Cuesta Pass consist mostly of sedimentary rocks deposited on top of the Franciscan complex during Tertiary time. Hills west of La Cuesta Pass contain a large body of intrusive serpentinite.

California 25

California 198—Hollister

64 miles

See map on page 177

Movement on the San Andreas system of faults carved the Coast Range into the long slices that make the series of narrow valleys and ridges between the Gabilan Range to the west and the southern Diablo Range to the east. All trend north to northwest. For most of its length, California 25 snakes its way from one valley to the next through this linear topography of fault slices. The main trace of the San Andreas fault is visible from the highway along much of the route.

Solid rocks are exposed in the hills on either side, but most of the road crosses soft Tertiary and Pleistocene sandstones and shales that have weathered into smooth slopes. The jagged rhyolite crags of Pinnacles National Monument punctuate the subdued topography of the Gabilan Range west of California 25.

Along the San Andreas Fault

Long stretches of the highway are almost on the San Andreas fault, the main boundary between the North American and Pacific plates. The Gabilan Range on the western skyline consists mostly of smooth, gently rolling hills eroded in granite of the Salinian block. The Salinian block is part of the Pacific plate and moving north. The lower ridges near the road nearly obscure the view of the considerably higher Diablo Range to the east, on the North American plate. Their rocks belong to the Franciscan complex and the Great Valley sequence.

The mountains east of the San Andreas fault are also moving north along other faults in the swarm but not as fast as the Gabilan Range. So the Gabilan Range is moving north relative to the Diablo Range in the same way that a passing car moves past another, even though both are going in the same direction.

Unlike the locked parts of the San Andreas fault farther north and south, this long section of the fault slips more or less continuously. Evidently, the friction along the fault is low enough to permit it to slip before high levels of stress can build. That almost certainly explains why this part of the fault has caused no major earthquakes during historic time. Seismographs do record numerous small earthquakes, most of them too weak to feel.

The creeping fault offsets roads, fences, and buildings, thus enabling geologists to measure how fast it is moving. For example, two fences and a driveway built in 1908 at the Flook Ranch now bend a little more than 9 feet where they cross the fault, an average displacement of almost 1.5 inches per year. Offset stream channels nearby indicate an average slip rate of a little more than 1 inch per year over the past 1,000 years. Continue that movement for a million years, and you have nearly 16 miles of slip.

Creeping Hollister

Hollister is famous among geologists as the town built on a stretch of the Calaveras fault that continuously slips. Here and there in parts of the older section of town, you can see cracked and buckled sidewalks, curbs, and driveways with obvious offset along the cracks.

Sidewalks laid in Hollister since the early 1900s were constructed straight, at known dates, so their offsets provide raw material for a detailed study of slip along the Calaveras fault. The fault began to creep sometime after 1929, at a rate of about 0.3 inch per year. It accelerated to twice that rate between 1961 and 1967, then slowed to an average rate of about 0.25 inch per year. It moves twice that fast less than 2 miles northwest of town, but not nearly as fast as the main trace of the San Andreas fault.

Of course the constant creeping means that the Calaveras fault is not stuck and that no strain is accumulating. The Hollister part of the fault seems unlikely to cause an earthquake. But the constant creep at Hollister does not protect against earthquakes on other segments of the Calaveras fault, or on other faults.

Pinnacles National Monument

Watch 30 or 35 miles south of Hollister for the jagged and craggy peaks of the Pinnacles. They rise abruptly from smooth grassy hills eroded in the granite basement of the Salinian block, at the southern end of the Gabilan Range. Andesite and rhyolite erupted during early Miocene time, about 23 million years ago. Geologists estimate that about 7 cubic miles of lava erupted from five main volcanic centers. At least half that volume

Rugged peaks of resistant Miocene rhyolite at Pinnacles National Monument.

Volcanic debris typical of that in Pinnacles National Monument. Maroon fragments in a green matrix. View is 12 inches across.

is now lost to erosion. What remains is in a dropped fault block in the crest of the Gabilan Range.

The Pinnacles volcanic pile is nearly identical in composition and age to the Neenach volcanic rocks west of Lancaster, in the western end of the Mojave Desert and on the opposite side of the San Andreas fault. It is hard to avoid concluding that the San Andreas fault snipped off the western part of the Neenach volcanic pile and carried it north some 180 miles.

The match between the Pinnacles and Neenach volcanic piles on opposite sides of the San Andreas fault was one of several items of evidence that convinced many skeptics of its enormous displacement. That eased their acceptance a few years later of plate tectonics theory, which requires large movements of entire continents, and was then considered revolutionary.

California 41

Morro Bay—Interstate 5

76 miles

The highway between Morro Bay and Atascadero crosses the southern, less rugged part of the Santa Lucia Range, part of the Western Franciscan complex. The road crosses the Sur-Nacimiento fault just east of Atascadero and crosses the Salinian block between there and Cholame. At Cholame, the road crosses a broad valley that marks the trace of the San Andreas fault zone. East of the fault zone lies the North American plate, where Franciscan and Great Valley sequence rocks dominate the high ridges of the southern Diablo Range. Folded ridges of Tertiary sedimentary rocks flank the western edge of the San Joaquin Valley.

Santa Lucia Range

The 16 miles of highway between Morro Bay and Atascadero wind through the southern part of the Santa Lucia Range. Franciscan mélanges, including some sizable bodies of pale green volcanic rocks and dark green serpentinite, dominate the western half of this segment. Watch for outcrops of serpentinite a few hundred yards east of the national forest boundary. That is probably the serpentinite that normally separates the Franciscan complex from the Coast Range ophiolite, although the relationship is not altogether clear.

Just west of the turn to the Cerro Alto Campground, the jumbled Franciscan mélanges give way to the Great Valley sequence—dark sandstones and shale still lying on the oceanic crust on which they were deposited.

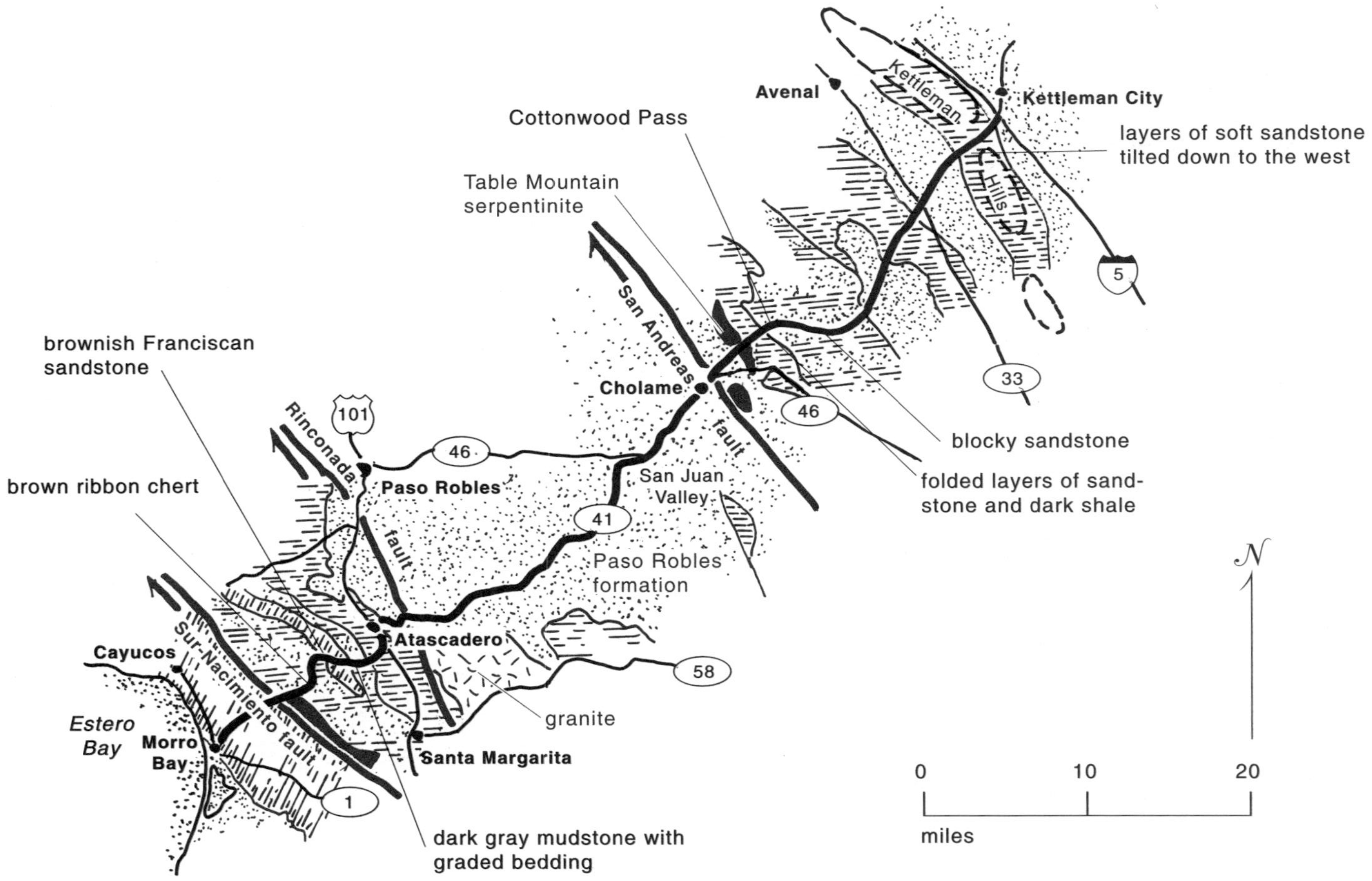

Rocks along California 41 between Morro Bay and Interstate 5.

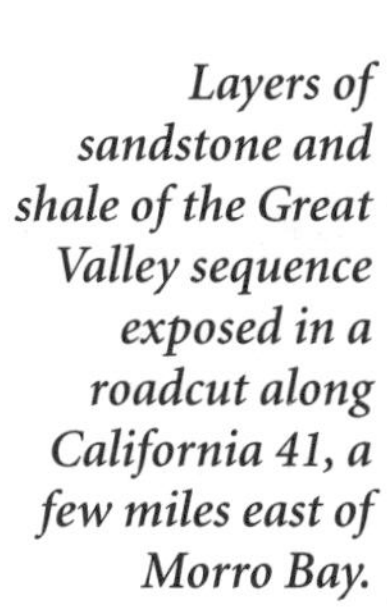

Layers of sandstone and shale of the Great Valley sequence exposed in a roadcut along California 41, a few miles east of Morro Bay.

Watch for them in the roadcuts between the campground turnoff and Atascadero.

Paso Robles Formation

The highway crosses the Salinian block between the Sur-Nacimiento fault at Atascadero Summit and the San Andreas fault at Cholame. Horizontal layers of tan silt, sand, and gravel of the Paso Robles formation completely cover the granite bedrock. A small component of vertical movement along the San Andreas fault dropped this central part of the Salinian block during Pliocene and perhaps the early part of Pleistocene time. Streams flowing from the Santa Lucia and La Panza Ranges washed sediment into the low area, depositing the Paso Robles formation. About 2 million years ago, the subsidence stopped and this part of the Salinian block rose, exposing its cover of soft sediments to erosion.

Cholame Valley

Watch about 1.5 miles west of the junction of California 41 and 46, near Cholame, for layers of tan sand and fine gravel that stand vertically. Movement on the San Andreas fault upended them. California 41 crosses the fault where California 46 branches to the southeast.

The trace of the San Andreas fault follows the southwest side of the Cholame Valley for some miles northwest, then disappears and reappears on the northeast flank of the valley, still heading northwest. The ends of the two fault segments overlap for several miles, although they are 1 mile

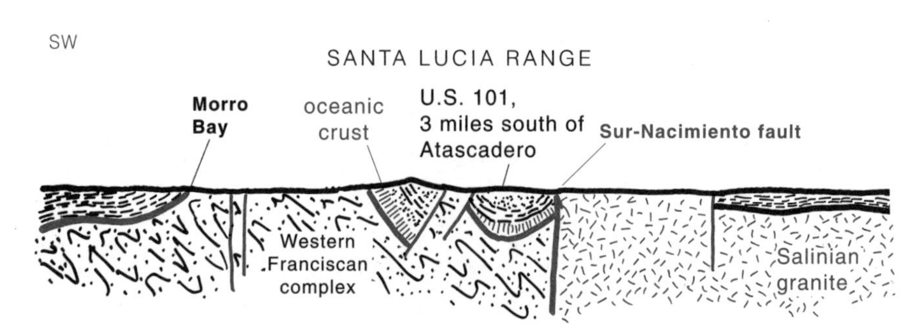

Geologic section drawn across the Coast Range south of Atascadero, across the Western Franciscan complex, the Salinian block, the Franciscan rocks east of the San Andreas fault, and into the San Joaquin Valley.

apart. Their separation opened the Cholame Valley as a basin in the Coast Range. Sediments washed in from the surrounding hills fill its floor.

Southern Diablo Range

East of the Cholame Valley, the highway crosses Cottonwood Pass and follows Cottonwood Canyon through the southern end of the Diablo Range. This is one of the longest, highest, and most inaccessible of the Coast Ranges, a formidable barrier along 100 miles of the west side of the San Joaquin Valley. The core of the range is high because it contains an assortment of resistant rocks: mélanges of the Franciscan complex, sandstones of the Great Valley sequence, and dark igneous rocks of the oceanic crust at the base of the Great Valley sequence.

Most of the roadcuts near Cottonwood Pass expose tilted beds of oceanic sandstone and shale laid down during Mesozoic time. They are interesting sandstones with graded beds, typical sediments dumped from great clouds of muddy water. Look carefully to see the individual layers grading from coarse sediment at the base to fine at the top.

Kettleman Hills

The low Kettleman Hills rise grassy brown from the flat plain of the western San Joaquin Valley. California 41 crosses their middle part just west of Interstate 5.

The Kettleman Hills are the eroded top of a fold, an anticlinal arch, in the Pliocene and Pleistocene sedimentary rocks of the San Joaquin Valley floor. Movement along the San Andreas fault a few miles to the west raised

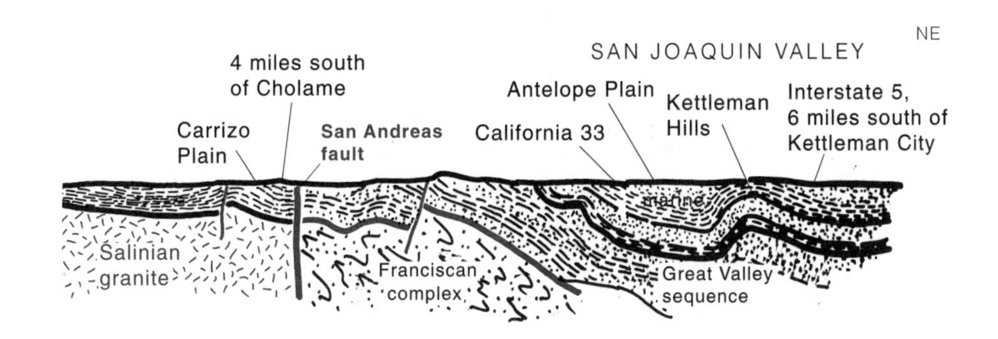

the arch as it compressed the sedimentary fill along the western edge of the San Joaquin Valley. The San Andreas fault is still moving, and the fold bends some of the youngest sediments in the valley floor, so it seems quite likely that the Kettleman Hills are still rising.

Roadcuts along California 41 offer glimpses of some of the sediments that lie beneath the Kettleman Hills and the farmland on the flat valley floor to the east. As recently as 3 to 4 million years ago, most of the San Joaquin Valley was still a bay filled with seawater, enclosed between the Sierra Nevada and the Coast Range. Rivers flowing east out of the Coast Range dumped sediment in numerous small deltas, laying down the fossiliferous sandstone and siltstone beds now visible along the highway. Watch for the brownish layers of soft sand and the conspicuous white layers that consist mostly of seashells.

The youngest sediments of the Kettleman Hills accumulated in fresh water during Pleistocene time, within the past 2 million years. Even those young rocks are folded. By this time, movements along the San Andreas and related faults had closed the formerly broad entrance to that bay, leaving an enormous lake flooding the San Joaquin Valley.

A huge oil field produces from the northern part of the Kettleman Hills. A much smaller one produces from the area south of the road. The oil comes from Miocene sediments deep within the fold. Oil collected in sandstones in the crest of the fold, trapped beneath layers of impermeable shale. Most of the production in the San Joaquin Valley comes from this and similar folds within sight of the Coast Range.

California 58

Santa Margarita—Interstate 5

87 miles

The eastern part of the route crosses the flat floor of the San Joaquin Valley, then winds through dry, rolling hills eroded in Franciscan rocks. The western part winds across the La Panza Range, passing exposures of granite in the Salinian block. The highway crosses the San Andreas fault where it separates the arid Carrizo plain from the equally barren slopes of the Temblor Range.

Salinian Granite in the La Panza Range

A short distance east of Santa Margarita, the highway enters the rolling hills of the northern La Panza Range, covered with forests of picturesque oak trees. You can actually see the basement rocks of the Salinian block, not just a cover of younger sedimentary rocks. About 20 miles of road passes numerous roadcuts in weathered granite, pale rock generously peppered with black crystals. The geologic map shows it exposed in more than 100 square miles of the northern La Panza Range. Numerous wells drilled through the sedimentary cover in the surrounding regions suggest that the same distinctive granite may exist at depth in an area several times that large, making it the most widely distributed rock in the Salinian block. Age dates show that this mass of uncommonly homogeneous granite crystallized about 80 million years ago, long before the Salinian block moved into the Coast Ranges.

San Andreas Fault in the Carrizo Plain

The Carrizo Plain is a broad depression in the dry eastern Coast Range, between the Temblor Range on the northeast and the Caliente Range on the southwest. No stream drains the Soda Lake playa in its floor.

The San Andreas fault sharply defines about 30 miles of the boundary between the eastern edge of the Carrizo Plain and the abrupt front of the Temblor Range. A small component of vertical motion on the fault raised the Temblor Range as the Carrizo Plain moved north. The unimproved Soda Lake Road follows the Carrizo Plain south past Soda Lake to California 166. It provides a magnificent view of a major fault conspicuously displayed in the landscape.

Franciscan rocks are the basic bedrock in the Temblor Range, and Mesozoic granite of the Salinian block is the bedrock in the Carrizo Plain and the Caliente Range. Thick accumulations of much younger sedimentary rocks deposited in seawater cover most of that bedrock.

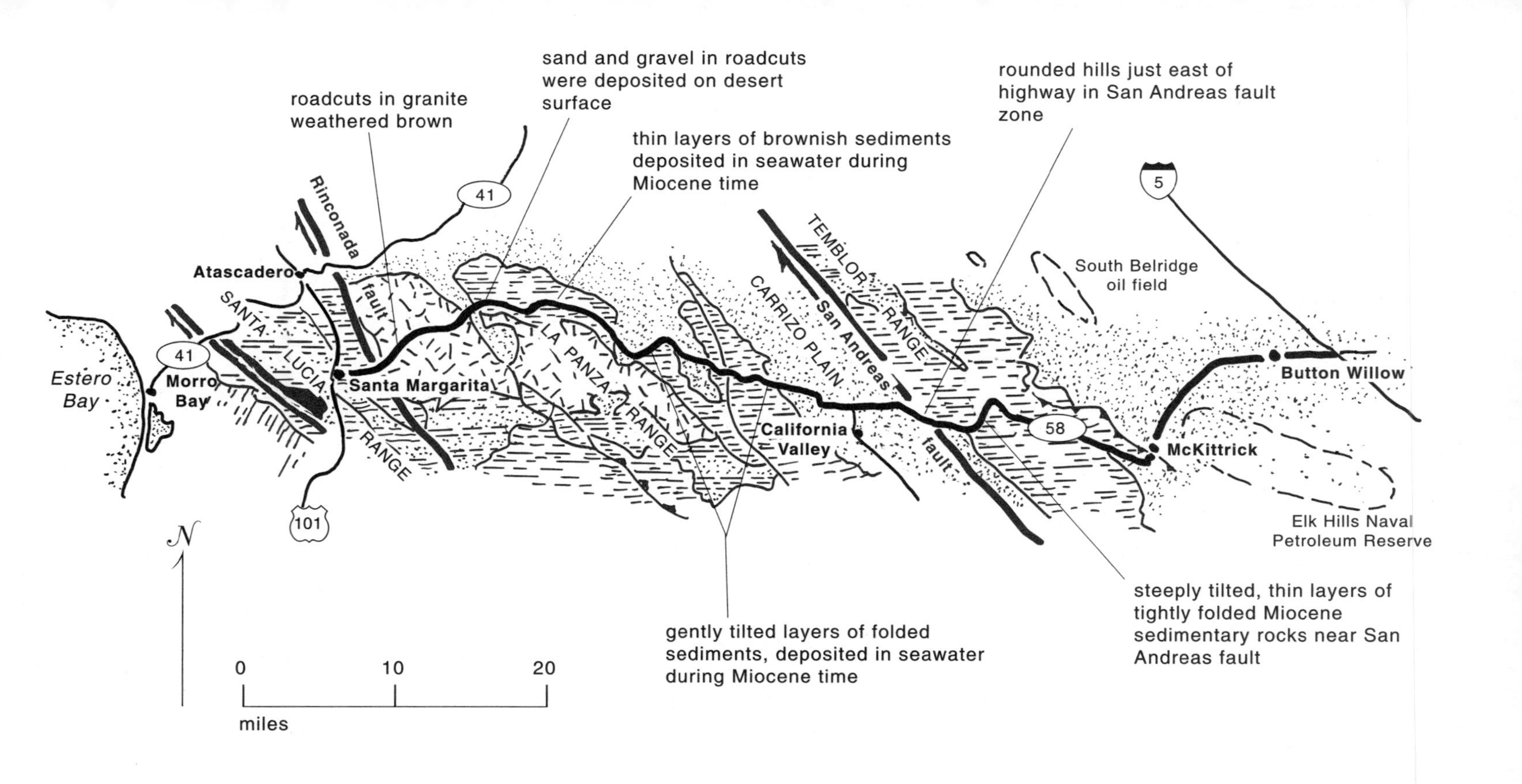

Rocks along California 58 between Santa Margarita and Interstate 5.

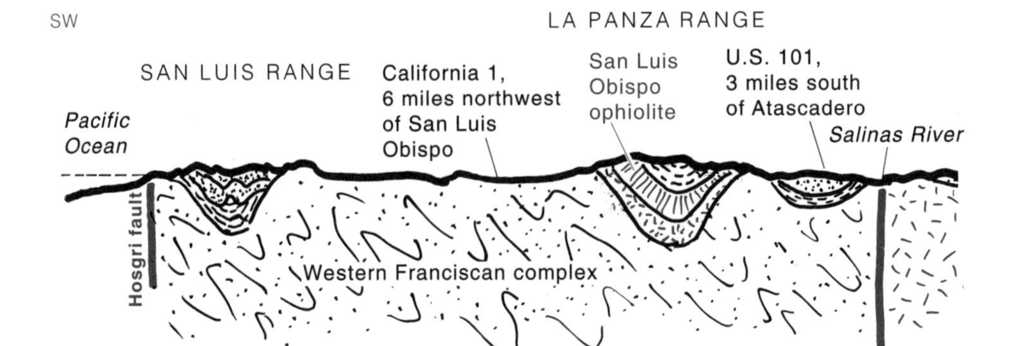

Section across the La Panza Range, San Andreas fault, and the Temblor Range. Drawn parallel to California 58 in the La Panza Range, north of the highway farther east.

The Carrizo Plain segment of the San Andreas fault moved during the great Fort Tejon earthquake of 1857. It has not slipped since, but the segment to the north has moved several times during moderate earthquakes near Parkfield, north of Cholame. The lack of recent movement, together with the sparse plant cover, has helped geologists study the record of fault movement.

Many small streams drain west off the Temblor Range, across the San Andreas fault onto the Carrizo Plain, and finally empty into Soda Lake playa. They attempt to adjust as slip along the fault offsets their channels but never quite keep up. The result is a series of bends where ridges and streams cross the fault. Many streams flow along the trace of the fault for distances of as much as a quarter mile before they break out of it to resume their normal course toward Soda Lake.

Geologists thoroughly studied one such creek about 4 miles south of California 58. It is offset about 300 feet, but an abandoned channel marks its old location, now moved about 1,000 feet north. The stream cut a new channel across the fault, rather than continuously erode a lengthening channel parallel to it. Trenches through the streambed revealed a detailed record of fault movement. Radiocarbon dates showed that this section of the fault moves 31 to 40 feet every 240 to 350 years, each time causing a great earthquake.

That movement accounts for about two-thirds of the total slip between the Pacific and North American plates. The other third may go into movement on other faults parallel to the San Andreas fault, or into folding the rocks in a direction at right angles to the direction of fault movement.

NE

TEMBLOR RANGE

5 miles north of Cholame

Monterey formation

Salinian granite

San Andreas fault

San Andreas fault looking north, circa 1965. The lowland on the left is the Carrizo Plain. The long furrow was eroded in the crushed and easily weathered rock along the fault.

—R. E. Wallace photo, U.S. Geological Survey

Movement on the San Andreas fault offset this small stream draining onto the Carrizo Plain. —R. E. Wallace photo, U.S. Geological Survey

Vertical beds of beige siliceous silt of Miocene age near the San Andreas fault, just west of the Kern County line.

If you lay a rug on a floor and push it from one edge, the rug may slide across the floor, or it may crumple into folds. The same behavior almost certainly explains the tight folds in the sedimentary layers in the Temblor Range, next to the San Andreas fault.

Temblor Range

California 58 negotiates the steep eastern flank of the Temblor Range in a series of tight curves west of McKittrick. *Temblor,* Spanish for "earthquake," seems a good name for a range with the San Andreas fault on its western flank.

During Miocene time, this area was part of the Cuyama basin, with the California coastline still somewhere to the east, probably near the present western edge of the Sierra Nevada. Sediment accumulated in the Cuyama basin to a thickness of nearly 3 miles. The San Andreas fault began to move during middle Miocene time and by several million years ago was crumpling the sediments in the Cuyama basin into folds that become tighter closer to the fault. Some are so tightly compressed that they broke and moved along small thrust faults. Folded layers of Miocene sandstone and shale appear in a nearly continuous roadcut between the San Joaquin Valley and the crest of the range.

Elk Hills Oil Field

East of McKittrick, California 58 skirts the northwest edge of the barren Elk Hills. Their smooth lump is the surface expression of a fold, an anticlinal dome broadly arched in the sedimentary formations. The rocks at the surface are Pleistocene sandstones. Any force that folds and lifts rocks deposited as recently as Pleistocene time must still be active. The smoothly bulging and uneroded surface of the Elk Hills also testifies to the extreme youth of the fold, which will probably continue to rise as long as the nearby part of the San Andreas fault continues to move.

Organic matter in sediments that were deposited in seawater during Miocene time baked into oil in the hot depths of the Cuyama basin. Then the folded arch of the Elk Hills anticline trapped the oil as it rose toward the surface. The discovery well was drilled in 1915, but commercial production did not begin until 1919, after another thirty-five mostly futile wells had been drilled. The oil field covers approximately 50,000 acres of the crest of the anticline, which is some 15 miles long. This phenomenal field produced more than 800 million barrels of high-quality oil between 1919 and 1986, when the reserves still underground were estimated at some 700 million barrels.

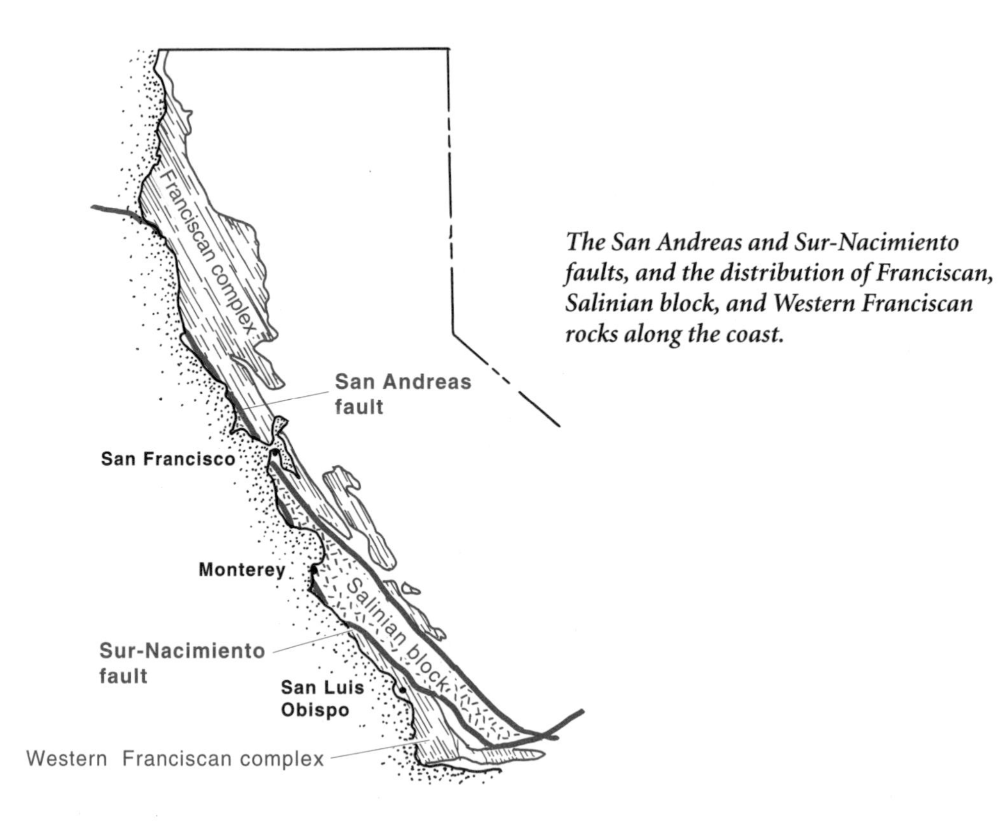

The San Andreas and Sur-Nacimiento faults, and the distribution of Franciscan, Salinian block, and Western Franciscan rocks along the coast.

Wavecut bench exposed below a sea cliff at low tide.

5

The Magnificent Coast

The coast is a creature of the breakers. They restlessly shape and reshape it, making the shorescape an artistic work in progress. From time to time, the land rises a bit and provides the breakers with fresh raw material. It is a splendid scene that changes considerably more rapidly than most of the many splendid scenes in California.

COASTAL BEDROCK

Franciscan rocks exposed along the coast north of Point Arena are much less folded and broken along faults than those in most of the Coast Range. They contain very little ribbon chert, limestone, basalt, and serpentinite. And they are not metamorphosed. They contain the fossils of animals that lived during early Tertiary time, 60 to 40 million years ago, so that must be when the rocks were deposited—the Franciscan trench got them sometime later. These rocks are less mangled than the older rocks to the east because they were last into the trench, perhaps not long before the Coast Range rose.

Coastal Franciscan rocks south of Point Arena, and in the Western Franciscan complex, are like those in the central part of the Coast Range. They are severely deformed and include the usual complement of ribbon chert, basalt, and serpentinite. Whatever fossils they may have contained are now battered and sheared beyond recognition.

Several stretches of the coast west of the San Andreas fault are on granite of the Salinian block. It is much less fractured than the Franciscan rocks, so better withstands the breakers to make bold shorescapes. Most of the Salinian granite lies beneath a cover of sedimentary rocks deposited since the Franciscan trench went out of business. Most of those sedimentary rocks are considerably folded and broken along faults, but not so severely as the Franciscan rocks.

WHERE WATER MEETS LAND

Waves rise in deep water as they take their energy from the wind, and they may travel across half an ocean or more to expend it in shaping the coast. There they become breakers, shaping the coast as they crash.

Throw a stick out of a boat bobbing in the waves in deep water offshore and watch it bob along with the boat. Neither the boat, nor the stick, nor the water travel with the waves. Only their energy travels.

Imagine tying one end of a rope to a post, then waggling the other end to send a train of waves down the rope. The rope would simply move up and down as the waves pass, but each wave would use its energy to wiggle the post. Do that long enough and you would eventually tear the post out of the ground. Let the wind drive enough breakers against a cliff and they eventually tear it apart.

Waves drag on the bottom as they enter water shallower than their length from crest to crest. As friction slows them, the waves coming fast from the deep water behind crowd in upon them. Their crests pile higher as their lower part drags on the bottom. Then the top of the wave outruns the dragging lower part, curls over, and crashes. That converts the original wave into a breaker. Large waves are longer from crest to crest than small waves, so they break in much deeper water. That is why the line of breakers shifts back and forth with the size of the swells.

Breakers approach the shore more or less directly, regardless of the wind and the direction of the waves offshore. That happens because the onshore end of a wave approaching the coast at an angle drags on the bottom, and slows, while the part that is still in deep water races ahead. The effect is similar to that of wheeling a band by having the marchers at one end of the line step along faster than those at the other. So the wave front pivots about the end in shallow water to approach the shore directly. That is wave refraction.

But wave refraction rarely works perfectly; most waves approach the coast at a slightly oblique angle. Then the swash of the incoming breaker sweeps sediment obliquely up the beach, and the momentum of the water carries into the backswash, continuing the sweep down the beach. Meanwhile, the oblique approach of the waves drives a current along the coast, carrying sediment offshore in the same direction the waves sweep it along the beach. This makes the beach a river of sand that flows along the coast. Sand that lies before you in the morning will be someplace down the beach

Sea stacks on the coast near Fort Bragg.

by afternoon. But the flow reverses as the wind changes. Sand moves one direction for a while, then back for a while when the wind changes—three steps forward and two steps back.

Sea Cliffs and Wavecut Benches

Breakers crashing against rocks drive pulses of compressed air into the cracks that all rocks contain. Those unrelenting pulses eventually pop blocks of rock out of the cliff, one after the other. That is why breakers preferentially attack heavily fractured rocks, leaving the more intact parts standing in the surf as little rocky islands, or sea stacks. Even those stalwarts eventually succumb to the unrelenting blows of the breakers.

Breakers attack the coast only in the narrow zone between high and low tide. Where they batter a sea cliff, they cut a notch into its base, undermining the rock above. Eventually the face of the cliff collapses, and the breakers wash its debris along the coast and out to sea. Meanwhile, they start cutting a new notch, preparing to drive the coast another step back into the land.

As the sea cliff retreats, step by step, it leaves a broad, gently sloping surface extending seaward, the wavecut bench. The breakers carve it right across the bedrock, almost regardless of variations in rock type. You see wavecut benches at dead low tide, especially the spring tides that come when the moon is full or new. They are fairly narrow along the coast of California, rarely more than a few hundred yards across. That is probably because the elevation of the land is not stable long enough to let a wavecut bench develop to any great width. Any change in sea level or land elevation sets the process back by raising or dropping the old wavecut bench out of reach of the breakers.

Headlands and Coves

Glaciers stacked enough water on the land to drop sea level more than 300 feet during the last ice age, moving the coast west to where the water is now that deep. As the last ice age ended, the melting ice raised sea level almost to its present stand by about 10,000 years ago—the exact timing is the subject of a debate that seems never to end. The rising sea level submerged broad reaches of former coastal land, flooding the valleys to make coves and leaving ridges projecting to sea as headlands. It flooded the lower courses of the larger streams to make deep tidewater estuaries.

Waves approaching a rocky headland wrap it in their embrace as they curl around it, focusing their energy on its tip. That explains why the breakers always crash so much higher on the tips of headlands than along nearby straight reaches of the coast. Meanwhile, the breakers sweep the debris of the headland along the coast. They dump some of it into sheltered coves and use some to build bars across the lower ends of the estuaries. So the breakers destroy the headlands and the harbors, converting an initially irregular coast into a series of smoothly sweeping curves. Some of the California coast has already reached that point, and the rest is well on its way.

Marine Terraces

Marine terraces, old strips of shallow seafloor now above sea level, make broad benches that fringe much of the California coast. View them

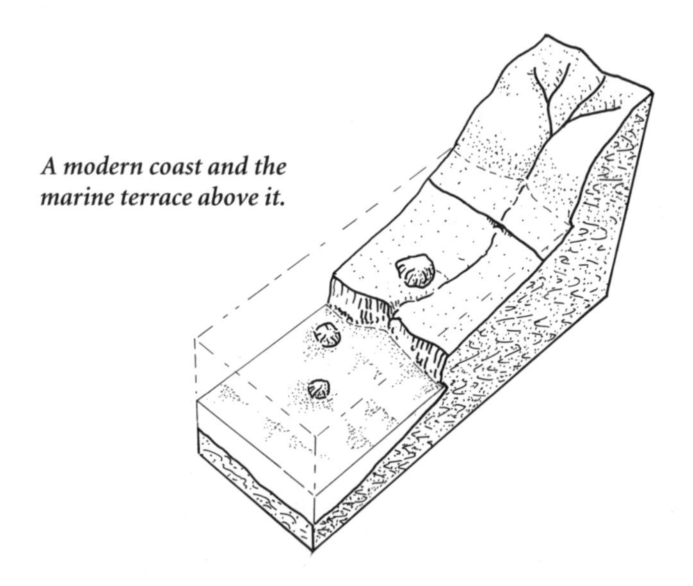

A modern coast and the marine terrace above it.

broadly and they look like giant steps that rise from the ocean to the hills—in some areas a single step, in others a short flight of steps. Long stretches of the coastal highway follow the smooth and nearly level surfaces of marine terraces.

Watch for smooth land surfaces that slope gently down toward the sea, then end abruptly at the top of a sea cliff. When they first emerged, the seaward parts of marine terraces were almost certainly deposits of sediment that the waves had worked seaward; their landward parts were wavecut benches planed across bedrock. Breaking waves have since eroded most of the sedimentary deposits, leaving a terrace that is mainly the old wavecut bench. Look on the marine terraces for old sea stacks that rise as little rocky hills. Poison oak now grows where sea lions once lounged.

No one has yet read the whole story of those terraces. The past 2 million years saw many ice ages come and go, perhaps as many as twenty. Each time, sea level fell as the great glaciers grew, then rose as they melted. Some of those interglacial high stands were above modern sea level. In some complex way, the marine terraces must tell of the long interaction between rising land and fluctuating sea level.

Coastal Dunes

The coast of California has more sand dunes than its deserts. Most form in places where rivers bring large quantities of sand to the coast, or where the breakers produce a supply of sand as they attack soft sandstones.

Ripples formed by wind in sand near Morro Rock.

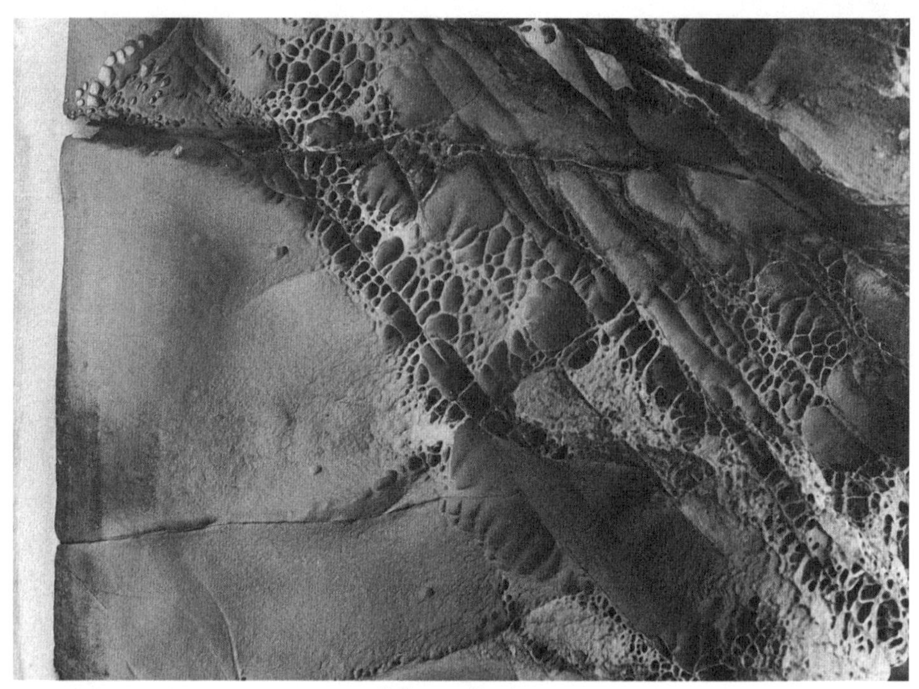

Salt weathering.

Waves wash sand high onto the beach when the tide is in, leave it there to dry in the sun when the tide is out. Strong sea breezes blowing across the beach at low tide pick up the dry sand and sweep it inland, creating dunes high above the reach of the waves. As the sea breeze drives the coastal dunes inland, beyond easy reach of the salt spray, an increasing variety of plants covers them, stabilizing the sand.

Salt Weathering

The salty sea air is at least as hard on the rocks as it is on house paint and car bodies. It also destroys rocks. Rocks within the reach of heavy salt spray weather differently and more rapidly than the same formations inland.

Seawater contains a large amount of common salt, along with much smaller amounts of many other dissolved substances. If seawater soaks into the pore spaces in a rock, then evaporates, the dissolved material precipitates within the pore spaces. The new crystals growing within the rock pry its mineral grains apart. Then some of the new crystals swell as they absorb water the next time rain or salt spray wets the rock.

Seawater is weakly alkaline. Most common rocks, except limestone, are far more soluble in weakly alkaline solutions than in water that is neutral

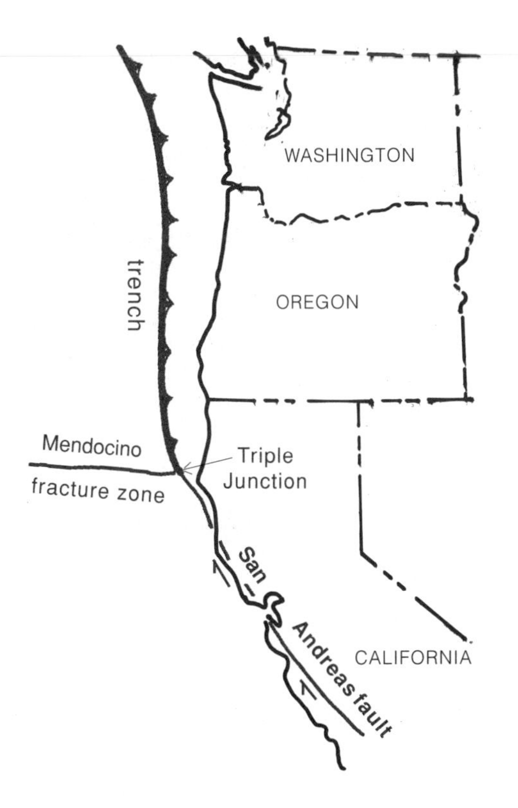

The modern trench off the coast of the Pacific Northwest is all that remains of the Franciscan trench.

or acidic. Seawater standing in a depression dissolves the rock, eventually creating a shallow pan ranging in size between a walnut and a bathtub. Solution pans are fairly common in the granites of the Salinian block, less so in the Franciscan rocks because they are too fractured to hold water.

GREAT EARTHQUAKES OF THE NORTH COAST

Geologists wondered and argued for a long time before they finally agreed that an oceanic trench really is swallowing ocean floor off the coast of the Pacific Northwest, north of Cape Mendocino. One of their main concerns was the apparent absence of the numerous earthquakes that typically come from the sinking ocean floor, from sources deep beneath the area landward of the trench. How could ocean floor sink without generating the usual earthquakes? Research since about 1980 has shown that occasional deep earthquakes do indeed strike the Pacific Northwest. They make up in ferocity what they lack in frequency.

The best evidence of large coastal earthquakes is in nondescript beds of sand a few inches to a foot or so thick. Geologists dig them up in the beds of coastal lagoons. These sand beds were laid down right across the sharply broken off stumps of trees and bushes. Tidal waves, more properly

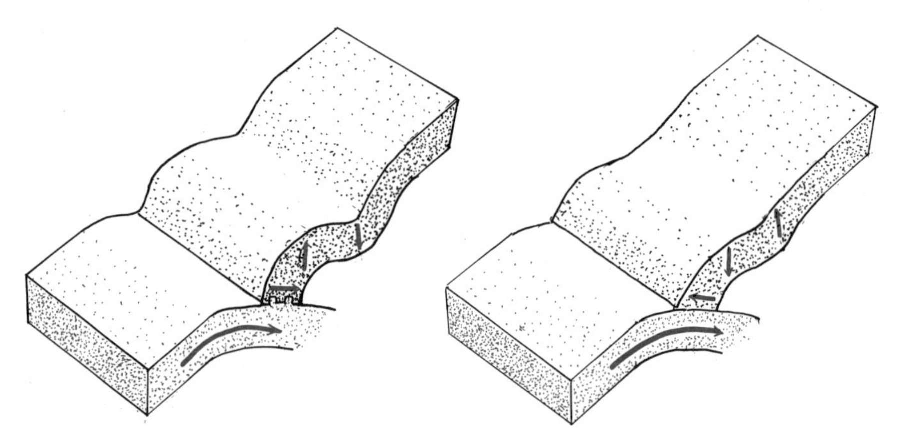

How a stuck, sinking slab of ocean floor would flex the coastal rocks onshore, and how they would rebound when it finally slips to release a great earthquake.

known as tsunamis or seismic sea waves, are known to deposit similar layers of sand, so it seems reasonable to suppose that prehistoric tidal waves deposited those along the coast. Radiocarbon dates on the wood incorporated in them show that they were deposited in several events that involved the entire section of coast from California north of Cape Mendocino to Puget Sound. Their estimated frequency is once every 300 to 600 years. Evidently, great tidal waves washed across the entire coast, probably because great earthquakes struck the entire coast.

The most recent great coastal earthquake apparently struck at nine o'clock in the evening of January 26, 1700, with an estimated magnitude of 9.0—an extraordinarily strong earthquake. That would explain otherwise mysterious Japanese records of strong tsunamis that struck their Pacific coast with no apparent preceding earthquake.

Evidently, the sinking ocean floor off the West Coast sticks against the rocks above. That drags the coast down, while raising the area a few miles inland. Precise surveys show that coastal Washington and southern British Columbia are moving in exactly that way. When the stuck slab finally snaps loose, the rocks along the coast suddenly rise, while those a few miles inland drop during a thundering great earthquake much larger than any of historic record in California. It now seems likely that the northern California coast occasionally experiences earthquakes much greater than those along the San Andreas fault.

Is it possible that some of those marine terraces along the coast north of Cape Mendocino rose from the surf in a single jump during a great earthquake? In 1992, the coastline just north of Cape Mendocino rose 4.6 feet during an earthquake of magnitude 7.1, suddenly making a new marine terrace. What geologists now know about the history of great earthquakes along the northern coast suggests that terraces rose about 800, 2,500, 3,300, and 6,300 years ago. Higher terraces south of Cape Mendocino probably rose for entirely different reasons.

POINT REYES NATIONAL SEASHORE

Point Reyes Peninsula is an afterthought of tectonics, a scrap of granite dabbed at the last moment onto the coast of northern California. According to most geologists, it lurched at least 270 miles north along the San Andreas and San Gregorio faults. It is still moving.

The fundamental bedrock of Point Reyes is granite. Watch for it in small, weathered outcrops beside the road and in the woods along the west side of Tomales Bay. Bold outcrops of granite make up the towering sea cliffs that buttress the extreme tip of the peninsula, around the Coast Guard station. You can also see granite at the north end of the peninsula, at Tomales Point and in the cliffs at Kehoe Beach. Look carefully to see that each of these areas has its own distinctive kind of granite. Age dates show that all crystallized about 100 million years ago, during middle Cretaceous time.

Elsewhere, the granite lies beneath sedimentary rocks. The oldest are sandstone and conglomerates of the Point Reyes formation that cover the

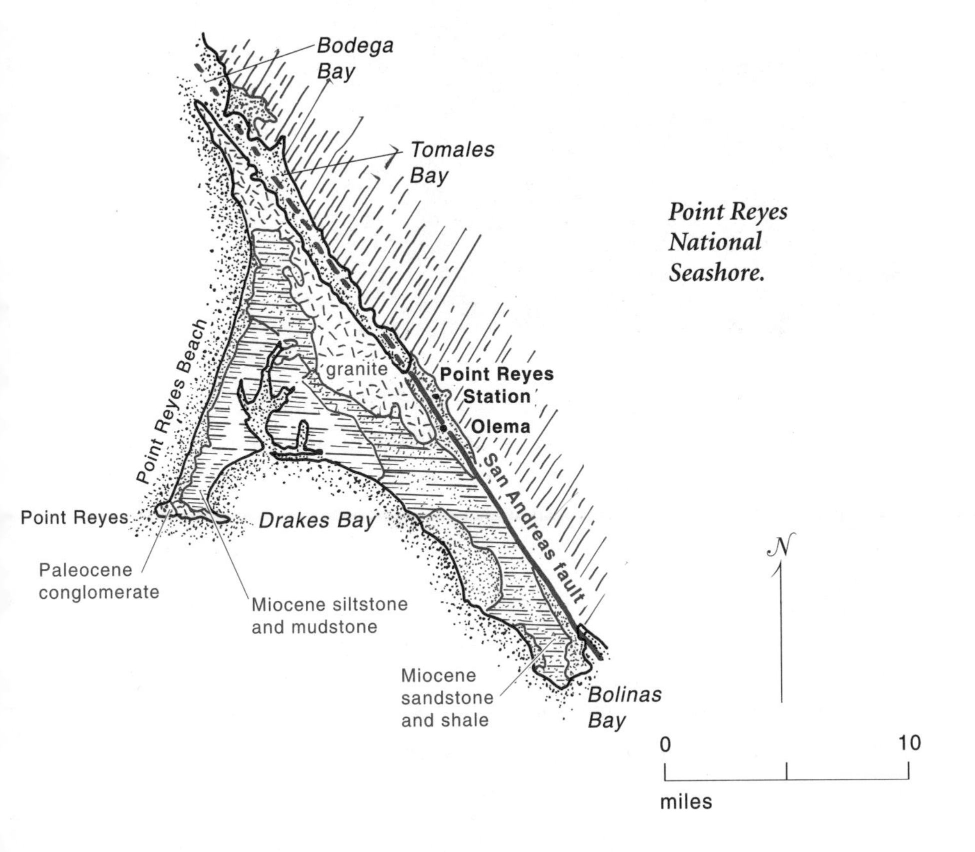

Point Reyes National Seashore.

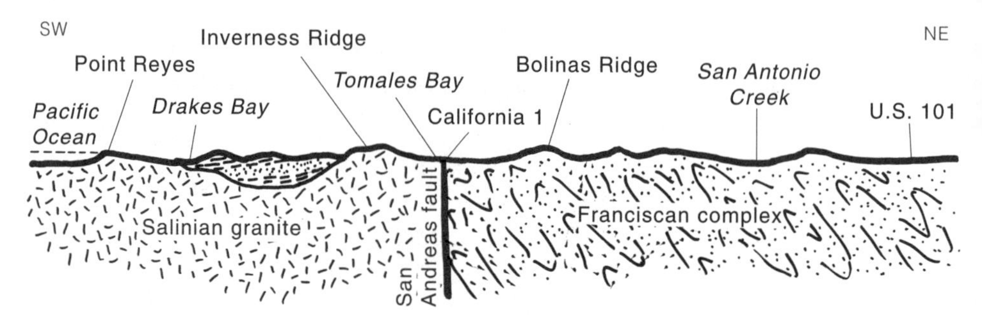

This section, drawn northeast from Point Reyes, shows the scrap of Salinian block west of the San Andreas fault and the Franciscan complex to the east.
—Adapted from Wagner and Bortugno, 1982

Fence displaced by 1906 movement on the San Andreas fault. Point Reyes earthquake trail.
—G. K. Gilbert photo, U.S. Geological Survey

granite at the tip of the peninsula. They were laid down during Paleocene time, about 60 million years ago, probably in deep water. Most of the others were deposited in shallow seawater between 20 and 5 million years ago, during Miocene and Pliocene time. They make up the long line of pale sea cliffs that face south across the sheltered waters of Drakes Bay.

Point Reyes Beach

The northwest wind builds the waves into heaping rollers as it drives them across thousands of miles of open ocean to burst onto Point Reyes Beach. They curl around the rugged granite cliffs at the extreme tip of the peninsula, focusing their energy into enormous breakers.

Pebble conglomerate and sandstone of the Point Reyes formation on Point Reyes.

The sand on Point Reyes Beach consists mostly of small red and green pebbles of chert, bits of Franciscan ribbon cherts. That is surprising because no Franciscan rocks exist on the peninsula, and Tomales Bay separates the beach from those east of the San Andreas fault. The waves may wash some of those colorful pebbles across the shallow north end of the bay, but it is easier to imagine that most of them washed down the coast and onto the beach at a time when Tomales Bay did not exist.

Drakes Bay

The same great waves that crash so heavily onto Point Reyes Beach stretch themselves thin as they wrap around the end of the peninsula to lap gently onto the long curve of shoreline along Drakes Bay on its south side. The soft beaches are made of sand eroded from the white cliffs that rise above them.

Long tradition maintains that Sir Francis Drake and his crew beached their ship in Drakes Bay to spend the winter of 1579 overhauling it before they started across the Pacific Ocean. The long row of white cliffs reminded Drake and his crew of the white cliffs of Dover. Those more famous cliffs are in chalky Cretaceous limestone; these are pale siltstone laid down in shallow seawater during late Miocene time. They contain fossil bones of seals and sea cows, among other water beasts. Many geologists correlate the pale siltstone in the cliffs at Drakes Bay with the Purisima formation of the Santa Cruz Mountains and along the coast between Martins Beach and Pescadero Point.

COAST ROADGUIDES

Long stretches of the coastal highways are within sight of the ocean, so close to the shore that salt spray dims the view through the windshield. Other long stretches dodge inland for a few miles, then return to the coast for another view of another seascape. Most of the hard rocks exposed in the sea cliffs belong to the Franciscan complex. A few sea cliffs provide views of granite and of much younger and mostly unconsolidated sediments deposited on the Franciscan complex and the granite.

U.S. 101

Eureka—Oregon

106 miles

The northern stretch of the California coast differs greatly from that south of Cape Mendocino. Hills clad in somber evergreen forests rise directly from the beach. In many places, steep cliffs and a scarcity of flat terraces along the coast force the road into the hills.

This is the northern end of the California Coast Range, where it narrows to an elongate panhandle of Franciscan rocks as little as 5 miles wide. They separate the older rocks of the Klamath block from the ocean. All the bedrock exposed along the road is Franciscan muds and sands, deposited far offshore on the deep floor of the Pacific Ocean and then stuffed into the oceanic trench about 100 million years ago. The coastal Franciscan rocks visible along the road are the least mangled and least complex of the whole lot. But a large area of severely deformed Franciscan rocks, including much blueschist, is exposed a few miles inland.

Humboldt Bay

Between the area several miles south of Eureka and that several miles north of Arcata, the highway follows the western edge of the hilly Coast Range, with Humboldt Bay and coastal lowlands to the west. Neither the bay nor the lowlands existed when sea level rose to its present stand at the end of the last ice age. Then waves washed large quantities of sediment down the coast from the north to build a large sand spit, the Samoa Peninsula, which is now the western edge of most of Humboldt Bay. Lesser quantities of sediment moving north built the much smaller sand spit that

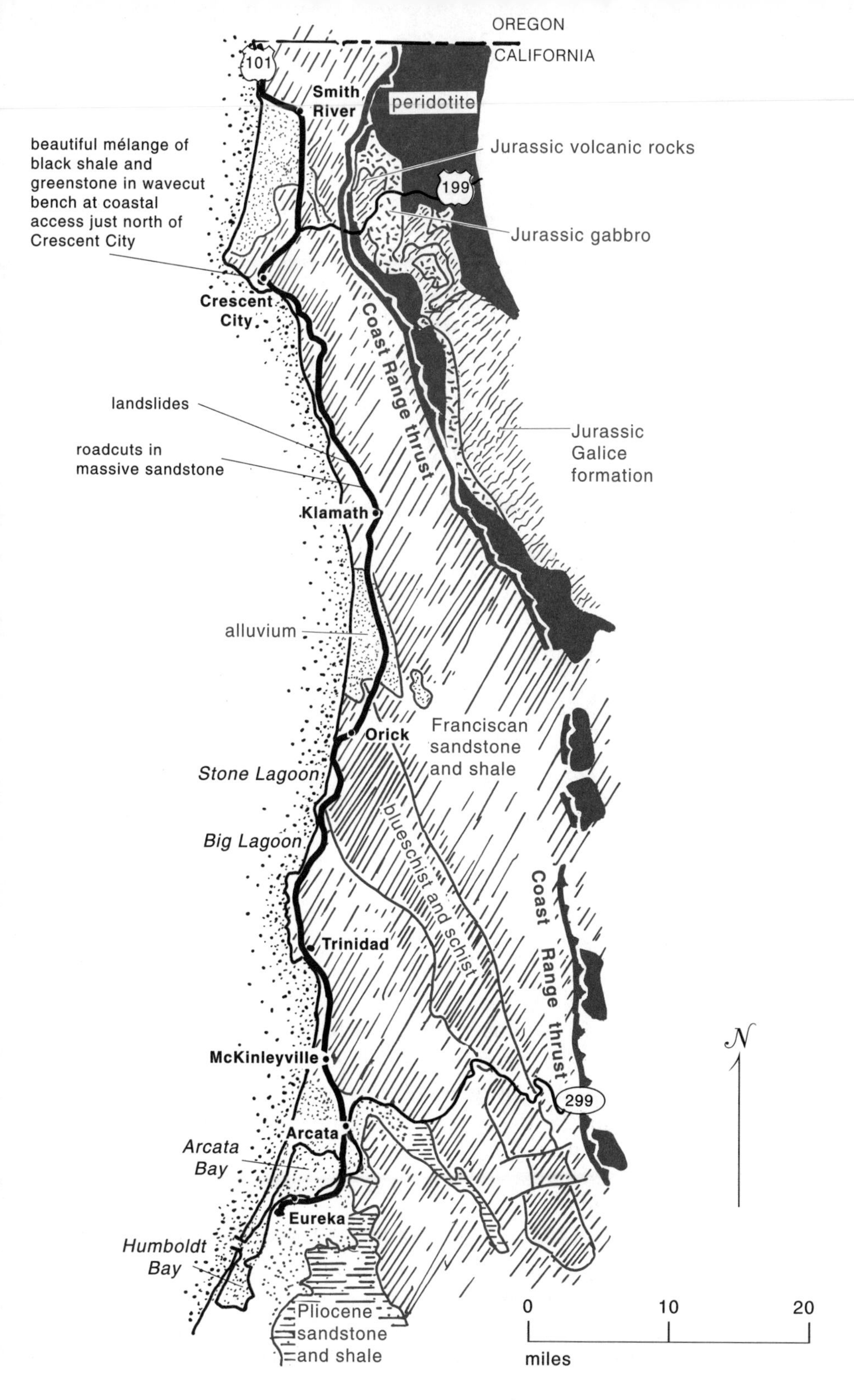

Rocks along U.S. 101 between Eureka and the Oregon line.

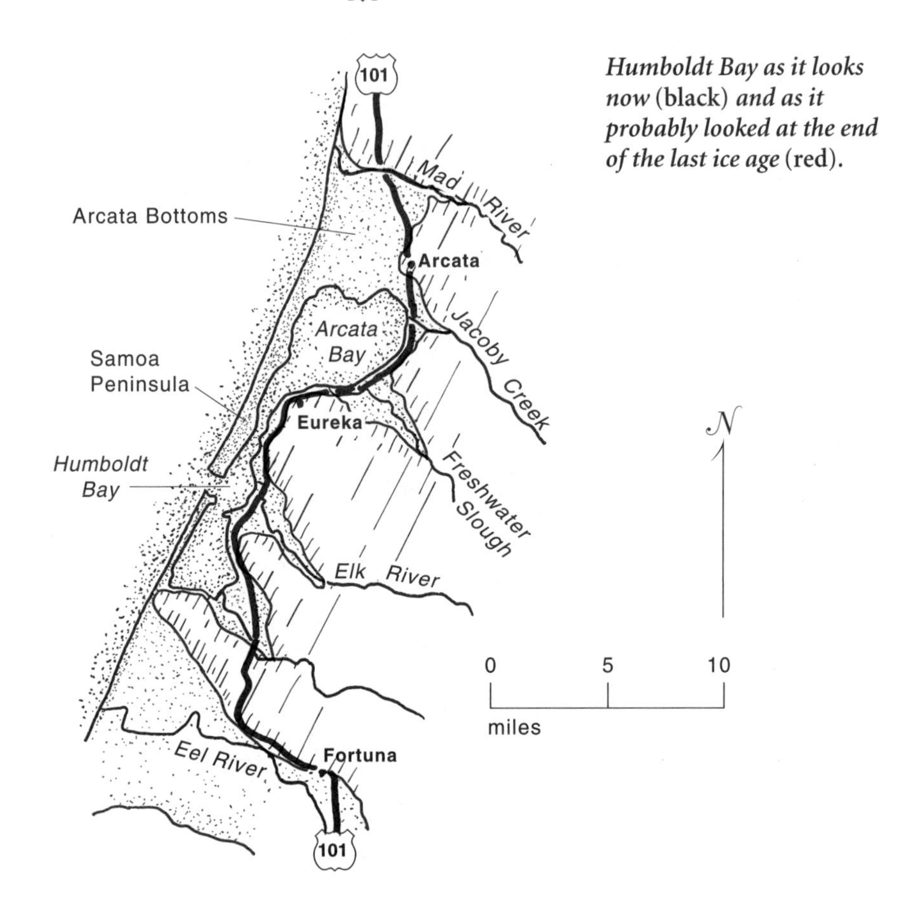

Humboldt Bay as it looks now (black) *and as it probably looked at the end of the last ice age* (red).

now bounds the southern end of Humboldt Bay. The result was a large bay, technically a lagoon, with an outlet to the ocean between the two sand spits.

Streams have since dumped enough sediment into that bay to fill large parts, converting them into coastal lowlands. The largest filled area is the Arcata Bottoms, in the north end of the original bay. The wind has also swept enough sand off the upper beach to build large tracts of dunes on top of the Samoa Peninsula.

Humboldt Lagoons

Big Lagoon, Stone Lagoon, and Freshwater Lagoon, from south to north, punctuate the steep coast between Trinidad and Orick. All were embayments in the rocky coast when sea level rose to its present stand, as the glaciers of the last ice age melted. The waves have since built sand spits across the mouths of all three, isolating them from the open ocean. The highway

Landslide beside U.S. 101.

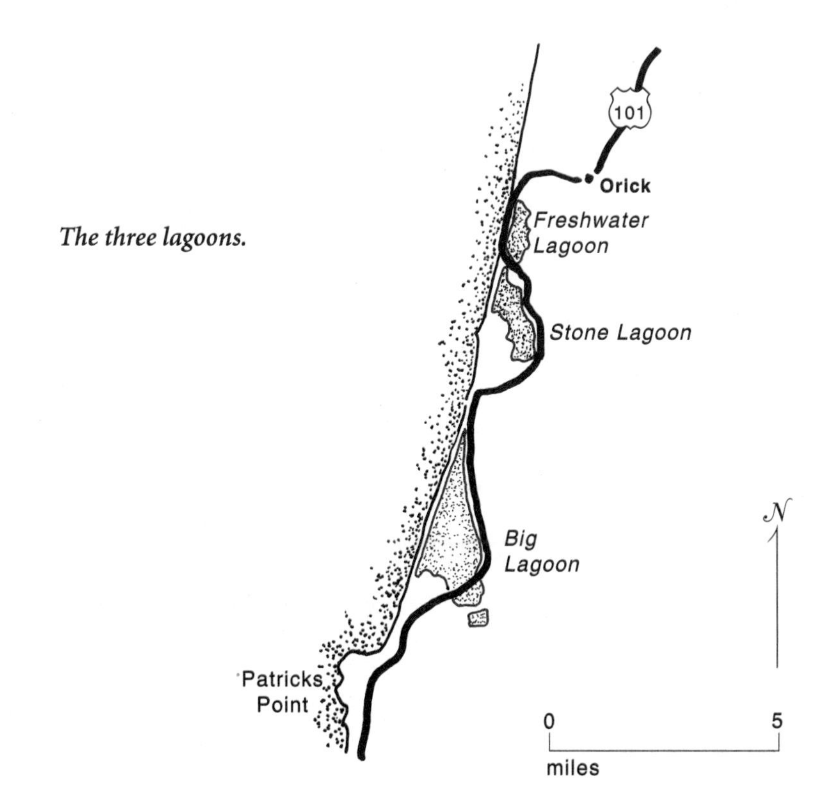

The three lagoons.

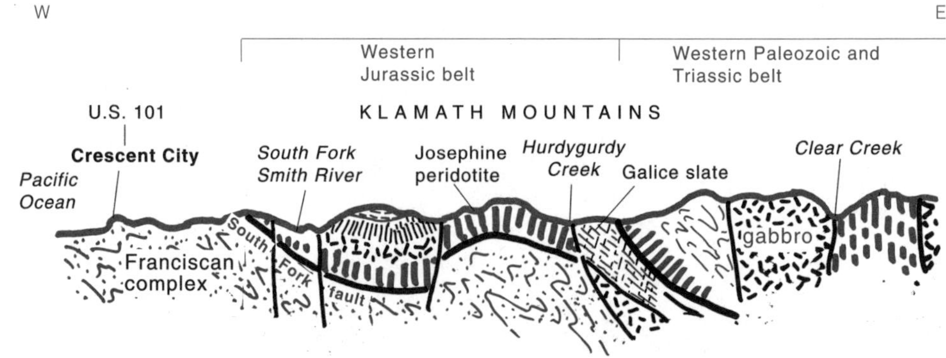

Geologic section through the Coast Range east of Crescent City. The Franciscan rocks were dragged beneath the older rocks of the Klamath Mountains. —Adapted from Wagner and Saucedo, 1987

follows the inland sides of Big and Stone Lagoons, with their impounding sand spits visible across the water; it follows the sand spit on the seaward side of Freshwater Lagoon. None of the three lagoons has an outlet to the ocean because no streams of any consequence drain into them. The small amounts of freshwater they receive seep into the ocean through the permeable sand spits.

U.S. 101 follows a strip of redwood parks between Orick and Crescent City. It loops inland in the 11 miles north of Orick to follow the floodplain of Prairie Creek, wind through the hills, and cross the Klamath River. Black shale is exposed at the base of many roadcuts. The road between Klamath and Crescent City passes occasional exposures of massive Franciscan sandstone in shades of dark gray and brown. Landslides abound.

Crescent City Coast

Crescent City faces south from low hills that must have been a small headland when sea level rose at the end of the last ice age. A shallow embayment extended north from those hills to the Smith River. Then the waves built 12 miles of sand spit and barrier island that connected Point St. George, at the tip of the headland, to the mouth of the Smith River.

The sand spit and barrier island isolated the embayment from the ocean as a large coastal lagoon. The Smith River has since dumped enough sediment into that lagoon to fill its entire northern part, leaving Lake Earl as the only watery remnant. The river now flows through the filled portion of the lagoon all the way to the ocean, carrying its load of sediment along. So Lake Earl receives too little water from the land to maintain an outlet to the ocean, and very little sediment. That explains why it is not filled.

Sediments of the filled lagoon cover a wavecut bench eroded on Franciscan rocks. The Franciscan rocks are spectacularly exposed on the

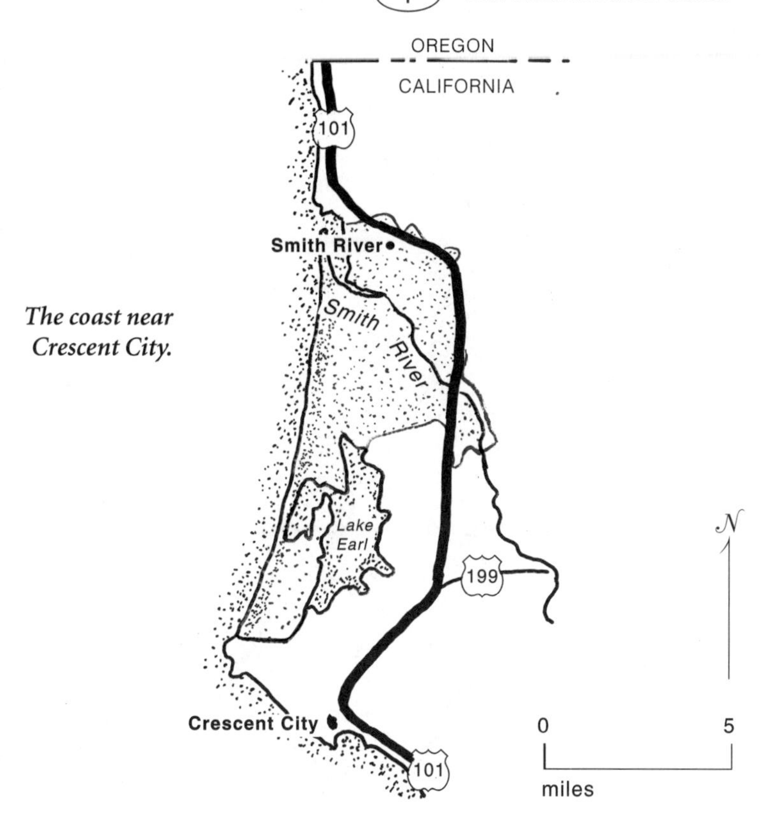

The coast near Crescent City.

broad wavecut bench at the coastal access from the west edge of Crescent City. Most of the Franciscan rocks are a mélange of thoroughly sheared greenstone full of white quartz veins. Scattered blocks of black shale, tens of feet long, are enclosed in the greenstone.

California 1
San Francisco—Monterey
125 miles

California 1 clings tightly to the coast all the way between Pacifica and Half Moon Bay. The geologic map shows that its northernmost part is east of the San Andreas fault and on the North American plate. It crosses the San Andreas fault onto the Salinian block and the Pacific plate in the south end of Daly City, near its border with Pacifica.

Franciscan rocks appear along the northern part of the road. The granite basement of the Salinian block is exposed near Montara Point and in various places around Monterey and Carmel. Elsewhere, the Salinian basement lies

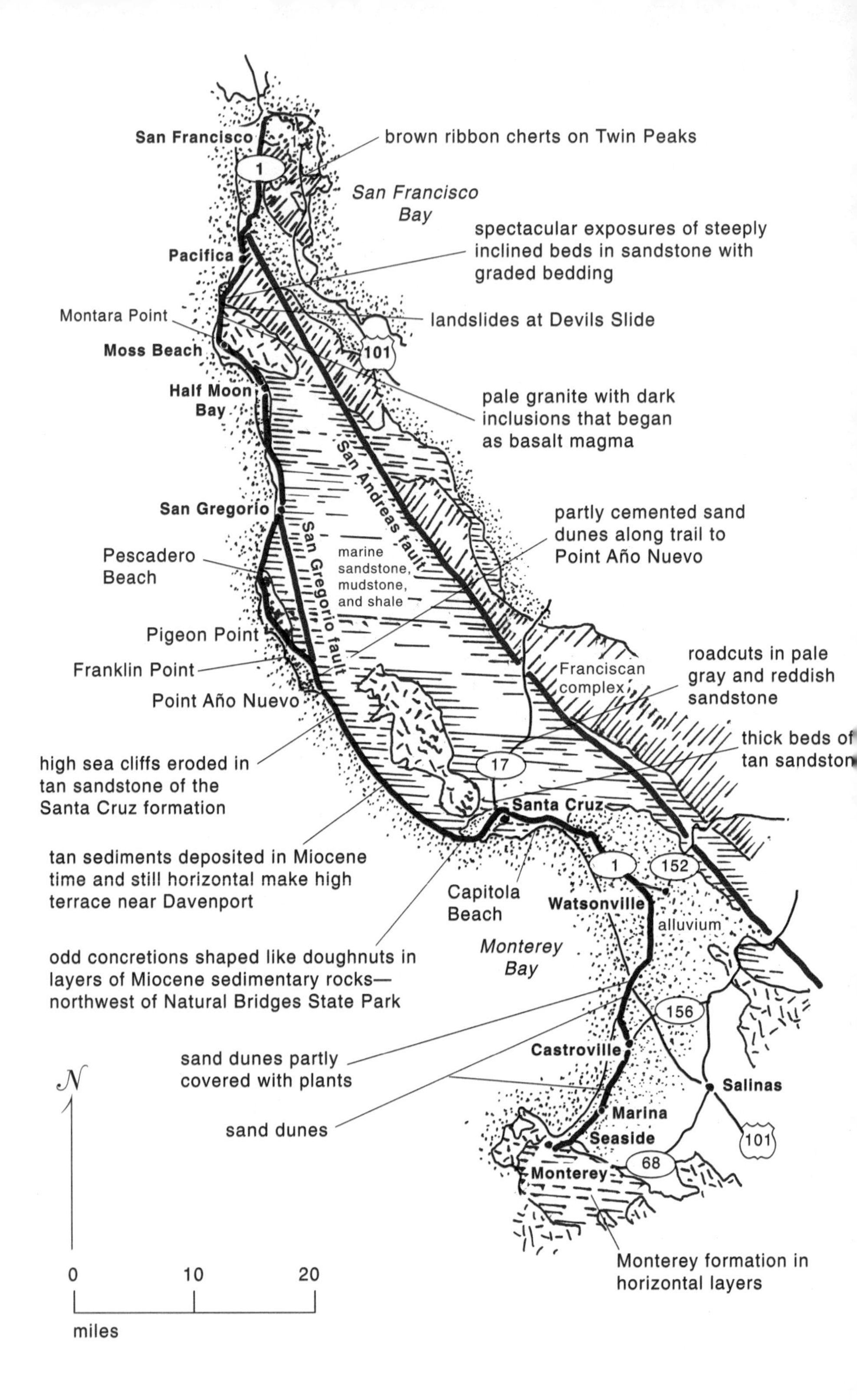

Rocks along California 1 between San Francisco and Monterey.

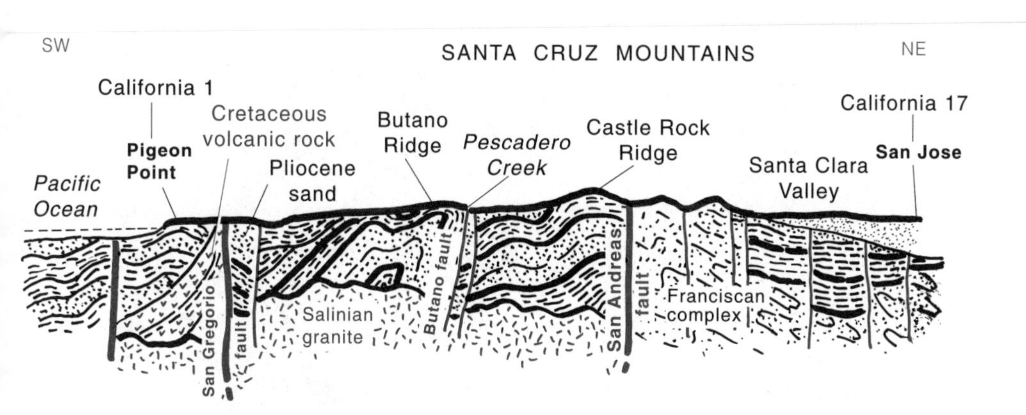

Geologic section between Pigeon Point and the Santa Clara Valley. Drag along the San Andreas fault probably caused most of the strong deformation of the younger sedimentary rocks on the Salinian granite. —Adapted from Wagner and others, 1991

beneath a cover of sedimentary formations, most of them soft, almost all of them deposited in shallow seawater during Tertiary time.

Between Half Moon Bay and Santa Cruz, the highway follows a long stretch of nearly undeveloped coast, where fields and pastures stretch from the road to the edge of the sea cliff, almost as though the calendar had slipped a century. The road between Santa Cruz and Monterey passes through a varied series of coastal landscapes.

San Francisco—Montara Point

The San Andreas fault goes to sea along the straight escarpment about a half mile north of Mussel Rock. The fault moved a few feet in 1957 during a moderate earthquake of magnitude 5.5. Rocks exposed in the cliffs are sediments of the Merced formation, which were laid down during Pleistocene time. They are soft, weak, and sliding. A large landslide that opened a deep cove in the sea cliff many years ago now threatens the residential developments perched on the edge of the cliffs around the slide scar.

Between the south end of Pacifica and Montara Point, the highway passes through a magnificent coastscape eroded into the resistant granite. Roadcuts expose pale gray rock deeply weathered to shades of reddish brown. Age dates show that the granite is 88 million years old, a late Cretaceous age. Many geologists think it was probably detached from a former southern extension of the Sierra Nevada, but they differ in their ideas of the exact source.

Tilted Purisima formation sandstone makes resistant ribs on a wavecut bench at James Fitzgerald Marine Preserve, just north of Half Moon Bay.

Breakwater at Half Moon Bay

The map view of Half Moon Bay shows a beautifully simple outline that spirals smoothly from a tight curve at the north end to an open curve at the south. A similar curve appears in the expanding spirals of seashells. Waves coming from the northwest and bending around Pillar Point shaped it as they attacked easily erodible sedimentary formations along the coast to the south. The spiral curve of the coast is typical of those in which an exact equilibrium exists between the rate at which waves erode sand from the sea cliffs and the rate at which they move it off down the beach. Such coasts are normally fairly stable.

The breakwater that now protects Pillar Point Harbor from the southwest swell was built in 1961. It destabilized the coast. Before then, those waves dissipated their energy in the north end of Half Moon Bay; since then, they reflect off the breakwater and on to the beaches farther south. The increased wave action rapidly eroded sand off the beaches, while the new breakwater caught the southbound sand that might have replaced it.

The beach has shriveled since 1961, so the sand no longer absorbs the force of the breakers as it once did. Waves have since eroded the unprotected sea cliff at about four times the previous rate. Large boulders dumped along the cliff to protect it disappear into the sand during heavy winter storms. Meanwhile, sand has accumulated in the protected area behind the breakwater.

Purisima Formation

Many sea cliffs at the north end of Half Moon Bay and in the area between Martins Beach and Pescadero Beach expose the Purisima formation, which consists mostly of mudstone, sandstone, and siltstone. It was deposited in shallow seawater during Pliocene time, several million years ago. Some geologists believe that this formation correlates with the rocks exposed in the white sea cliffs that line Drakes Bay, on the south side of Point Reyes. If they really are the same, movement on the San Gregorio fault has put about 75 miles between the two parts of the formation within the past few million years. Movement at an average rate of about 2 inches per year causes a displacement of some 30 miles per million years.

Movement on the Seal Cove fault explains why this segment of the coast is so straight and steep. It also explains why the Purisima formation is so sheared and folded. Certainly the waves find it easy to undercut the sea cliffs, collapsing them into the surf.

Pigeon Point Formation

Along most of the way between Pescadero Beach and Franklin Point, the road crosses the Pigeon Point formation, a thin fringe of sedimentary rocks along the coast. It is well exposed in the sea cliffs and at low tide in the wavecut benches.

The Pigeon Point formation is a sequence of about 8,000 feet of mudstone, sandstone, and gravel conglomerates, all deposited during late Cretaceous time, about 75 to 65 million years ago. The great thickness of the deposit and the rocks it contains both suggest that it was laid down in a deep submarine canyon. Much of the formation consists of dark gray mudstones that contain layers of muddy sandstone—stuff with a distinctly oceanic look. The sandstone layers are coarsest at the base and become progressively finer upward. That is typical of deposits laid down from clouds of muddy water draining down an undersea slope.

The Pigeon Point formation was deposited on the granites of the Salinian block while they were still somewhere far south of where we now see them. Although the formation is on the coast now, it was probably deposited some distance off some vanished shore. The base of the formation is not exposed, so its exact relationship to the Salinian granite is unclear.

San Gregorio Fault Zone

The San Gregorio fault zone branches west off the San Andreas fault at Bolinas Bay and trends south along the coast to Monterey Bay. Some parts of the fault are offshore, some onshore. Those onshore have been studied rather closely. Like all members of the San Andreas family of faults, they slip mainly horizontally, with the side west of the fault moving north.

Monterey formation on the marine terrace at Point Año Nuevo.

Some estimates place the total displacement along the San Gregorio fault zone at about 95 miles. That estimate is based partly upon the close similarity of some of the granite on the Point Reyes Peninsula to the Hobnail granite on the south side of Monterey Bay and on the opposite side of the San Gregorio fault.

Faults at Point Año Nuevo

At least five faults cut across Point Año Nuevo. The sea cliffs along its south shore provide good exposures of several, and the marine terraces reveal their recent movements with special clarity. Terraces are typically smooth and nearly flat. Their nicely planed surfaces provide a perfect setting for studying the movements of faults that break them.

The Año Nuevo thrust fault cuts across the seaward end of the marine terrace on a northerly trend. Every time the fault moves, the landward side jams up over the seaward side, raising a low scarp across the marine terrace. You can see that scarp on the terrace surface and the fault in the sea cliffs along the south side of the point. The fault appears to have moved at least four times since the terrace rose, displacing its surface about 17 feet. Fossils in the deposits on the terrace surface are about 100,000 years old, so the terrace rose since then. The sea cliff farther east along the south side of Point Año Nuevo provides good exposures of the Frijoles fault.

Doomed Dune Field of Point Año Nuevo

Strong northwest winds blow sand from the upper beach on the north side of Point Año Nuevo into the dunes that march southeast across the point and over the sea cliff on its south side. Aerial photographs taken over a period of years show that the dunes move about 40 to 80 feet per year—across the whole point in a century. Old maps and photographs show that the dune field between the two beaches shriveled considerably during the twentieth century and that plants spread into large areas. Evidently, the supply of sand has diminished, and the plant cover has increased. The rapid spread of plants began here during the 1940s but not in any other coastal dune field in California, so the cause was local. It seems likely that irrigation water from the farm fields immediately northeast of the dune field has been soaking into the sand, raising the water table. Fresh water seeps from the sea cliffs.

The journal of Sebastian Vizcaino, who sailed past in 1603, contains the earliest account of Point Año Nuevo. He did not describe an island off the point, perhaps because none then existed. Maps from the nineteenth century show a much larger beach on the north side of the point than now exists. Some geologists argue that sand deposits connected Año Nuevo Island to the mainland until sometime in the 1700s, when accelerated beach erosion

Natural Bridges State Beach at Santa Cruz in 1977. The bridge collapsed a few years later.

separated the island from the point. If movement on the San Gregorio fault zone dropped Point Año Nuevo, that would have exposed the sand deposits to more wave action. Soon after the waves isolated Año Nuevo Island, they would have carried most of the sand off the beaches along the north side of the point, thus cutting the supply of sand to the dune field.

Capitola Beach

Capitola Beach, just south of Santa Cruz, disappeared after the Santa Cruz Small Craft Harbor was built in 1965. Evidently, the jetty trapped sand that the waves would otherwise have washed down the coast and onto Capitola Beach. Like most beaches, that at Capitola had been in balance, with the same amount of sand washing in and washing out. When the jetty reduced the flow of incoming sand, the beach eroded for lack of nourishment. Studies made before the small craft harbor was built predicted that it would cause beach erosion to the east.

After its beach disappeared, Capitola built a groin, a wall across the beach that extended some 200 feet into the surf, and hauled in some 2,000 truckloads of sand to restore the beach. The restored beach eroded during the stormy winter of 1977–78. Since then, the groin has trapped enough sand to restore much of the original beach.

Loma Prieta Earthquake

October 1989. Baseball fans were just settling in front of their television sets to watch the third game of the World Series when they saw the light towers in San Francisco's Candlestick Park begin to sway. Television brought the Loma Prieta earthquake to the entire country.

The earthquake happened as the San Andreas fault slipped at a depth of about 6 miles beneath an epicenter at Loma Prieta, about 5 miles north of Santa Cruz, some 75 miles south of San Francisco. The movement did not break the surface. Its magnitude of 6.7 qualifies it as a strong earthquake but not by any means the Big One that lurks so ominously in the future of the Bay Area. It damaged the same parts of San Francisco that had shaken most violently in 1906, areas developed on watery bay mud and unconsolidated fill. It also devastated Watsonville and several nearby communities.

Monterey Bay

Monterey Bay is at the head of Monterey Canyon, a deep undersea valley that cuts seaward across the continental shelf. We suspect that the canyon must somehow explain the existence of the bay, but we do not understand the exact relationship. In any case, this was one of the first submarine canyons discovered and is one of the most thoroughly investigated.

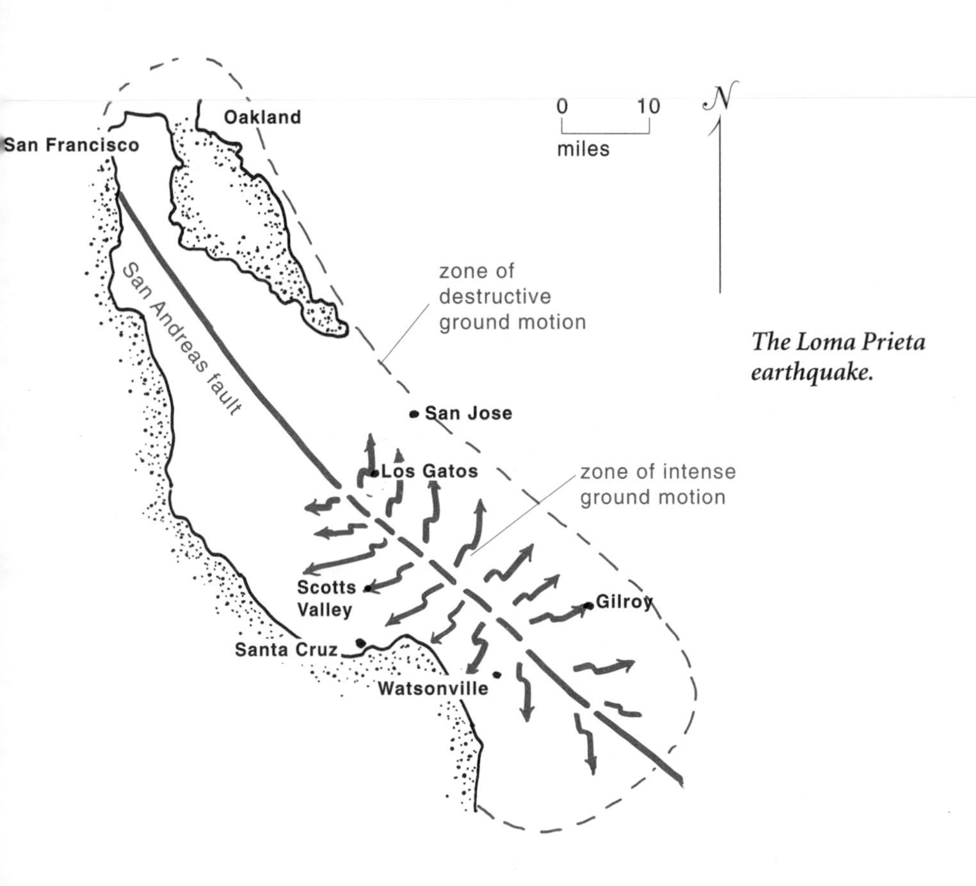

The Loma Prieta earthquake.

These gaping cracks opened near the epicenter north of Santa Cruz during the Loma Prieta earthquake of 1989.
—G. Plafker photo, U.S. Geological Survey

Submarine canyons were a mystery when geologists first learned of them during the 1930s. Everyone then assumed that nothing resembling stream flow existed anywhere on the ocean floor. How then to explain deep canyons that start offshore and continue to the deep ocean floor? Monterey Canyon was an especially troublesome case because it is eroded in granite, a resistant rock.

Theories abounded and proliferated for decades. Many geologists argued that submarine canyons were eroded above sea level, then submerged as the land sank or sea level rose. Others suggested that they are ancient features, preserved underwater for billions of years. Again, Monterey Canyon was troublesome because it is in Cretaceous granite—not all that old, as granites go.

As the fog of confusion eventually burned off, it became clear that submarine streams erode submarine canyons in basically the same way that subaerial streams erode subaerial canyons. Water with sediment suspended in it flows downslope along the bottom because it is denser than the clear water above. Most of those flows appear to happen episodically, when great clouds of muddy water pour down the slope—think of them as the submarine equivalent of floods on land. The hard particles of sediment suspended in the flowing water abrade bedrock exactly as sediment suspended in ordinary streams abrades bedrock. Submarine canyons are really not all that different from subaerial canyons.

Flatlands and Vegetables

The broad stretches of flat ground between Watsonville and Marina are the filled estuaries of the Pajaro and Salinas Rivers. When sea level rose as the glaciers of the last ice age melted, these were deep embayments in the coast. Then waves built sand spits across their mouths. The rivers filled them with sediment. Now they produce enormous crops of succulent vegetables.

Monterey Formation

The thin layers of soft, tan sediments in the eastern part of Monterey belong to the Monterey formation, which was deposited during middle Miocene time. It appears here and there in the Coast Range, from Point Reyes south.

The Monterey formation is, by any reckoning, the strangest and most valuable rock formation in California. It contains a thick interval of dark shale generously loaded with organic matter, phosphate rock, diatomite, and other commodities. The Monterey formation is the source of most of the oil production in southern California. Few rock formations anywhere have yielded so much oil from such small areas.

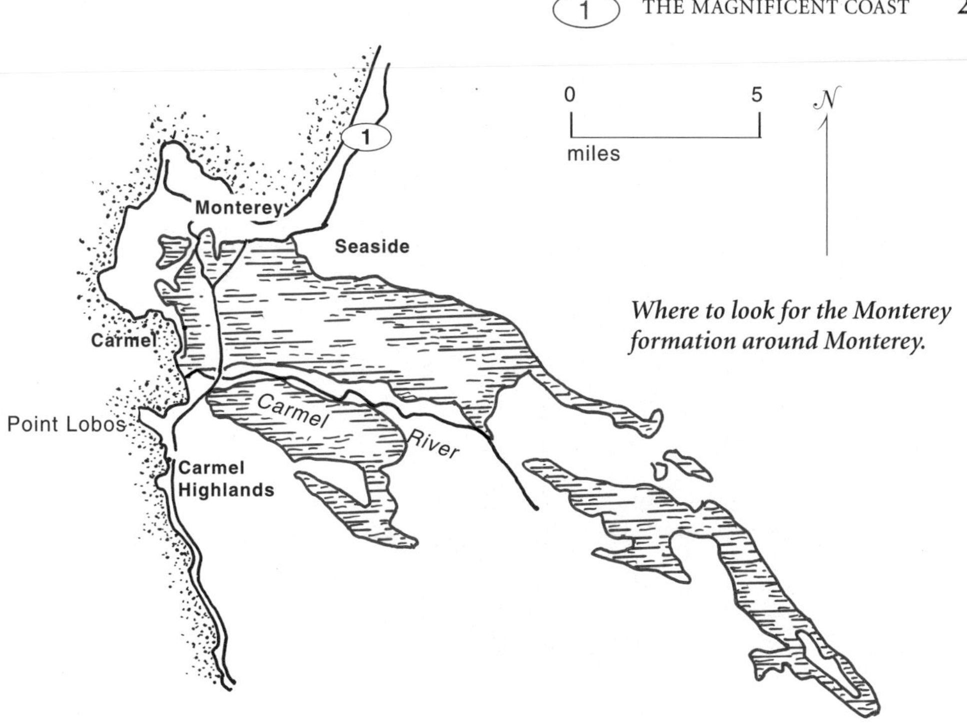

Where to look for the Monterey formation around Monterey.

So far, the Monterey formation has yielded very little phosphate, but the reserves are large and the potential considerable. Phosphate rock is used in making fertilizer, soap, and ammunition. Demand for all three should increase with the world population.

Mines around Lompoc, in southern California, produce large quantities of diatomite from the Monterey formation. Diatomite is a snowy white sediment composed mostly of the skeletal remains of diatoms, one-celled algae that live in shallow seawater. Through a microscope, diatoms look almost like fine lace. A large number of those makes a filter fine enough for such purposes as filtering yeast out of beer. If you touch your tongue to a freshly broken surface of diatomite, it will stick because capillary action pulls the moisture from your tongue into the diatomite.

Age dates show that the dark shale interval in the Monterey formation is between about 17 and 15 million years old. That is the age of the great flood basalt lava flows that cover much of the Pacific Northwest, including the Modoc Plateau in northeastern California. The association does not appear to be coincidental.

Abundant evidence of various kinds and from many places shows that the climate was very warm between 17 and 15 million years ago. We believe that the flood basalt eruptions released enough carbon dioxide into the atmosphere to cause a severe global greenhouse effect.

Diatomite of the Monterey formation along U.S. 101 at Monterey.

Plants use atmospheric carbon dioxide as a source of carbon for their tissues. Algae, including diatoms, could have deposited the enormous amount of organic matter in the dark shales of the Monterey formation. That was what eventually cooked into oil. And algae are certainly the source of the diatomite.

Basalt lava flows also release substantial quantities of phosphorous dioxide; this reacts with water vapor in the atmosphere to make phosphoric acid, which comes down in the rain. The enormous flood basalt flows must have added large amounts of dissolved phosphate to the ocean. Phosphate rock is rare. Most of the world's minable supply was deposited during times when flood basalt lava flows were erupting somewhere.

California 1

Monterey—San Luis Obispo

157 miles

Long stretches of California 1 cling precariously to the steep slopes of the rugged Santa Lucia Range, between roadcuts on the east and long drops of fresh air on the west. Other long stretches cross flat marine terraces that emerged from the sea with the rising Santa Lucia Range. Beaches, many of them nearly inaccessible, softly line the coves. Rocky headlands resist the crashing surf. The few small towns and many visitors hardly diminish

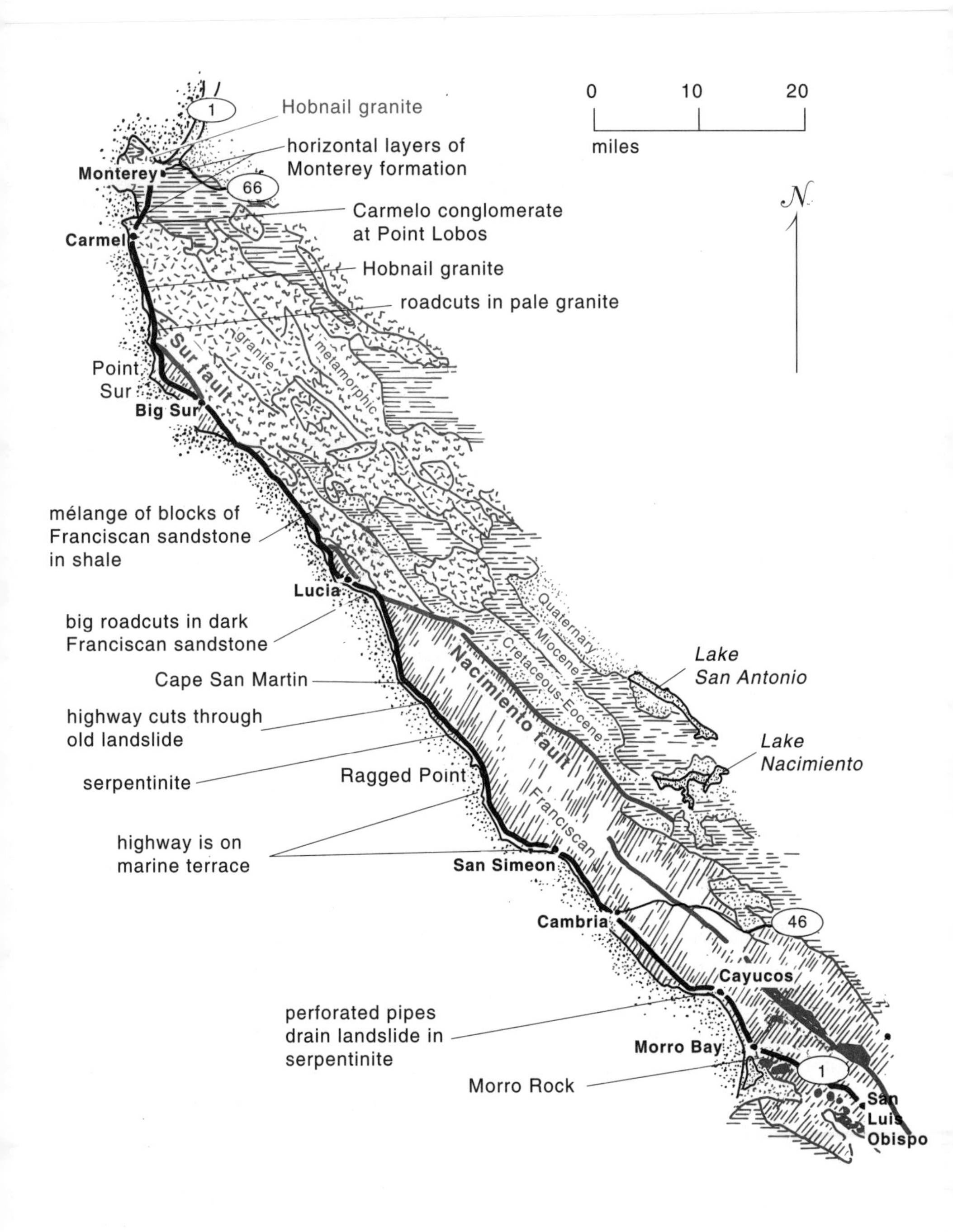

Rocks along California 1 between Monterey and San Luis Obispo.

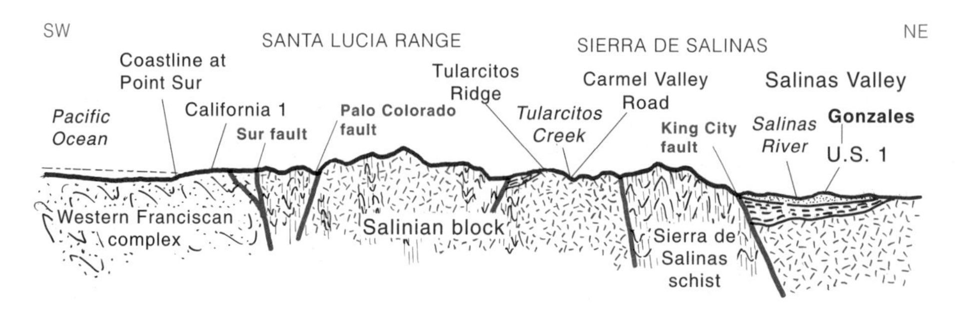

Geologic section across the Coast Range northeast from Point Sur. The Sur fault separates the Western Franciscan complex rocks from the complex rocks in the Salinian block. —Adapted from Ross and McCulloch, 1979

the sense of abiding wildness. Soil and brush cover most of the bedrock, except along the coast.

Franciscan rocks in somber shades of gray appear along the southern two-thirds of the route. The geologic map shows that they are west of the Sur-Nacimiento fault, part of the Western Franciscan complex. Farther north, the Sur-Nacimiento fault, the western boundary of the Salinian block, plays hide-and-seek with the coastline. Pale outcrops of Salinian granite dominate where the fault lies just offshore; dark Franciscan rocks appear where it lies onshore. Other outstanding rocks include the towering volcanic monolith of Morro Rock and the beautiful sandstones at Point Lobos.

Hobnail Granite

The Hobnail granite underlies much of Monterey and a broad area immediately to the south. The road crosses it between Point Lobos State Park and Motleys Landing, halfway south to Big Sur. The best exposures are in the coastal outcrops of the Monterey Peninsula and at Carmel River State Beach, south of Carmel. The Hobnail granite is a distinctive rock full of blocky crystals of orthoclase feldspar that resist weathering better than the other minerals, and so tend to stand out in relief on weathered outcrops.

Some geologists believe the Hobnail granite matches a similarly peculiar granite that appears on the other side of the San Andreas fault, in the La Panza and Gabilan Ranges. And some contend that the Hobnail granite matches some of the granite on the Point Reyes Peninsula north of San Francisco, 95 miles away in the other direction, and also on the other side of the San Andreas fault.

Hobnail granite at Carmel River State Beach, south of Carmel.

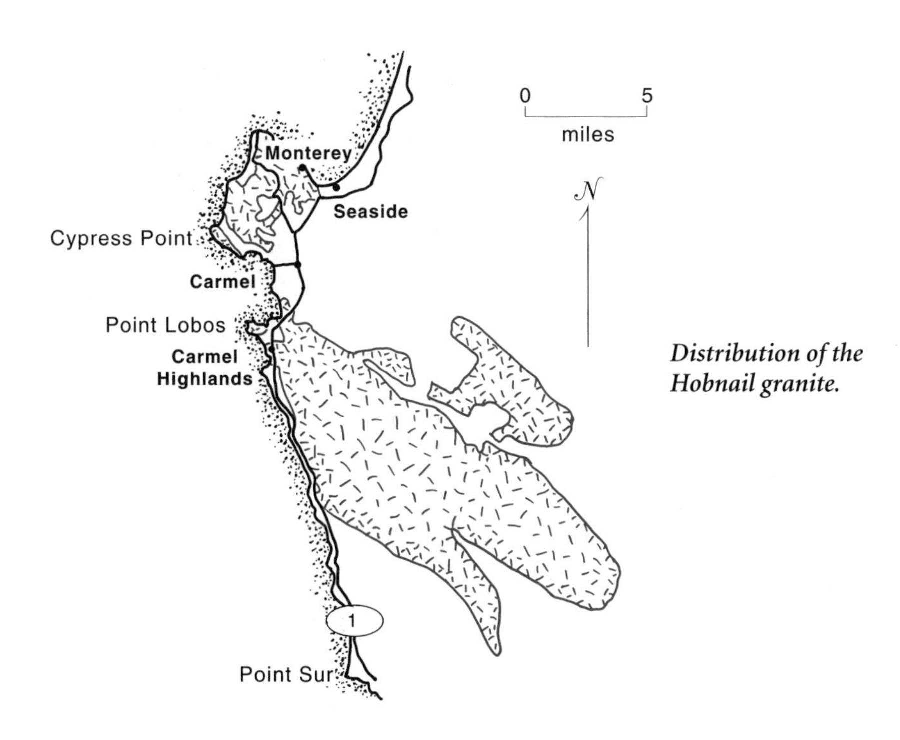

Distribution of the Hobnail granite.

Crystals of orthoclase feldspar 3 inches across stud this exposure of the Hobnail granite in the Monterey Peninsula. It is part of the Salinian block.

Point Lobos

Point Lobos State Reserve encompasses the rocky peninsula at the south end of Carmel Bay, about 8 miles south of Monterey. Paved roads and good trails lead to spectacular headlands that expose a fascinating sequence of muddy sandstones and shales in the Carmelo formation, which may correlate with the conglomerate at the seaward end of Point Reyes.

The Carmelo formation probably accumulated within a narrow submarine canyon eroded into granite of the Salinian block. Turbidity currents, dense mixtures of debris and water, eroded the canyon that holds the Carmelo formation and also deposited it. Look for the graded beds, coarse at the base and grading upward into finer sediment. They formed as particles settled from clouds of muddy water, the larger particles first, the smaller ones last. Look also for fossil burrows and strange marks or tracks on bedding surfaces. Those are probably traces of animals with soft bodies that did not become fossils, phantoms of the rocks.

Sur-Nacimiento Fault

In the 35 miles between Point Sur and Lucia, the highway parallels the complex zone of faults at the western boundary of the Salinian block. This zone includes the Sur fault toward the north end of the Salinian block and the Nacimiento fault toward the south end; so it is the Sur-Nacimiento

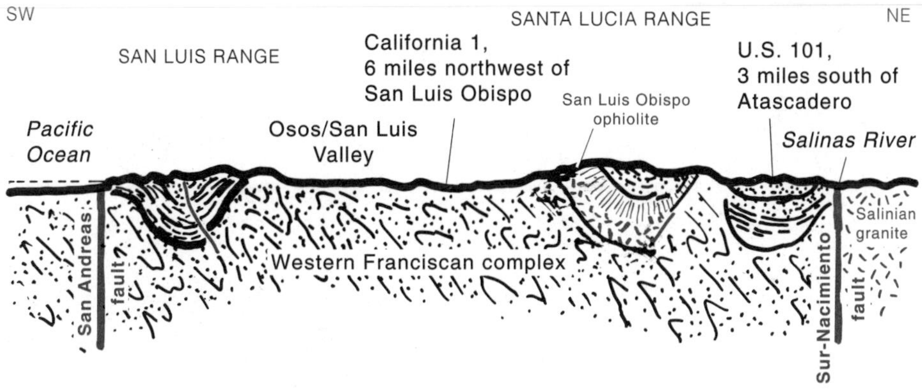

Geologic section across the Western Franciscan complex west of the Sur-Nacimiento fault. —Adapted from Page and others, 1979

fault. Rocks west of the Sur-Nacimiento fault belong to the Western Franciscan complex. Those to the east are the granites and metamorphic rocks of the Salinian block. Farther east, the Salinian block ends against the San Andreas fault, with more Franciscan rocks beyond.

Spot the Sur-Nacimiento fault by watching for the dramatic change between pale granite of the Salinian block to the east and the dark Franciscan rocks to the west. The Sur-Nacimiento fault lies a short distance offshore along the north half of the drive between Monterey and Point Sur, and along the 24 miles between Big Sur and Lucia. White granite and streaky gneiss of the Salinian block appear in the coastal outcrops and roadcuts in both areas.

Franciscan Rocks

Spectacular sea cliffs and rugged headlands along most of the coast between Point Sur and Morro Bay provide the choicest views of Franciscan rocks in central California. Point Sur is a large sea stack of Franciscan volcanic rocks that are on magnificent display in the sea cliffs.

A few miles south of Big Sur, California 1 crosses several miles of grungy granite and associated metamorphic rocks of the Salinian block. Farther south are more Franciscan rocks. The McWay thrust fault moved the Salinian rocks over the Franciscan rocks. The Salinian rocks resist erosion more successfully than the Franciscan rocks beneath them, so the fault makes a cliff. McWay Creek tumbles 70 feet over it to reach the ocean at McWay Cove.

Watch in the Franciscan complex for muddy sandstones in dark shades of gray and greenish gray, their original sedimentary layering sheared almost beyond recognition. Reaction with water altered the dark minerals in most of the basalts to the greenish mineral chlorite, which makes the basalt blackish green. A few basalts still show the distinctive pillows typical of

Crude pillow structures in Franciscan basalt exposed in a sea stack north of Morro Bay. View is about 10 feet across.

Thinly bedded sandstone 8 miles south of Pacific Valley.

lava flows that erupted on the ocean floor. Occasional outcrops of crumpled ribbon cherts in shades of red, green, and white brighten the somber Franciscan scene. Serpentinites make outcrops of broken rocks with slick surfaces in various shades of green.

Marine Terraces—Present and Absent

Along several parts of the route, the highway follows smooth marine terraces that slope gently seaward about 100 feet above the beaches. The best are between Point Lobos and Point Sur, and north of Cambria.

The steep slopes of the Santa Lucia Range drop smoothly into the ocean, without stepping down any marine terraces, along the coast between Point Sur and the area about 20 miles north of San Simeon. Marine terraces form when the coast rises quickly, perhaps during a great earthquake, after having been stable for a long time while the waves eroded a broad wavecut bench. Is it possible that the Santa Lucia Range is rising so smoothly and continuously that no marine terraces can form? Or did erosion wipe them off those steep slopes, as some geologists suggest?

Pretty Pebbles

Moonstone Beach, at the north edge of Cambria, and the beaches north along the coast to San Simeon are widely noted for the excellence of their pebbles. Look for beautifully rounded and smooth pebbles in all possible translucent shades of green, as well as gray, brown, and red. Most of those lovely pebbles are Franciscan ribbon chert. A few are other Franciscan rocks, such as blueschist or eclogite.

The beaches farther north, near Cape San Martin, contain jade. Placer mines there produced several tons of jade, most of it poor quality. The quality of jade depends upon its color and on whether it is nicely translucent.

Morro Rock

Morro Rock rises 576 feet above the entrance to Morro Bay, a bold and shapely sentinel visible for miles from land and sea. It is a solid mass of rhyolite. Age dates show that the mass of molten, but stiffly viscous, magma rose into the interior of a volcano some 24 million years ago. Erosion has since stripped the rubbly pile of volcanic ash and lava that formed the bulk of the volcano, leaving the solid plug of Morro Rock.

Morro Rock is the western end of a chain of fourteen neatly aligned volcanic plugs that extends southeast from Morro Bay through San Luis Obispo. Cerro San Luis Obispo is just east of the highway in San Luis Obispo. Watch southwest of the highway within a few miles of Morro Rock for the rocky summits and grassy lower slopes of the others. The old volcanoes probably lie along the Huasna fault, which continues many miles to the

Hollister Peak, near the northwestern end of a chain of volcanic plugs, southeast of Morro Rock.

Morro Rock.

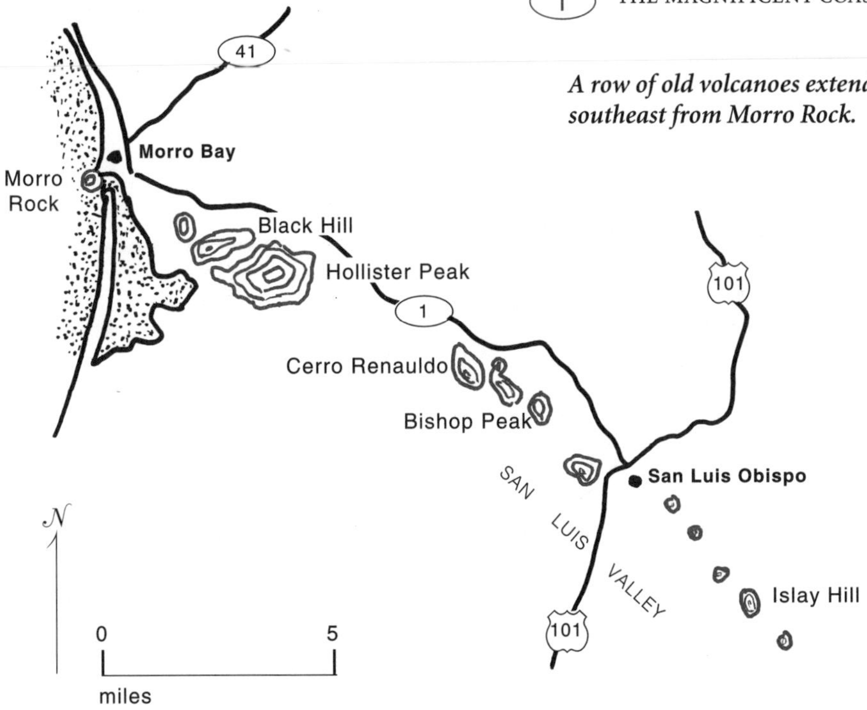

A row of old volcanoes extends southeast from Morro Rock.

southeast. Volcanoes commonly do align along faults, presumably because their fracture zones provide an easy pathway for the rising magma.

Morro Rock was once a small island standing in the surf just offshore. But it interfered with the waves, so sand accumulated in the sheltered area behind it to build the narrow causeway that now connects it to the mainland. Westerly winds blow part of that sand off the upper beach and into the field of low dunes that stretches inland under a partial cover of plants. Waves sweeping sand north along the coast built the long sandbar that isolates Morro Bay from the open ocean. Streams supply enough water to the bay to maintain the outlet between the northern tip of the bar and Morro Rock.

California 1

Golden Gate—Point Arena

127 miles

The road winds along the seashore, nosing into the stream valleys and venturing onto the headlands. It is a tedious drive for people in a hurry and a long series of scenic marvels for those who have the time to enjoy them.

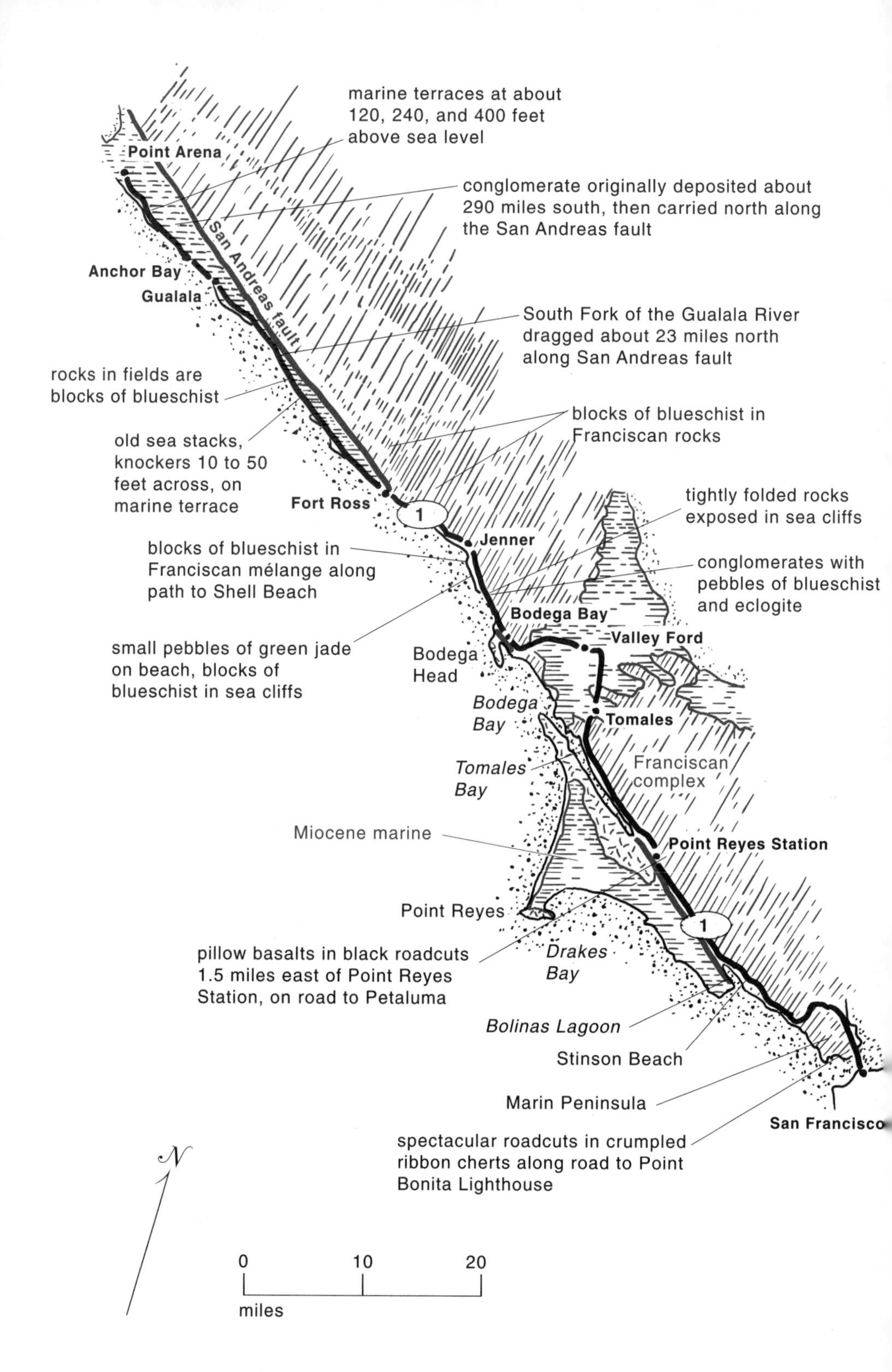

Rocks along California 1 between Golden Gate and Point Arena.

Black Franciscan pillow basalts a few miles east of Point Reyes Station. View is 8 feet wide.

San Andreas Fault

The road follows close to the trace of the San Andreas fault all the way between San Francisco and Point Arena. West of the fault, it is on the Salinian block, part of the Pacific plate. East of the fault, it is on Franciscan rocks of the North American plate, the stuff that was jammed into the oceanic trench while it defined the plate boundary. Most of the rocks along the coast are somber Franciscan sandstones, along with the occasional outcrops of colorful ribbon chert, greenish black pillow basalt, and serpentinite. The Franciscan sandstones are fairly dark where they are exposed in the sea cliffs. A crust of lichens paints a coat of pale gray on outcrops in the hillsides above the road.

The sandstone and chert were deposited at least 100 million years ago and then crushed into the Franciscan trench. Most of the exposures are too small to show how severely these rocks were deformed. But geologists who plot them on maps in an attempt to see the structures in the rocks find it extremely difficult, or even impossible, to determine exactly how they are folded and broken.

Some of the serpentinites contain chunks of dark blueschist, dull greenish eclogite, and even streaks of jade. Serpentinite has so little mechanical strength that it breaks up easily in streams and in the surf. Blueschist, eclogite, and jade all share the quality of toughness—they are hard to break. Tough rocks survive well in the surf. Watch for them along the beaches.

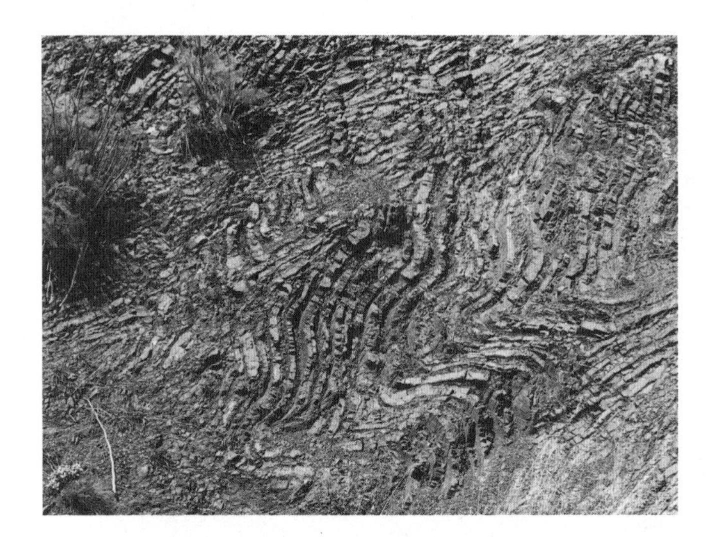

Tightly kinked ribbon chert just west of the north end of the Golden Gate Bridge.

The highway winds along the coast of the Marin Peninsula between the Golden Gate Bridge and Stinson Beach. Outcrops reveal murky Franciscan sandstones, occasional glimpses of colorful ribbon chert, and many small exposures of serpentinite in shades of dusky green. According to some connoisseurs of such matters, the exposures of ribbon chert in Marin Headlands State Park are the best in the world.

Stinson Beach perches precariously on a long spit of loose sand that the waves built right across the San Andreas fault, which passes under the western edge of the town of Stinson Beach, then north up Bolinas Lagoon. When the fault moves, as it someday must, that unconsolidated sand will shake like the proverbial bowl full of jelly. Stinson Beach faces a wobbly future.

The road north of Stinson Beach follows the east side of Bolinas Lagoon, a drowned valley developed along the San Andreas fault zone. The road stays just east of the fault, so all the outcrops near it expose Franciscan rocks. Their severe fracturing testifies to the harsh treatment they suffered, first as they were stuffed into the trench, then as movement along the San Andreas fault ground them finer. Rocks within the fault zone are so broken that they are easy prey to weathering and erosion. That explains why the fault zone is in a valley along most of its length.

The floor of the fault valley rises above sea level at the north end of Bolinas Lagoon and continues north with the road close to its eastern edge. Movements of about 20 feet, as great as those measured anywhere, occurred along most of this valley during the 1906 earthquake, offsetting roads, fences, and lines of trees.

About 2 miles south of Olema, an unfortunate cow fell into a crack that momentarily opened in the crest of a passing ground wave during the 1906 earthquake. Then she was buried, except for her tail, when the crack slammed shut as the wave crest passed. Macabre news photos of the tail sticking out of the ground help explain why so many people fear that the ground may swallow them up during an earthquake. In fact, the risk is minuscule.

A short distance north of Point Reyes Station the fault valley drops below sea level to become Tomales Bay. The road continues along the eastern side of the fault zone almost to the mouth of the bay. It winds far inland through the hills between Tomales and Bodega Bay, passing few outcrops of the Franciscan bedrock.

Bodega Head

Bodega Head, like the Point Reyes Peninsula, is a scrap of granite that is moving north along the San Andreas fault and is now perched temporarily on its bit of coast. It is the northernmost granite scrap on the Salinian block. Technically, the rock is diorite, which is much darker than granite, and contains little or no quartz. Age dates show that it is 94 million years old, a middle Cretaceous age. It presumably started its northward journey somewhere in southern California, or possibly Mexico.

Heavy surf rolling in from the northwest deposited the broad northern sand spit that connects the granite at the tip of Bodega Head to the mainland. The wind blew some of the sand off the beach into a tract of dunes, adding bulk to the connection. Much smaller waves, coming from the southwest, swept sand north along the shore of Bodega Bay to make the much smaller spit that nearly encloses Bodega Harbor. Part of the community of Bodega Bay stands on loose sand either directly above the San Andreas fault, or nearly so. It will shake with great violence when the fault moves. The San Andreas fault lies offshore between Bodega Bay and Fort Ross. That length of coastline is entirely in Franciscan rocks, mostly muddy sandstone and some serpentinite.

Blueschist

Roadcuts near Jenner expose serpentinite that contains large chunks of unusually elegant blueschist. Watch in the dark green serpentinite for the chunks of dark blue rock. Blueschist occurs only in the older Franciscan rocks that were stuffed into the trench deep under the Coast Range ophiolite during Jurassic and Cretaceous time. None exists north of Fort Ross, where the highway crosses younger Franciscan rocks.

The 1906 Earthquake at Fort Ross

The San Andreas fault comes ashore about 2 miles south of Fort Ross. Crushed, broken Franciscan rocks exposed in roadcuts mark where it crosses

Mélange of Franciscan sandstone and shale near Jenner. View is about 4 feet across.

Modern sea stacks stand in the surf and old sea stacks stand on a marine terrace near the road to Goat Rock State Beach south of Jenner.

the coastline and continues into the wooded hills to the north. The 1906 earthquake split several large redwood trees that were growing directly over the fault in the woods north of Fort Ross. A study of their growth rings showed that those trees had survived similar damage during previous earthquakes.

After the great earthquake of 1906, geologists of the California Earthquake Commission found a picket fence near Fort Ross that was offset 12 feet, the side west of the fault having moved north. Nearby, the fault offset the old Russian Road, which dates from 1812, by about the same amount. That is important because it shows that the fault moved only that one time since the road was built.

On the Pacific Plate

The road follows the coast along most of the distance between Fort Ross and Point Arena, in places following a marine terrace about 30 feet above sea level. This part of the coast is west of the fault, so the rocks are on the Salinian block, even though they are tilted and sheared.

The gentle and almost perfectly straight valley that trends northwest 2 or 3 miles inland is the topographic swale along the San Andreas fault. It contains the community of Plantation. The Gualala and Garcia Rivers follow the fault, flowing parallel to the coast for some miles before they finally turn west to enter the ocean.

Rocks exposed along the coast between Fort Ross and Gualala are folded layers of sandstone and mudstone deposited offshore between about 70 and 57 million years ago. They probably lie on oceanic crust, instead of on the granite that makes up the Salinian block farther south. Where were these rocks before they started moving north along the San Andreas fault?

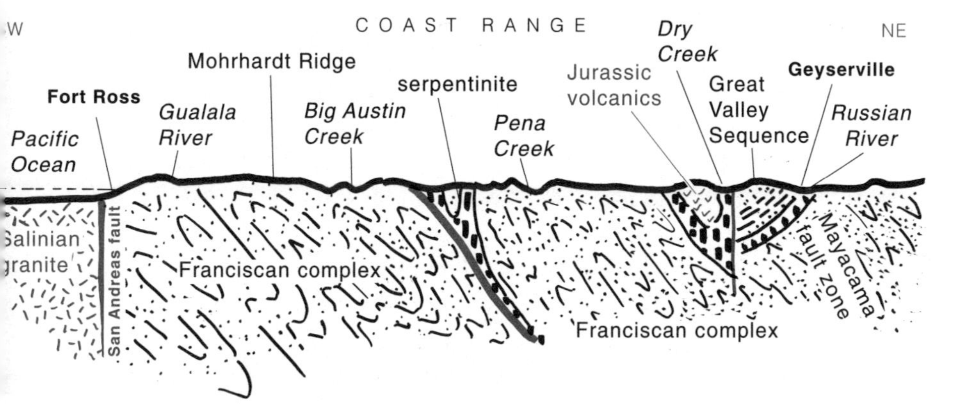

Geologic section across California 1 at Fort Ross, where the San Andreas fault comes ashore and the Salinian block is just offshore.

Distinctive dark pebbles in some of the layers exposed near Anchor Bay closely resemble rocks in part of the Temblor Range east of San Luis Obispo. If the correlation is valid, it means that this part of the Salinian block was near the Temblor Range when these rocks were deposited and has since moved north about 350 miles. A conglomerate that contains pebbles of an unusual granite flecked with crystals of red garnet appears to have moved about 290 miles north along the San Andreas fault. Several other ties across the San Andreas fault in other areas give about the same result.

Approximately 100,000 repetitions of the displacement that accompanied the earthquake of 1906 would move the Salinian block north about 350 miles. One earthquake like that of 1906 every century or two for 10 to 20 million years would do the job. The arithmetic does seem feasible.

California 1

Point Arena—Leggett

85 miles

The San Andreas fault passes out to sea at the mouth of Alder Creek, about 12 miles north of Point Arena, and the coast becomes much more sinuous and irregular than it is farther south. The fault follows the valley immediately east of Manchester Beach State Park and from there north closely parallels the coast for some miles, edging farther out to sea as the coast trends away from its path. It finally meets the Mendocino fracture in the triple junction some miles off Cape Mendocino and disappears westward in the floor of the Pacific Ocean.

Most of the Franciscan rocks along the coast are dark sandstones, but ribbon cherts appear here and there in shades of pale gray, yellow, red, and green. Roadcuts in ribbon chert are especially prominent within 1 or 2 miles north and south of Elk. Gravelly beach sediments cover the steeply tilted Franciscan rocks along many long stretches of this coast.

The road stays within sight of the waves along most of the way between Point Arena and Rockport. The view seaward looks down the gently sloping surface of a marine terrace to the edge of the sea cliff. Near Elk and Albion, the terrace is about 100 feet above the ocean, and sea stack remnants dot its surface. The view landward shows the old sea cliff, now softened to a steep slope in the row of hills that fringe the marine terrace. The marine terrace at Jug Handle State Reserve, 5 miles south of Fort Bragg, is the most recent of several that have been studied in the hills to the east.

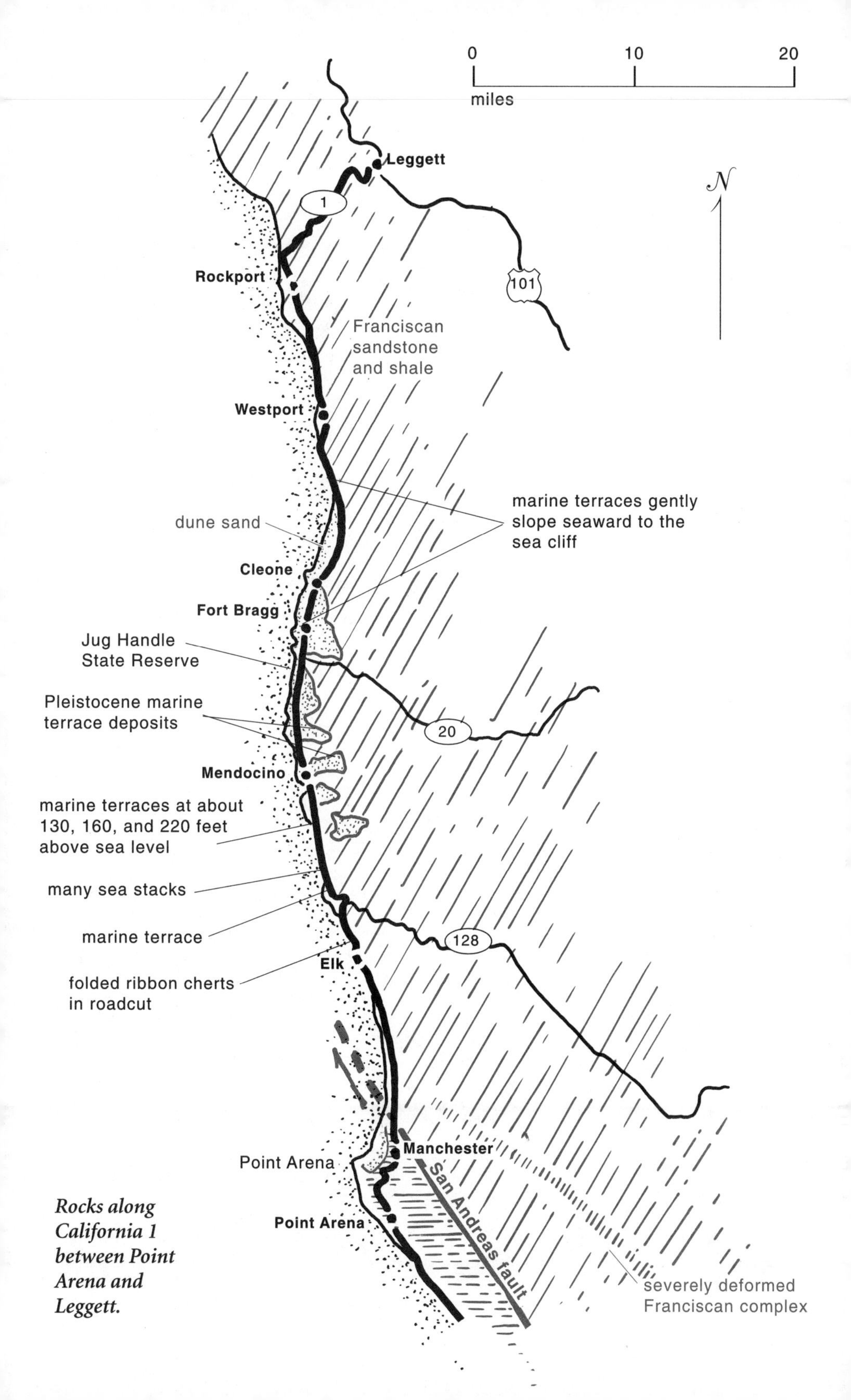

Rocks along California 1 between Point Arena and Leggett.

Beach cliffs and a sea stack in the making, about 10 miles north of Point Arena.

Big dune field, mined for sand, just north of Fort Bragg.

Mendocino

Trendy little Mendocino stands on a broad terrace remnant that makes a small cape just north of Mendocino Bay, at the mouth of the Big River. Do not confuse this with the big Cape Mendocino farther north. Any number of vantage points along the sea cliffs provide splendid views of the waves carving the land, isolating stacks as they exploit fracture zones in the rock and hollowing caves and arches in the narrow vertical zone between high and low tide.

MacKerricher Sand Dunes

MacKerricher State Park, about 3 miles north of Fort Bragg, contains one of the most spectacular dune fields on the California coast. It fills the mile that separates the highway from the beach. The Tenmile River appears to be the source of the sand. It empties into the ocean at the northern edge of this dune field, dumping its load of sand on the coast. Then the heavy winter storms sweep the sand south down the coast, maintaining a wide beach. Waves spread sand across the upper beach at high tide, then the sea breeze blows it into the dunes as the upper beach dries at low tide.

Winding through the Coast Range

Remnants of raised marine terraces become narrower and fewer farther north. Westport huddles on one of the northernmost. About 3 miles north of Rockport, the highway turns inland toward Leggett. The last views north along the coast toward Cape Vizcaino reveal steep mountain slopes descending directly into the foaming surf, unbroken even by the slightest hint of a marine terrace—no easy route for road construction.

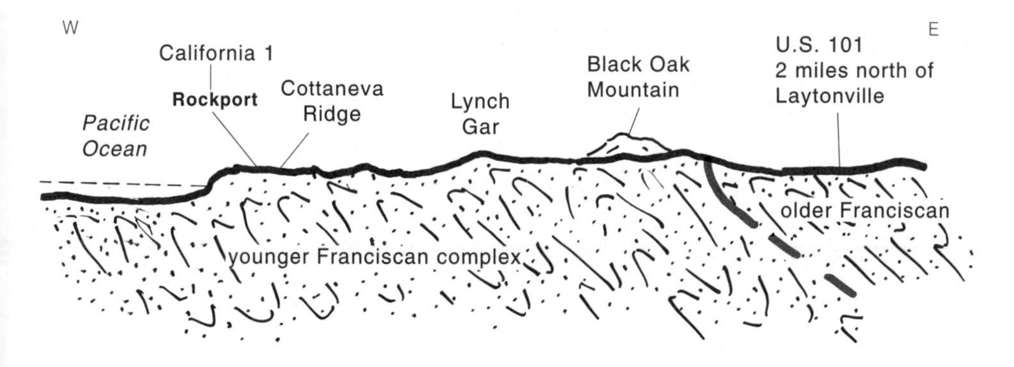

Section drawn east from Rockport shows mainly Franciscan rocks. —Adapted from Etter and others, 1981

The road between Rockport and Leggett winds a corkscrew route through the steep, corrugated landscape of the Coast Range, an endless series of sharp ridges and deep ravines. Most of the roadcuts expose the deeply weathered, reddish brown soils of the Coast Range. A few reveal Franciscan bedrock, at various stages of weathering into that reddish soil. Watch for thick layers of sandstone, thinly layered ribbon cherts, and occasional glimpses of greenish serpentinite. The few larger roadcuts expose layers of rock steeply tilted, twisted, and broken along faults. Nowhere can you follow an individual layer of rock from one roadcut to the next.

Layers of Franciscan sandstone steeply tilted in a roadcut west of Leggett.

6

The Great Valley

The Great Valley is one of the world's most productive agricultural regions. Early geologists marveled that the rocks in the Sierra Nevada to the east and the Coast Range to the west are tightly folded and much broken along faults. But those in the Great Valley are nearly undeformed, even though some of them are as old as those in the enclosing mountains. How could that be possible?

The Great Valley.

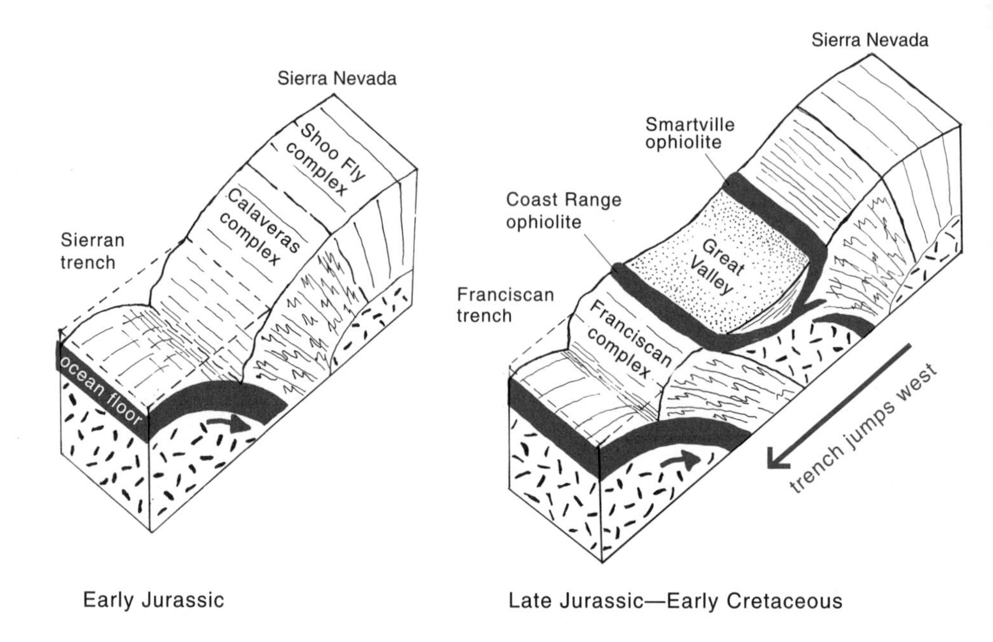

The trench shifts west, probably during late Jurassic time.

Sinking ocean floor first swept rocks into the Sierran trench to make the mangled accumulation in the Sierra Nevada. Then the trench jumped west to the line of the Franciscan trench and stuffed it with the equally mangled Franciscan rocks of the Coast Range. That probably happened during late Jurassic time, perhaps about 150 million years ago. The rocks that fill the Great Valley began to accumulate shortly after the trench jumped west, and continued until long after the Coast Range had risen from the depths of the oceanic trench.

THE TRENCH SHIFTS WEST

When the Klamath block moved west, the active oceanic trench shifted west to the Franciscan trench. A strip of ocean floor approximately 60 miles wide and at least as long as the Great Valley was trapped between the positions of the old Sierran and new Franciscan trenches, unable to sink through either.

The long stretch of ocean floor trapped between two trenches is now the deep bedrock floor of the Great Valley. You see its upturned western edge lying on the Franciscan rocks along the eastern edge of the Coast Range, tilted steeply down to the east. Other remnants probably exist, in more

Oceanic sedimentary rocks of the Great Valley sequence in the hills along the west edge of the valley.

mangled form, in the Smartville complex on the western flank of the Sierra Nevada. The oceanic crust that lies on the Franciscan rocks dates from late Jurassic time, so the oceanic ridge where it grew could not have been far offshore.

The Incipient Great Valley

The future Great Valley may have begun to separate from the open ocean as early as late Jurassic or early Cretaceous time. The water was still deep then but nowhere near as deep as an oceanic trench. The oceanic crust sinking through the trench was already stuffing the Franciscan formation against and beneath the western edge of the old ocean floor lying below the Great Valley, now the Coast Range ophiolite. The sediments stuffed under the ocean floor jacked up its western edge, raising a deeply submerged barrier to the movement of sediments. That was the beginning of the Great Valley as a separate sedimentary basin.

Almost all of the sediments that filled the Great Valley eroded from the Sierra Nevada. The oldest are full of fragments of volcanic rocks eroded from its early volcanoes. As erosion stripped the cover of volcanic rocks from the granites of the Sierra Nevada batholith, their detritus of pale quartz and feldspar sand began to wash into the Great Valley.

Filling the Basin

Rock cuttings brought up from oil and gas wells contain microscopic fossils from deep within the Great Valley. The fossils tell part of the story. Those from the older formations are the remains of animals that lived in the open ocean during Cretaceous time. By early Tertiary time, about 50 million or so years ago, enough sediment had accumulated to make the water in the Sacramento Valley quite shallow, but deep water still existed in part of the San Joaquin Valley. During most of Tertiary time, large parts of the Great Valley were filled to sea level with sediment and were receiving muds deposited on generally dry land. Other parts were still a bit below sea level and were receiving sediments deposited in shallow seawater. Finally, terrestrial sediments spread across most of the valley floor. The geophysicists suggest that the sedimentary fill in the western Great Valley is as much as 8 or 9 miles thick—wells can not penetrate to such depths. The fill thins to the east, where its feather edge laps onto the Sierra Nevada.

Petroleum

Sediments deposited in shallow seawater are likely to contain petroleum. An enormously thick wedge of such sediments exists within the Great Valley, so it is no surprise to find that it contains an abundance of oil and gas. Wellheads are part of the scene almost from one end of the Great Valley to the other.

Oil and gas start as organic matter disseminated through sediment, the sort of stuff that makes some mud black and smelly. Gas forms easily—it consists mostly of methane, a simple molecule that forms wherever organic matter decomposes. Oil is more difficult—it forms only when the sediments are within a narrow temperature window. Too cool and only gas forms. Too hot and the organic matter bakes into inert black carbon.

The Sacramento Valley produces gas in great quantity but no oil. Evidently, the sediments within it have never been hot enough to cook their organic matter into oil. The San Joaquin Valley, where the sedimentary fill is deeper, produces both gas and oil.

The original organic matter was concentrated mainly in shales deposited in seawater during Eocene and Miocene time. Gas and oil leaking from those rocks seeped up into the younger sandstones above, then stopped where it could rise no higher. Most of those traps in the western and southern part of the San Joaquin Valley are anticlines—layers of impervious shales draped over arches folded in the rocks. Along the eastern flank of the valley, the petroleum stopped where the pore spaces in the rocks became so small that it could travel no farther. Those are called stratigraphic traps.

GREAT VALLEY ROADGUIDES

The geologic experience of driving in the Great Valley is rather similar everywhere—a great flatness with no rocks and with mountains on the eastern and western horizons. Variety depends mainly upon the view on either side of the valley and to a minor extent upon details of drainage within it. The agricultural view is lush in all seasons.

Interstate 5

Sacramento—Redding

167 miles

Interstate 5 rises only a few feet above sea level as it follows the flat floor of the Sacramento Valley between Sacramento and Red Bluff. On a clear day, you can see the full width of the valley, from the ragged profile of the Coast Range along the western horizon to the Sierra Nevada along the eastern skyline. Shasta looms on the northern horizon in the northern part of the route. The road between Red Bluff and Redding passes through low hills eroded mostly in gravel weathered to a reddish color.

All of the road crosses loose sediments washed into the valley from the surrounding mountains during the past few hundred thousand years. Except in Sutter Buttes, no rocks worthy of a hammer exist anywhere on the floor of the Sacramento Valley. The lack of hard rocks is also a lack of raw material for road metal and construction aggregate.

Sutter Buttes looking east from Interstate 5.

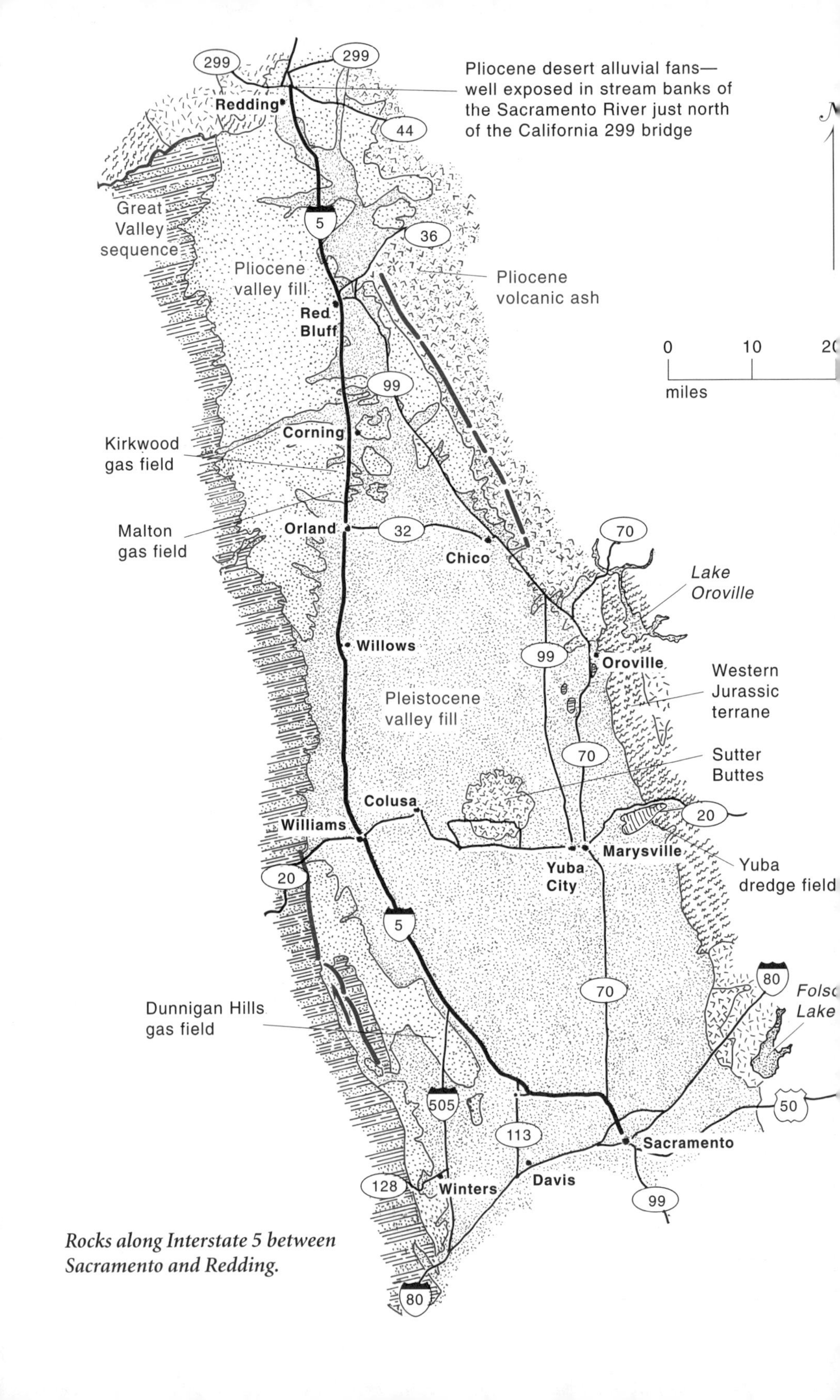

Rocks along Interstate 5 between Sacramento and Redding.

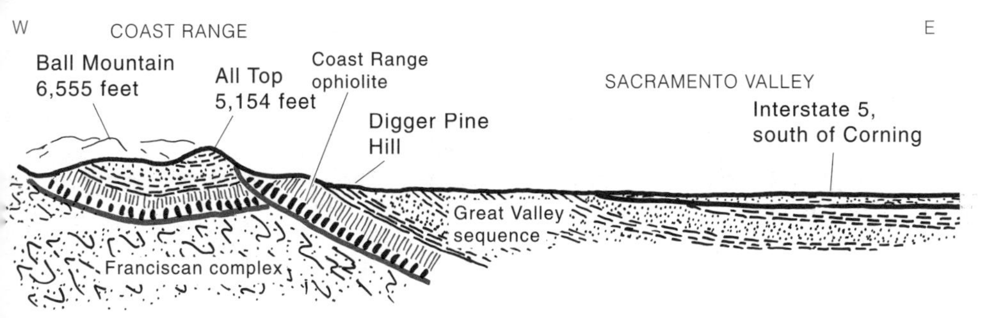

Geologic section across the Great Valley near Corning.
—Adapted from Blake and others, 1985

The Sierra Nevada abruptly ends almost directly east of Red Bluff. Mountains north of Redding are the southern edge of the Klamath block, the continuation of the Sierra Nevada, offset about 60 miles west during early Cretaceous time.

Sutter Buttes, the Misplaced Volcano

The scattered hills of Sutter Buttes rise above the dead flat floor of the southern Sacramento Valley in the area about 6 miles northwest of Yuba City. They are the ruins of a thoroughly defunct volcano that was active during early Pleistocene time, probably between about 1.6 and 1.4 million years ago. What is a dead volcano doing out there on the valley floor, all by itself? No one knows. None of the various theories now available accommodates all the evidence.

Many geologists argue that the volcano is a distant southern outpost of the High Cascades. It does seem suspicious that a line drawn through the active volcanic peaks of the High Cascades and extended south passes through Sutter Buttes. And the rocks in Sutter Buttes are certainly similar to those in the High Cascades. But why the wide gap between Sutter Buttes and Lassen? And why is the Sutter Buttes volcano so far south of the modern oceanic trench, which ends off Cape Mendocino? The trench could not have extended as far south as Sutter Buttes as recently as 1.6 million years ago.

Other geologists contend that Sutter Buttes and the volcanoes in the Coast Range erupted as the Mendocino triple junction passed offshore in its northward migration. But it is hard to imagine how the triple junction could have passed the latitude of the Sutter Buttes volcano as recently as the eruptions there. And why should the passage of the triple junction leave just one volcano isolated in the floor of the Great Valley? Why not a row of them that become younger northward?

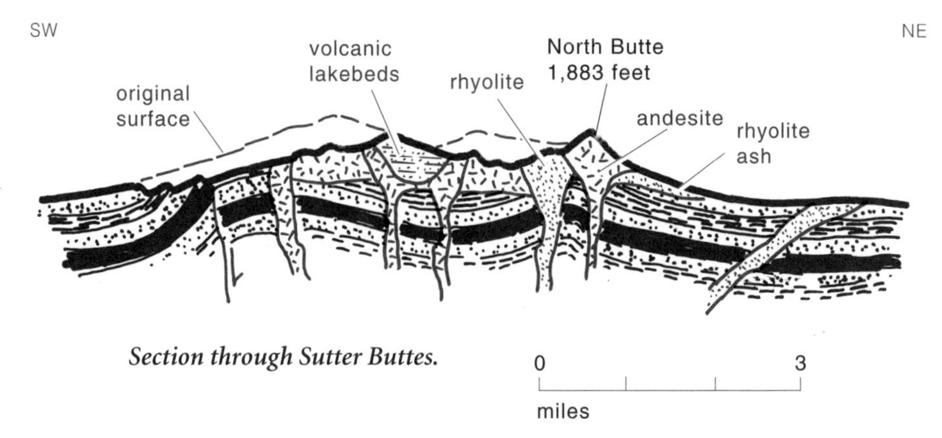

Section through Sutter Buttes.

Whatever its cause, the Sutter Buttes volcano consists basically of a swarm of andesite and rhyolite domes. Many of them quietly intruded the soft sediments in the floor of the Sacramento Valley. Others erupted in great sheets of ash and angular rubble, no doubt to a thundering accompaniment of steam explosions. Erosion has since carved the volcanic rocks into a distinctive landscape of pinnacles and steep hummocks.

Intrusion of the igneous rocks tilted the older sedimentary rocks in the valley floor and broke them along an intricate pattern of faults. All that tilting and breaking created structures that trapped large reservoirs of natural gas, which would otherwise have continued rising through the rocks.

Alluvial Fans, a Record of Climates Past

Watch for the gravelly roadcuts between Orland and Redding, where the highway follows a route near the western side of the Sacramento Valley. The roadcuts expose deposits of sand, mud, and gravel washed in from the Coast Range within the past few million years. These deposits make enormous alluvial fans, almost too big to perceive from the ground. But they are easily visible on aerial photographs, topographic maps, and from a few good vantage points. The pebbly sediments exposed in roadcuts along the 10 miles south of Redding are alluvial fan gravels, weathered red in

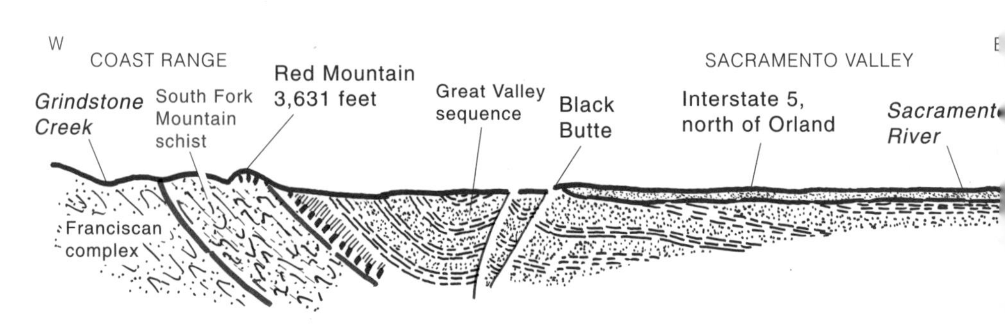

Geologic section across the Great Valley just north of Orland.
—Adapted from Etter and others, 1981

Gravel in roadcut south of Redding.

Gravels on bank of the Sacramento River at east edge of Redding.

the wetter climates that have prevailed since they were deposited. They extend west to the Coast Range.

Alluvial fans typically form where extremely muddy streams flow out of the hills and onto more open ground—out of the Coast Range or Sierra Nevada and onto the floor of the Great Valley. You generally see active fans in arid regions, such as Nevada, where the plant cover is too sparse to protect the soil from erosion during occasional heavy rainstorms. Most alluvial fans are creatures of the desert. Others may grow in places where a watershed supplies an overload of sediment to its streams.

The alluvial fans dissected by gullies along the edge of the Sacramento Valley tell a story of at least two climates. At some time in the past several million years, the climate was much drier than now. Then the wetter climates of the ice ages followed during the past 2 million years. The more vigorous streams of the last ice age carved deep channels into the old fans, probably because a good plant cover in the watershed greatly reduced the rate of soil erosion, therefore reducing the sediment load of the streams.

Great Valley Sequence

Look west from the northern part of the Sacramento Valley to see the long ridge of the Great Valley sequence, which defines the boundary between the flat valley floor and the gaunt ruggedness of the Coast Range. Its sharp crest is about 15 miles west of the highway. Its smooth slopes are nearly treeless.

Many of the side roads such as those that lead west from Williams and Willows pass through excellent roadcuts in the Great Valley sequence. They provide splendid views of alternating layers of pale sandstone and black shale that were laid down somewhere on the deep floor of the Pacific Ocean. The layers tilt down to the east and lie on the Coast Range ophiolite, which lies on serpentinite.

Volcanic Rocks

Great sheets of volcanic rock, mostly rhyolite ash, cover large areas of the Sacramento Valley east of Red Bluff. They are souvenirs of violent eruptions during the past few million years. Roadcuts just north of Red Bluff expose tan ash with prominent horizontal layers several inches thick. The ash is interbedded in the fan gravels. Watch for the several small volcanic cones visible from the highway. These are the westernmost edge and the southernmost end of the High Cascades.

Farmers Versus Hydraulickers

In the several miles south of Red Bluff, Interstate 5 follows the broad floodplain of the Sacramento River. Early farmers in the valley worked mostly along the floodplain, where irrigation water was plentiful. Like most streams that meander on broad floodplains, the Sacramento River has natural levees made of sediment deposits along its banks. The deposits are laid down when muddy floodwaters rise onto the floodplain, dumping part of their sediment load as they spread into the dense vegetation along the stream banks. Natural levees are hard to see because they are normally no more than a few feet high—perhaps only a few inches—and several hundred feet wide. But they are important because they control drainage on the floodplain.

Hydraulic gold miners operating in the northern Sierra Nevada between 1852 and 1884 dumped large quantities of mud and sand into the headwaters of the Sacramento River. The river responded, as rivers will, by depositing more sediment in its bed and on its natural levees. That raised the stream as much as 15 feet above large areas of its floodplain. In some places, the sediment deposit covered all but the tops of orchard trees. Boats, which had been the best means of travel, could no longer navigate the river.

The rising natural levees made it impossible for the floodplain to drain into the river, thus converting productive farmlands into marshes and swamps. So the hydraulic miners in the remote streams of the Sierra Nevada ruined good farmland in the valley. Now that the Sacramento River again carries its normal sediment load, it has largely repaired the damage and restored its floodplain to productive agricultural use.

It took most of 30 years for the great slug of sediment to work its way downstream to San Francisco Bay. That slug filled almost half the previous area of the bay—a high price for the gold hydraulically mined in the Sierra Nevada.

Interstate 5

Sacramento—U.S. 99

246 miles

The drive along the flat floor of the San Joaquin Valley provides a wonderful view of one of the world's most agriculturally productive regions. Lush crops grow on sediments that were deposited on the valley floor within the past 1 or 2 million years, many of them within the past few thousand years.

The Coast Range rises on the western horizon along the entire route. The high southern end of the more distant Sierra Nevada defines the eastern horizon. The valley comes to an abrupt southern end at the White Wolf fault, where the road enters Grapevine Canyon and the hard granite of the Tehachapi Mountains.

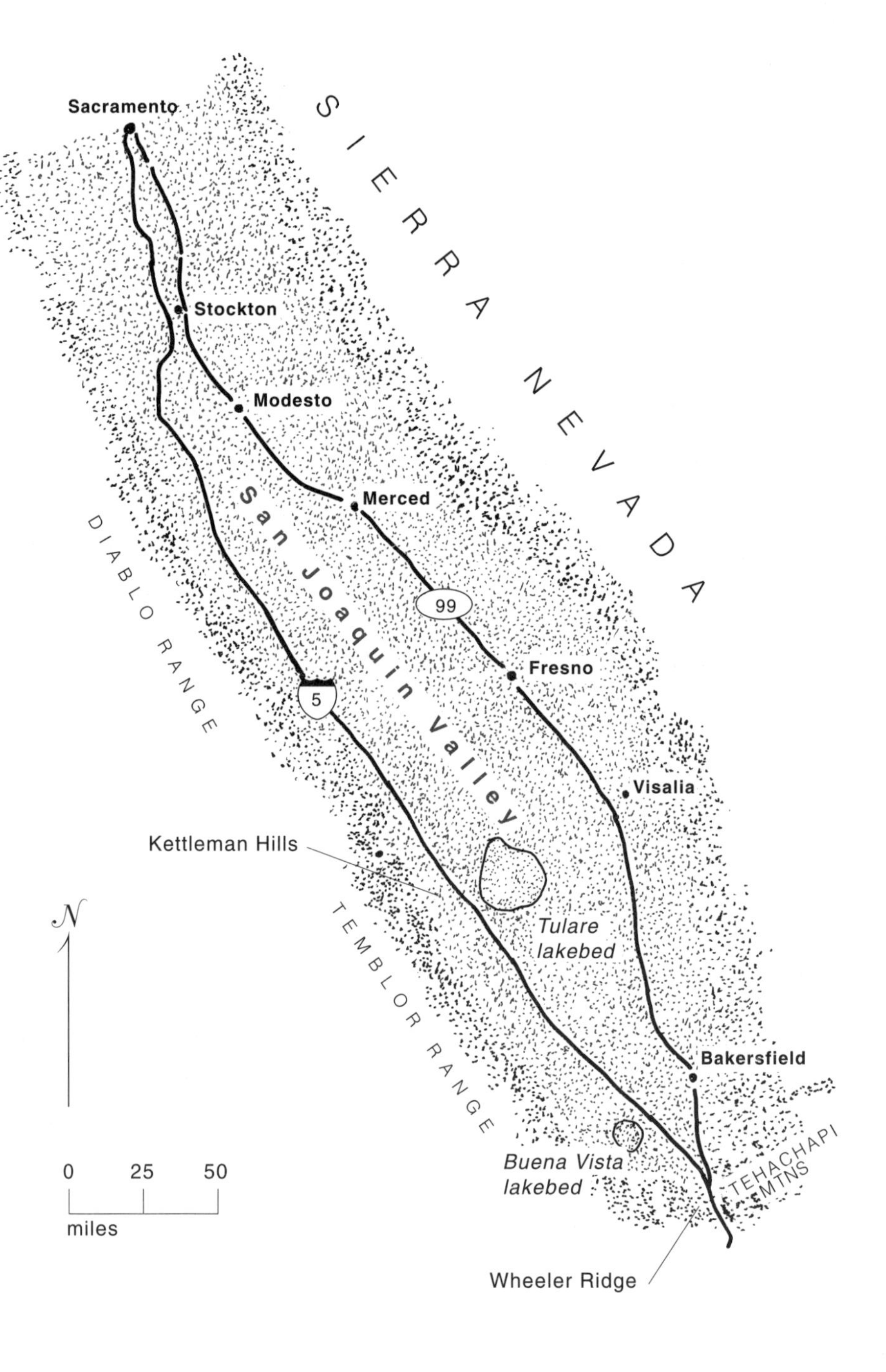

The San Joaquin Valley.

View west across the San Joaquin Valley to the low hills of the Coast Range. Taken from Interstate 5 north of Coalinga.

Wind Erosion

Along nearly the entire length of its route through the San Joaquin Valley, the highway passes just west of large areas of valley floor scoured by wind. Old blowout depressions and tracts of overgrown sand dunes tell of a time when the wind eroded the valley floor and blew sand across it. No one knows when that happened, but the appearance of the surface suggests that it was many thousands of years ago.

Some geologists suggest that the wind erosion happened during one or more of the ice ages, when strong winds drained off the great ice fields in the high Sierra Nevada. But the ice ages were also times of heavy rainfall, which probably supported a thriving plant cover in the San Joaquin Valley. It seems more likely that the wind erosion happened during interglacial periods, some of which may have brought dry climates and an extremely sparse plant cover.

Lakes on the Valley Floor

Great alluvial fans, gently spreading from the mouths of the Kings and Kern River canyons, give the floor of the southern San Joaquin Valley a gentle slope down to the west. That explains why water pouring out of those canyons flows almost to the western edge of the San Joaquin Valley, then ponds in the low areas between fans, where it evaporates and soaks into the ground. If the climate were wetter, those low areas would fill and overflow north to San Francisco Bay.

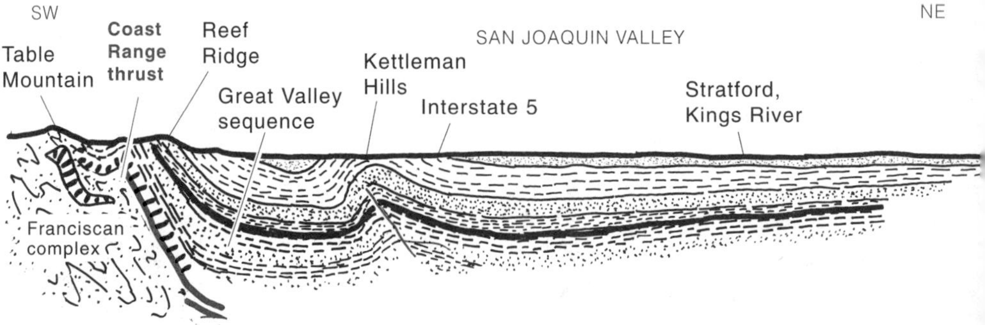

Geologic section along the San Joaquin Valley between the Coast Range in the southwest and the area south of Fresno. —Adapted from Page and others, 1979

The enormous expanse of almost perfectly flat land east of the highway near Kettleman City is the dry bed of Tulare Lake, directly east of the Kettleman Hills. Under natural conditions, Kings River and several lesser streams flowed out of the Sierra Nevada and into Tulare Lake. Before about 1920, Tulare Lake had the largest surface area of any lake in California. Then most of its water supply was diverted for irrigation. Tulare Lake dried up, and its flat bed was cultivated. Parts of it still flood during unusually wet years.

The Kern River originally flowed into Buena Vista Lake, the low area southwest of the highway where it passes west of Bakersfield. The area was a perennial wetland of shallow lakes and marshes. During uncommonly wet seasons, water flowed north across the low divide into Tulare Lake. Then the Isabella Dam stored most of that water, and the rest was diverted into irrigation ditches. Buena Vista Lake also dried up, and like Tulare Lake, most of its flat bed is now under cultivation. Buena Vista Lake also floods in wet years.

Wheeler Ridge

People traveling south get a good view of Wheeler Ridge as they approach the south end of the San Joaquin Valley. Look west of the highway. Wheeler Ridge is an anticline, an arch in the sediments of the valley floor. It is so young and so uneroded that its shape is almost exactly the shape of the anticline.

The big rocks that litter the surface of Wheeler Ridge are weathering out of the Tulare formation, which was deposited as recently as 250,000 years ago. They are granite from the Tehachapi Mountains, just to the south. Seeing a formation so young warped into the fold is clear evidence that the fold is even younger and strongly suggests that it may still be rising.

Oil wells at Kettleman City.

Notches in the crest of Wheeler Ridge are the abandoned valleys of streams that were flowing in those courses before Wheeler Ridge began to rise. They were displaced from their channels because they were unable to erode their beds fast enough to keep pace. Had the climate been wet enough to provide them with a stronger flow, they probably would have succeeded.

You can see the four great pipes of the California Aqueduct passing through the deepest of the old stream valleys, near the east end of Wheeler Ridge. The pumping station on the north side of the old valley raises the water about 2,000 feet, to boost it over the Tehachapi Mountains.

Oil and Gas

The highway passes east of a nearly continuous series of oil and gas fields between the area north of Coalinga and the south end of the San Joaquin Valley. All are basically similar. The apparent source of most of the oil is in Miocene rocks deposited while the San Joaquin Valley was still full of seawater. The oil is trapped in the crests of young anticlinal arches, which are probably still rising. The arches appear as low hills and groups of low hills on the valley floor. Early petroleum geologists discovered most oil fields simply by drilling on hilltops. The southern part of the San Joaquin Valley was one of the few places where that strategy ever worked, and it ran out of undrilled hilltops before 1920.

Movement on the San Andreas system of faults in the Coast Range is probably dragging the western edge of the San Joaquin Valley north, wrinkling its deep fill of sedimentary rocks into folds. Repeated surveys show that some of the folds are observably rising, so they must be extremely young.

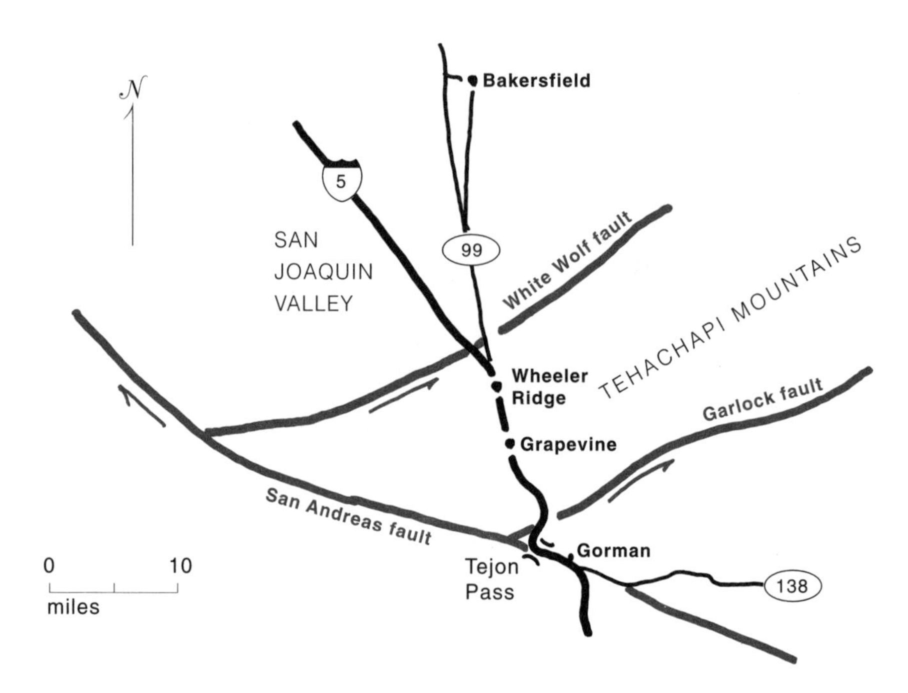

The White Wolf fault ends the San Joaquin Valley where it bluntly abuts against the granite of the Tehachapi Mountains.

Grapevine Canyon

The landscape along Interstate 5 changes abruptly at Grapevine Canyon, a narrow defile deeply eroded into pale gray granite. The San Joaquin Valley ends against a wall of this granite at the White Wolf fault. The Tehachapi and San Emigdio Mountains rise southeast of the fault along a remarkably straight mountain front that trends southwest. The Tehachapi are a western projection of the Sierra Nevada. The San Emigdio Mountains, west of the highway, are more of the same, masquerading under a different name. The White Wolf fault moves obliquely, south side up, and to the northeast. The total vertical movement during the past 2 million or so years appears to have been between 2 and 3 miles, the horizontal movement much larger.

The epicenter of a magnitude 7.5 earthquake that shook Kern County in 1952 was on the White Wolf fault near Wheeler Ridge, south of Bakersfield. That was the largest earthquake to rattle California since 1906. Fault movement raised the Tehachapi Mountains a bit and moved them a bit northeast.

Interstate 80
San Francisco—Sacramento
85 miles

The route between San Francisco and Sacramento crosses an eastern portion of the Coast Range and a portion of the Sacramento Valley. The western part of the route crosses slices of the Coast Range cut along faults related to the San Andreas fault. Rocks in them include the usual Franciscan rocks of the Coast Range, along with some that belong to the Great Valley sequence and some of the Sonoma volcanic rocks, andesite and rhyolite, that erupted within the past few million years. The eastern part of the route crosses flat ground, the floor of the Sacramento Valley, and part of the Sacramento Delta, which the river is steadily building downstream into San Francisco Bay.

Coast Range

Interstate 80 crosses the Hayward fault where it leaves the eastern limits of Richmond to climb into the low hills that face the southeast side of San Pablo Bay. The rocks in the hills between Richmond and Vallejo are sedimentary formations that contain fossils of animals that lived in seawater during Miocene time, about 20 million years ago. The poorly exposed rocks in the 3 miles north of Vallejo belong to the Great Valley sequence. The soft, pale gray to yellowish rocks exposed in the several miles northeast of them were deposited during Eocene time.

All these formations contain rather thin layers of a variety of sediments ranging from hard sandstone and volcanic ash to soft muds. Such rocks are rarely stable under the best of circumstances. Here they are tilted to fairly steep angles and are busily sliding on the layers of weak rock below them. Watch for the landslides and incipient landslides that wrinkle the

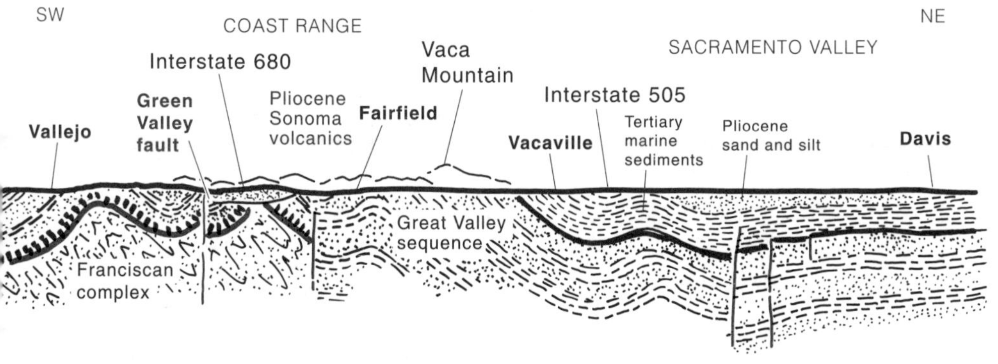

Geologic section drawn approximately along the line of Interstate 80 between Vallejo and Davis. —Adapted from Wagner and others, 1981

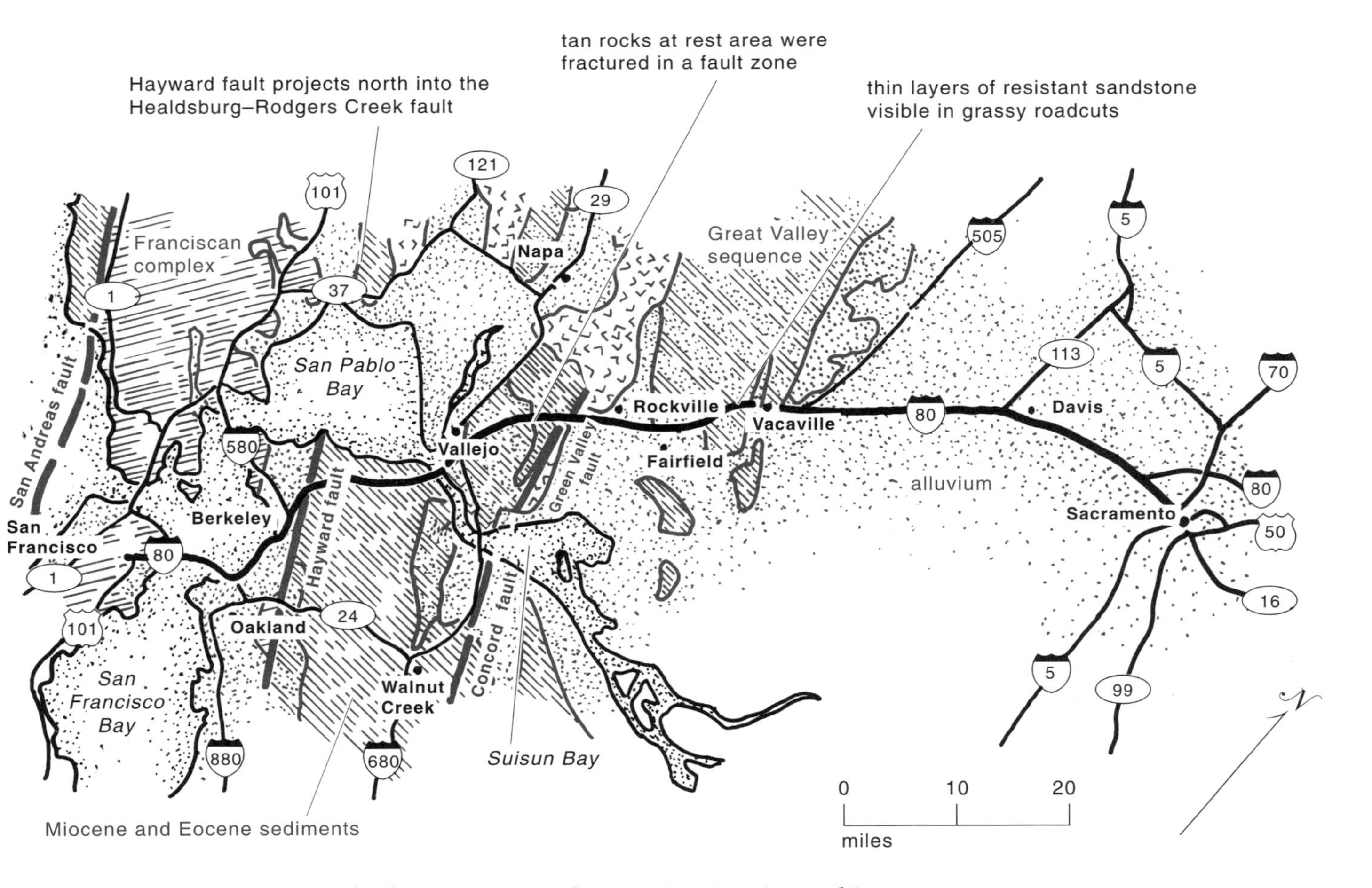

Rocks along Interstate 80 between San Francisco and Sacramento.

slopes of nearly every hill eroded into these unstable rocks. The future will bring a wider variety and growing number of catastrophes as developers build on those slopes.

The highway passes through a spectacular roadcut in Franciscan rocks at the top of Sulphur Springs Mountain, just east of Vallejo. Basically similar Franciscan rocks form the hills north of the highway between San Francisco and Vacaville, where the highway passes the southern ends of several long ridges and wide valleys that extend to the northern horizon. These are slices of the Coast Range that moved horizontally along faults parallel to the San Andreas fault.

Except for some knobs of volcanic rock north and south of the road near Rockville, bedrock in all these long ridges belongs to the Great Valley sequence. Most of it is dark sandstone, muddy stuff. Farther north, the Great Valley sequence is not so thoroughly sliced and makes a long ridge that more neatly defines the boundary between the Sacramento Valley and the Coast Range.

The low ridge north of Vacaville is the easternmost outpost of the Coast Range. Between there and Sacramento, the highway crosses stream sediments laid down on the flat floor of the Sacramento Valley since the last ice age. On clear afternoons, you can see the Sierra Nevada looking like cardboard stage scenery pasted onto the eastern horizon. The Coast Range looks more convincing, a long succession of ridges fading into the blue distance of the hazy coastal air. John Muir called it "the range of light."

The Sacramento Valley

The Sacramento and San Joaquin Rivers meet at the east end of Suisun Bay, which they are rapidly filling with sediment as they build their combined delta. They probably began building it when sea level was near its present stand during earlier interglacial times. If present trends continue, they will eventually fill all of San Francisco Bay, even though it is on a subsiding fault block.

The flat stretch of Interstate 80 between the junction with Interstate 680 and Fairfield follows the north edge of the Sacramento Delta, which the river built into the eastern end of San Francisco Bay since the end of the last ice age. Interstate 680 follows the delta between that junction and Benicia. The nearly perfect flatness of the delta top makes it ideal ground for growing rice.

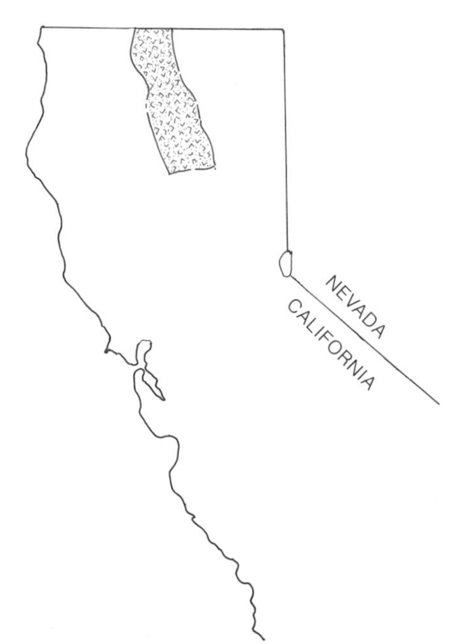

The High Cascades.

Lassen from Manzanita Lake.

7

High Cascades

The Western Cascades started erupting about 40 million years ago, then snuffed out about 17 million years ago, just when the great flood basalt eruptions of the Pacific Northwest began. Most of those rocks are in Oregon and Washington. They exist in only a few small areas in northern California.

The modern High Cascades began to erupt within a few million years after the flood basalt eruptions ceased, and they are still erupting. They include such monsters as Rainier in Washington, Hood in Oregon, and Shasta in California, as well as dozens of lesser eminence. They all erupt above the slab of ocean floor now sinking through the modern trench off the coast north of Cape Mendocino.

Shasta and Lassen are the most notable High Cascades volcanoes in California. Both have erupted within historic time and probably will again. Some of the many lesser volcanoes may also erupt again. And it is perfectly possible that brand new volcanoes will appear, as cinder cones, rhyolite domes, or even andesite cones.

VOLCANIC ROCKS

The High Cascades volcanoes erupt black basalt, white rhyolite, and all the gray varieties of andesite in between. That is the complete spectrum of common volcanic rock types. Many produce basalt early in their careers, andesite in the middle, and rhyolite as they near their ends—but that is just a broad tendency with too many exceptions to qualify as a predictably general rule.

Basalt is black, or almost so, and consists mostly of mineral grains too small to see without a microscope. If it is in volcano country and is black, hard, and nondescript, call it basalt.

Basalt full of gas bubbles, 15 miles north of Weed.

Rhyolite comes in white and in various pale shades of gray, yellow, pink, and lavender. It quite commonly contains visible grains of quartz and feldspar. The irregular crystals of quartz are nearly transparent but tend to look dark gray because you see through them into the dark shadows of the holes they fill. The feldspar crystals are blocky and white, with flat cleavage surfaces that glitter in the light like little mirrors. You will at least be decently close if you call an extremely pale volcanic rock rhyolite. The exception is the rhyolite glass, obsidian. It is intensely black.

Andesites cover the broad spectrum of volcanic rocks between basalt and rhyolite. They come in all shades of gray. Dark gray andesites are close to the basalt end of the spectrum, pale gray varieties are near the rhyolite end. Most andesites contain pale crystals of feldspar large enough to see without a magnifier. If the rock is some shade of gray, perhaps speckled with white grains, call it andesite. Some andesites, most commonly those near the basalt end of the spectrum, erupt quietly as lava flows to make reasonably solid rocks; most erupt with enough violence to make ash and ugly deposits of volcanic rubble called agglomerates.

Types of High Cascades Volcanoes

Most basalt eruptions build cinder cones. A cone begins when a fissure opens in the ground and blows off hot steam, carbon dioxide, and other

Close view of andesite agglomerate.

gases that carry shreds of molten basalt. The finer shreds drift on the wind in a dark cloud of volcanic ash, while the coarser pieces settle around the vent to build a cinder cone. It continues to grow as long as steam blows out of the vent, typically during a period of several days to several weeks.

When most of the steam is gone, a lava flow bursts out through the base of the cinder cone. That happens for the same reason that milk flows out through the base of a pile of corn flakes. A cinder cone is just a loose pile of debris without the strength to hold and confine a column of heavy basalt lava.

Basalt lava is fairly fluid, so the flow runs thin and covers a large area. Many basalt flows raft large chunks of their cinder cone along with them. Most flows are less than 100 feet thick and cover several square miles. The eruption ends after the lava flow appears, and the cinder cone never erupts again. Future eruptions in the neighborhood will build new cinder cones.

A long series of basalt eruptions within a small area builds a deep pile of thin lava flows that geologists call a shield volcano. Those come in all sizes, ranging up to genuine monsters, such as the Medicine Lake volcano near the Oregon border.

Andesite lava is not as fluid as molten basalt, so it does not run out into such thin lava flows. It may also contain a heavy charge of steam that explodes as it erupts, tearing the lava into clouds of ash and chunks of rubble. So andesite volcanoes consist largely of loose rubble standing at its angle of repose, like a talus slope. Most contain just enough lava flows to knit the pile into an edifice solid enough to contain a column of lava. So andesite volcanoes tend to erupt repeatedly from a summit

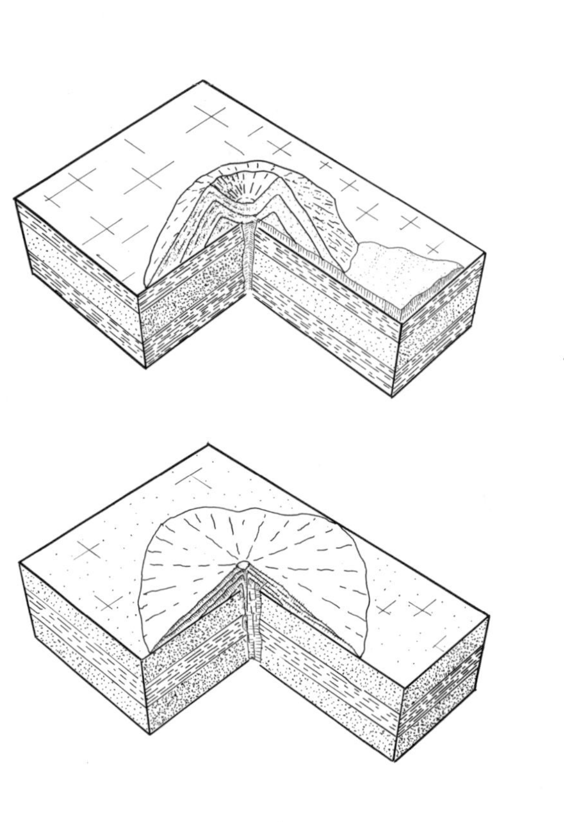

A cutaway view of a cinder cone with a basalt lava flow erupted from its base.

Cutaway view of a basalt shield volcano.

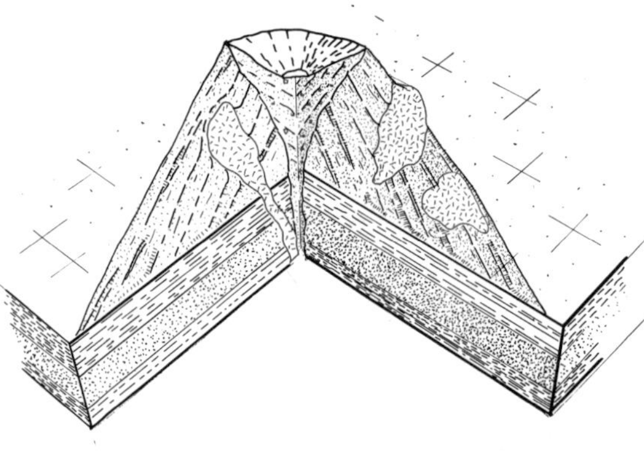

Cutaway view of an andesite volcano. They commonly have complex internal structures that record a complex eruptive history

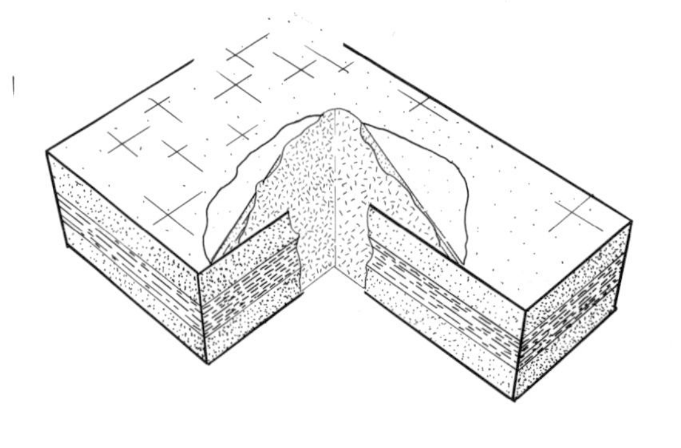

Cutaway view of a rhyolite dome Many of them sag at the base into thick and sluggish lava flows.

Lassen from the south.

crater and tend to build a tall cone that grows with every eruption. Those are the raw material for travel posters. Andesite volcanoes may also erupt from their flanks.

Rhyolite lava is so viscous that it seems almost solid. What happens when it erupts depends mainly on how much steam it contains. If it contains little or no steam, rhyolite erupts as great bulges of lava that rise ever so slowly from the ground, like giant mushrooms. Many spread sluggishly down the slope as enormously swollen lava flows. Rhyolite eruptions may continue for years in such tediously slow motion that it is hard to be sure that anything is happening.

If rhyolite lava erupts with a heavy charge of steam, it explodes into great clouds of volcanic ash and pumice. The lighter part of the eruption cloud drifts downwind, dropping volcanic ash like winter snow. The part that contains enough ash to make it denser than air spreads across the ground as an ash flow that instantly ignites and buries everything in its path. Most ash flows are still so hot when they finally settle that their central parts fuse together into a solid rock called welded ash.

Rhyolite tends to erupt in large volumes, catastrophically if it contains much steam. Many large andesite volcanoes end their careers in a great debacle of steaming rhyolite that destroys the volcano and leaves a large

Shasta from the north. Shastina on its right.

collapse basin, a caldera, where it once stood. We have no historic record of a rhyolite eruption on the largest scale, but the rocks leave no doubt that they happen with a ferocity far beyond what most of us would expect from a volcano. The recorded human experience with volcanoes does not nearly encompass their full potential.

SHASTA

So far as can be told, Shasta began its career sometime around 600,000 years ago. The mountain appears to have grown during four major eruptive periods, each of which produced a large andesite volcano. So the massive bulk of Shasta is really a huddled cluster of volcanic cones.

Although the southern flank of Shasta is deeply eroded, most of the mountain is not. Obviously, it has been active enough to repair most of the scars of ice age glaciation. Radiocarbon dates suggest that Shasta has erupted, on average, once every 600 to 800 years since the last ice age. It erupted ash flows about 1,100 and 750 years ago, a hot mudflow about 800 years ago, and modest quantities of ash and hot mudflows in 1786.

Shastina, the satellite cone on the west flank of Shasta, is a substantial volcano in its own right, one of the larger of the High Cascades. Its smooth slopes show no sign of glacial erosion. A detailed series of radiocarbon dates on charcoal associated with its lavas show that the entire cone grew during an episode of vigorous activity between 9,700 and 9,400 years ago. Ash flows poured down the west side of Shastina as it was growing, and across more than 40 square miles of countryside west of the mountain, including the sites of Weed and Mount Shasta city. It has not erupted since.

The Hotlum Cone on the summit of Shasta appears to have been the source of most of the activity since Shastina went out of business. It apparently erupts at least once every thousand years, probably more frequently. Steam vents in the Hotlum Cone show that hot rock still exists within the volcano.

Gravity surveys reveal evidence of a large volume of abnormally light rock at shallow depth beneath Shasta and the surrounding area. Earthquake waves passing beneath the area of the gravity anomaly lose much of their shear wave motion, which means that they are passing through a liquid. It must be magma. No one actually knows what kind of magma, but pale andesite or rhyolite seem a good bet because they commonly come in large volumes. Also, the small eruption of 1786 produced pale andesite. If that magma lurking beneath Shasta really is a large mass of rhyolite or pale andesite, and if it contains steam, it could cause a horrifying eruption. The potential may exist for a catastrophe of the kind that quickly converted Oregon's Mazama, a volcano on the scale of Shasta, into Crater Lake.

LASSEN VOLCANIC NATIONAL PARK

The geologic map looks at first glance like a chaotic assemblage of splotches. That is exactly what it is. Geologic maps of areas that have seen a long and varied history of volcanic activity do look that way.

Lassen stands within a large collapse caldera, which lies in turn within an older and much larger caldera. The larger caldera opened some hundreds of thousands of years ago when an enormous volcano that geologists call Maidu sank into a magma chamber that was emptying during a great eruption. Then Tehama, the smaller successor to Maidu, sank into the second caldera during another great eruption.

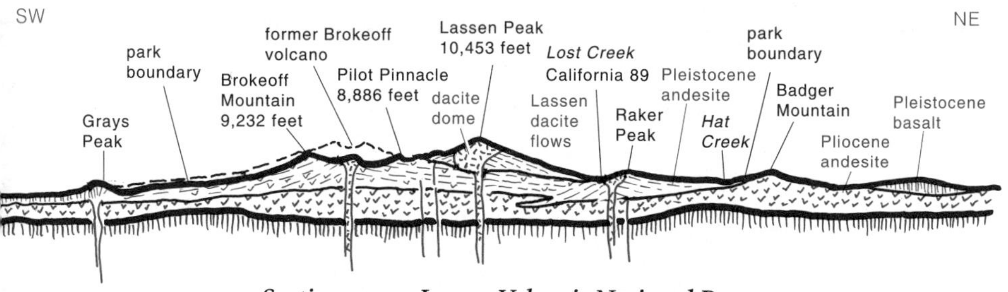

Section across Lassen Volcanic National Park.
—Adapted from Williams, 1932

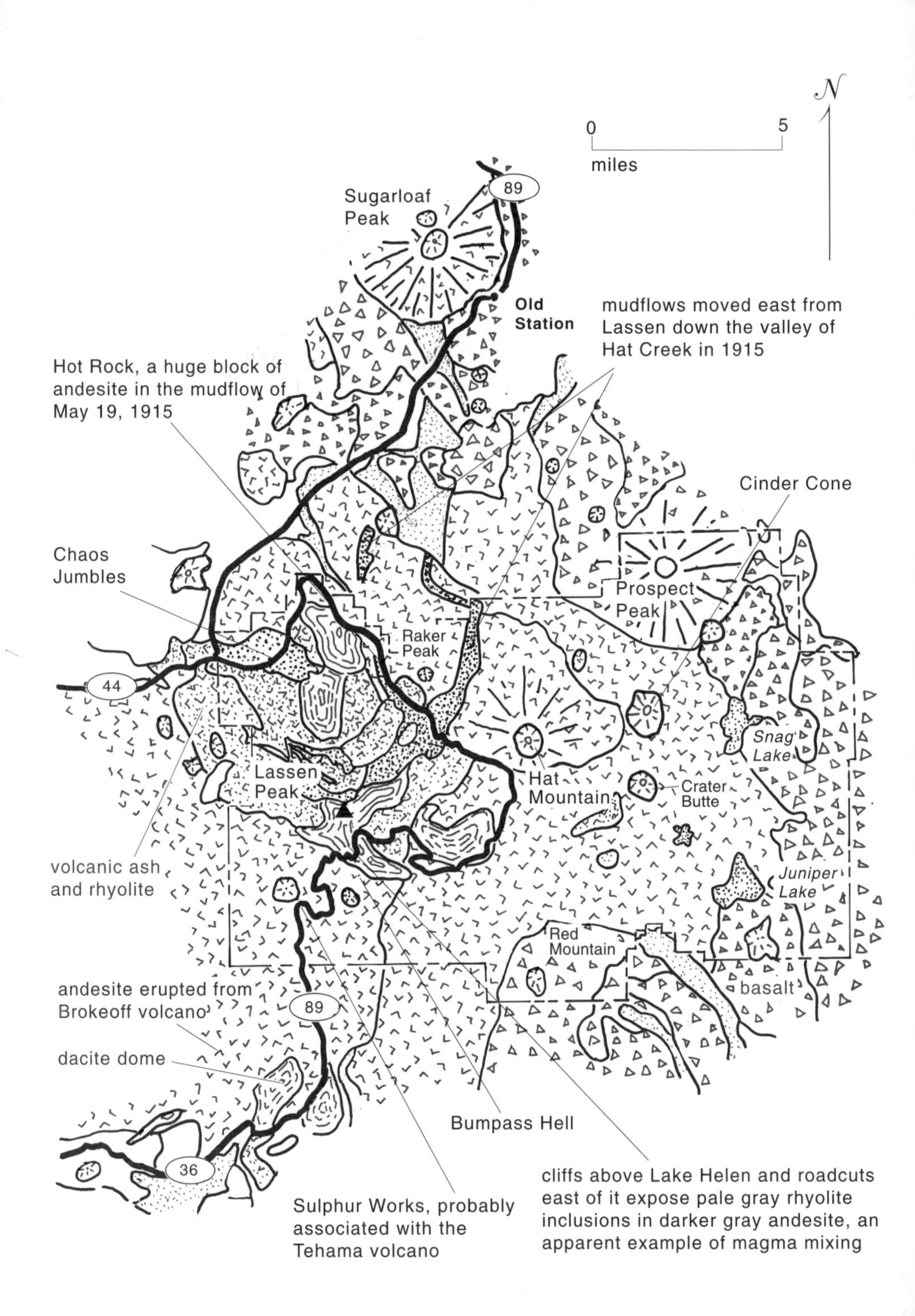

Lassen Volcanic National Park.

Lassen blowing off a cloud of steam and rhyolite ash, 1915.
—J. S. Diller photo, U.S. Geological Survey

Geologists mentally reconstruct Tehama by projecting its remnant lower slopes up at the usual angle for an andesite volcano, then arbitrarily lopping off part of the top for a crater. The result of that exercise shows that Tehama must have been the volcanic centerpiece of the region, a giant fully comparable to Shasta, about 11,000 feet above the surrounding countryside. Tehama first erupted about 450,000 years ago and continued at intervals until it sank into its caldera during a great rhyolite eruption about 350,000 years ago. All that remains of Tehama is the smaller caldera, where the peak once stood, and a few remnants of its outer flanks, of which Brokeoff Mountain and Mount Diller are the largest.

Lassen later rose in the caldera that opened as Tehama destroyed itself. Most large volcanoes grow in a series of eruptions from a central vent, but Lassen is an enormously oversize dome that rose full grown from the earth. Technically, the rock is dacite, a pale variety of andesite that lies close to the rhyolite end of the volcanic spectrum. Dome volcanoes typically grow within a few years, then solidify to become ragged lumps of solid lava clothed

Large dacite boulder rests against a ponderosa pine that was debarked on its upstream side by the 1915 mudflow.

in sliding screes of broken rock. Lassen is one of the largest of its type and is one of the few to erupt again after it stopped growing. It erupted four ash flows from its north flank about 1,000 years ago, three debris avalanches about 350 years ago, a basalt lava flow in 1850, and steam in 1854.

Then, in May 1914, Lassen suddenly blew off an immense cloud of steam that was dark with volcanic ash, apparently without the customary prelude of swarms of small earthquakes. More dark clouds of steam and ash appeared at irregular intervals throughout the next year, with progressively hotter steam and increasing violence. The steam explosions blasted a crater in the summit, which enlarged as the season continued. In May 1915, observers saw the crater glowing brightly at night. Lava had finally reached the surface.

A few days later, viscous lava began to overflow the crater and pour slowly down Lassen's western slope, eventually making a flow about 1,000 feet long. It looks like a long, black tongue lapping down the mountain. Its composition is that of a pale andesite, near the rhyolite end of the compositional spectrum, but it is black because it consists largely of rhyolite glass—obsidian. If it were crystalline, it would be pale gray.

Meanwhile, another flow oozed over the northeastern wall of the crater and poured down the slope into a deep snowfield, where it burst into fragments as though it were a casserole dish dropped hot into ice water.

The meltwater picked up a load of old volcanic ash, and the whole mess poured some 20 miles down the valley of Lost Creek as a hot mudflow. Part of the mud spilled over the divide into the valley of Hat Creek, where it buried several ranches, littering them with boulders weighing as much as 20 tons. Some of the boulders were red-hot lava, which puffed steam out of the mudflow for days after it stopped moving. Mudflows typically accompany volcanic eruptions. Usually, as at Lassen, they are more dangerous and destructive than the lava flows.

All that was just preamble. The main event came two days later when an enormous steam explosion blew a dark mushroom cloud 7 miles into the air, casting a pall over the surrounding countryside. The cloud caused panic in the streets of Redding. People watched, horrified, from the cupola of the state capitol in Sacramento. Ash fell like snow across a wide area of the northern Sacramento Valley and dusted a wide swath of northern California and Nevada. Meanwhile, a searing cloud of hot steam and fragments of lava swept down the valleys of Lost and Hat Creeks, which the mudflow had recently devastated. The heat of the newly erupted steam and ash melted more snow and started several more large mudflows.

After that superb effort, Lassen subsided into occasional small steam explosions that mainly ended in 1917 but continued on and off until 1921. Some fumaroles still blow hot gases, but they steadily diminish in number and decline in temperature. The action is over, at least for now. The big explosion left Lassen with a gaping crater on its northeastern flank. Old photographs show the volcano looking like a battered wreck, its sides streaked with the dark stripes of the big mudflows. Plants have now covered those so completely that the black lava flow is the only easily visible trace.

It is hard to form a clear impression of the surrounding rocks and their stories from the road through Lassen Park. The park contains a group of volcanoes, most of which were active in various ways for a long time. The complex pattern of late activity superimposed upon the wreckage of a long volcanic past is typical of large volcanoes.

The rocks come in a variety of colors, generally some medium shade of gray, brown, or reddish brown. They also come in a variety of disguises, depending upon how they erupted. Andesite agglomerates are chaotic mixtures of large and small fragments set in a matrix of ash. Most are mudflow deposits but some settled directly from an eruption cloud onto the volcano. Ash is simply a deposit of small fragments of lava that blew from the vent in a blast of steam. Some ash beds were dropped directly on the volcano; others drifted downwind and settled like snow to make beds of air-fall ash; still others were deposited from running water.

Mixed Magmas

Lassen's lava flow of 1915 is famous among interested geologists because it contains dark and light rock in streaks and blotches. It looks like an attempt at a volcanic marble cake. Look for the mixed rock in the blocks around the Devastated Area parking lot.

Such mixed lavas quite dependably cause arguments among geologists. The arguers invariably agree that the question of how to interpret them is important, but they never agree on the answer. Many believe that these streaky rocks formed through incomplete mixing of two different magmas. Others believe that basalt magma rising from the mantle melted chunks of continental crust to make the andesite. Still others argue that the rocks formed as an originally homogeneous magma began to separate into two different kinds of magma, like oil separates from water. The question has been such a playground for so long that it will almost seem a loss when someone finally proposes an acceptable answer.

Raker Peak

Raker Peak began as a shield volcano, a pile of lava flows of dark andesite, on the thin edge of being basalt. After those flows erupted, a large mass of extremely viscous dacite magma erupted to make a lava dome that rose above the older shield volcano. So Raker Peak is a combination of two kinds of volcano, made of two kinds of lava.

Cinder Cone

Cinder Cone, in the northeast corner of Lassen Park, erupted the Fantastic Lava Beds, the flows that impounded Snag Lake and partially filled Butte Lake. Cinders cover the earlier flows, apparently because they erupted while Cinder Cone was still building. Hot water, probably rain steaming off the flow, altered those cinders to a startling variety of earth colors, all due to iron oxides. No cinders lie on the last flow, which has a black surface.

Some of the basalt that erupted from Cinder Cone is peculiar in containing visible grains of quartz. Some common kinds of basalt contain microscopic quartz grains, but visibly large grains are exceedingly rare and contrary to all theories. The problem is that basalt magma contains so little silica that none is left to make quartz after the usual plagioclase, pyroxene, and olivine have crystallized. It is much easier to suppose that the magma at Cinder Cone picked up some contaminants on its way up than to suggest that quartz crystallized from it in the normal sequence of magmatic events.

Most estimates have Cinder Cone erupting sometime around 2,000 years ago. Several contemporary accounts, all from observers who were miles away from the action, suggest that some kind of eruption also occurred there during the winter of 1850–51. But it is not clear just what, if anything,

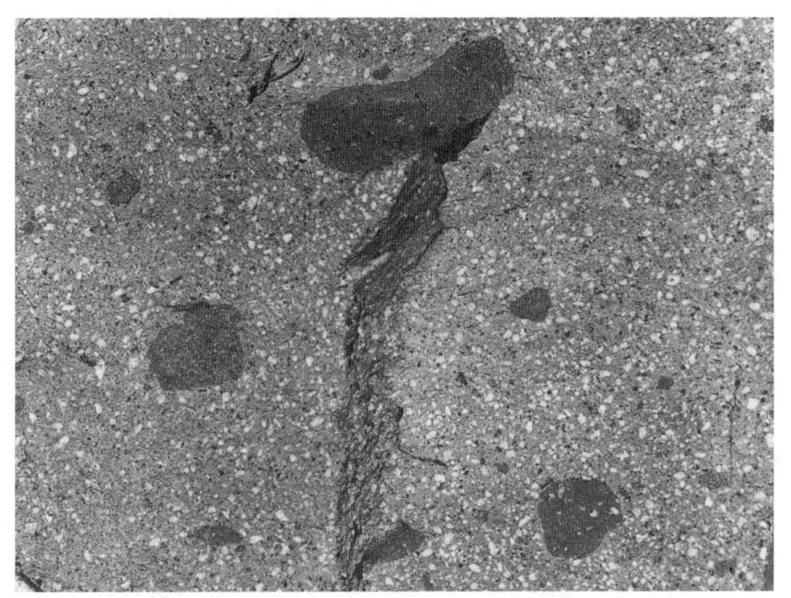

Inclusions of gray andesite in pink dacite in Chaos Jumbles.

Inclusions of pale gray rhyolite in massive gray andesite east of Lake Helen.

Chaos Jumbles below Chaos Crags.

actually happened. Perhaps that was when the last lava flow erupted. Nineteenth-century accounts of volcanic eruptions in the High Cascades are notoriously creative and unreliable.

Chaos Crags and Chaos Jumbles

Chaos Crags, a group of ragged hills near the north side of the park, is a lava dome of pale andesite, or dacite if you prefer, that rose about 1,800 feet. The rock is a pinkish shade of brown and has a glossy surface that befits a rock composed largely of glass. Look carefully to see the scattered white crystals of plagioclase feldspar.

The lava dome of Chaos Crags is basically similar to Lassen but is much smaller and much younger. Geologists estimate that it appeared sometime around 1700. Gold rush prospectors reported that it was still steaming during the 1850s. Future eruptions in Lassen Park will likely involve the appearance of more such domes.

About the time Chaos Crags reached its present height, blasts of steam blew a crater into the base of the mountain, undermining its northwestern side. The slope collapsed, spilling a mass of broken rock to create Chaos Jumbles. This rolling sea of angular chunks of rock covered nearly 1 square mile and splashed 400 feet up the side of Table Mountain, probably in a matter of a minute or two. It was a catastrophic rockfall. It fell so recently that few lichens grow on the blocks, little shrubbery among them.

The rockfall of Chaos Jumbles impounded Manzanita Lake, near the north entrance station to the park. As the lake filled, the water table rose and flooded low points in the rockfall, creating Reflection Lake and the Lily Pond, among others.

Bumpass Hell and Supan Springs

Most park visitors see the volcanic hot springs and hot gas vents at Bumpass Hell, as well as those at Supan Springs at the Sulphur Works. Both groups are associated with Brokeoff Mountain, the remnant of Tehama, rather than with Lassen.

Surface water seeps down through cracks to the vicinity of large bodies of hot rock. The water absorbs heat from the hot volcanic rocks and reacts with them, then boils back to the surface as mineralized hot water or steam. The water deposits much of its dissolved mineral matter around the springs as it begins to cool. The deposits grow into fantastic structures of siliceous sinter, composed mostly of silica, and of white travertine, composed mostly of calcite.

Mining geologists find hot springs especially fascinating because many ore bodies owe their origin to circulating hot water depositing metallic minerals. The hot springs and fumaroles in Lassen Park are the top of a deep circulation that may now be depositing new metallic ore bodies thousands of feet below the surface.

Gas vents at Sulphur Works.

Glacially scoured bedrock near Lake Helen.

Emerald Lake

Emerald Lake, like the other small lakes in the higher parts of the park, is the result of glacial erosion, not of volcanic activity. Snow and ice covered much of what is now Lassen Park during the last ice age. Large glaciers poured down the peaks to scour the valleys below. The scraping glaciers left beautiful exposures of solid lava in the highest part of the park.

HIGH CASCADES ROADGUIDES

The High Cascades cover only a small part of northern California, and few roads pass through them. Those few are richly rewarding.

Interstate 5

Dunsmuir—Oregon

69 miles

Most of the route between Dunsmuir and Grenada closely follows the boundary between the High Cascades to the east and the mountains of the Klamath block to the west. Most of the landscape east of the road is volcanic and very young. The area west of the road is entirely erosional, carved in the ancient rocks of the Klamath block. The volcanic rocks lap onto the much older mountains of the Klamath block approximately along the line of the road.

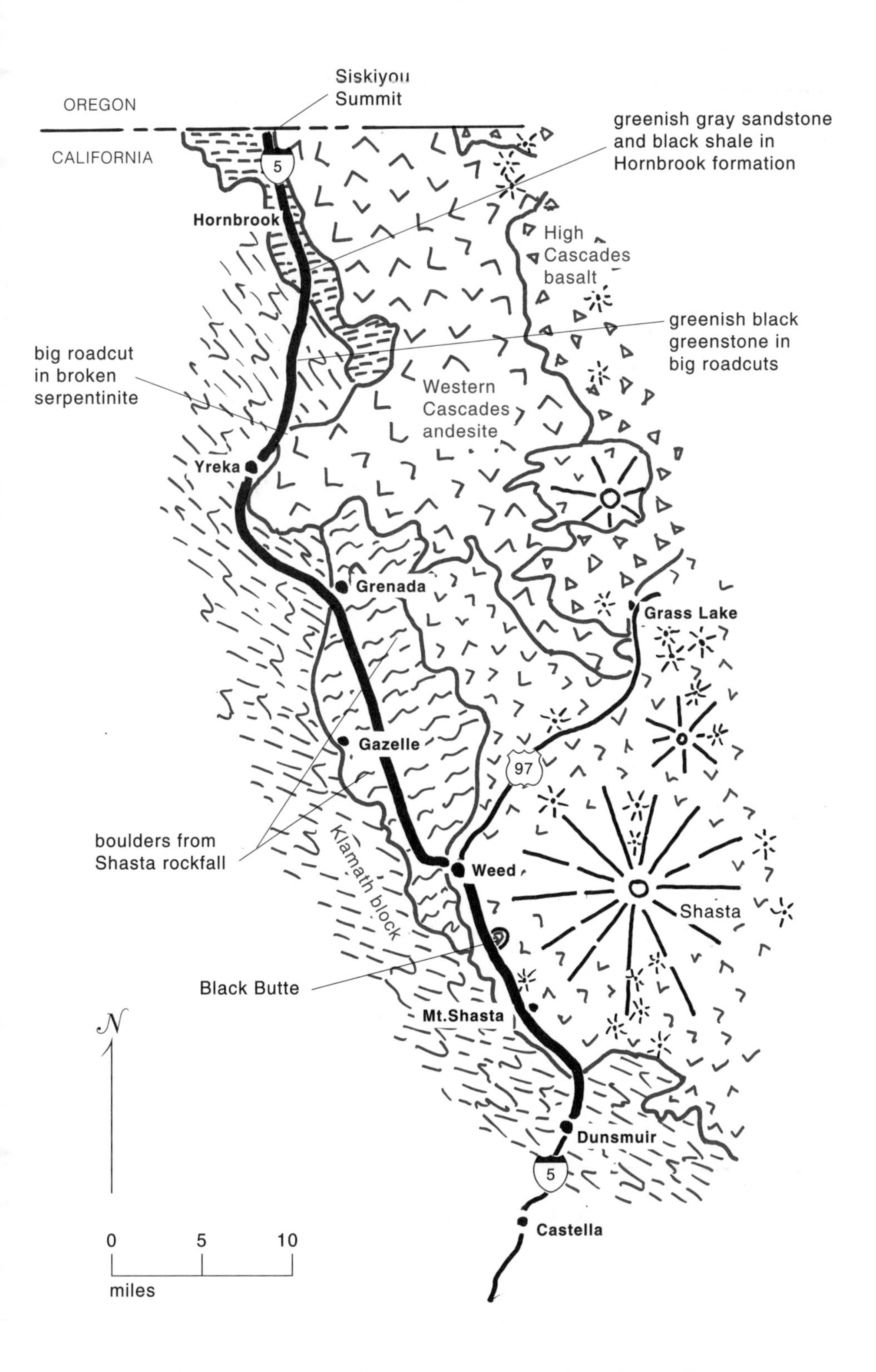

Rocks along Interstate 5 between Dunsmuir and the Oregon line.

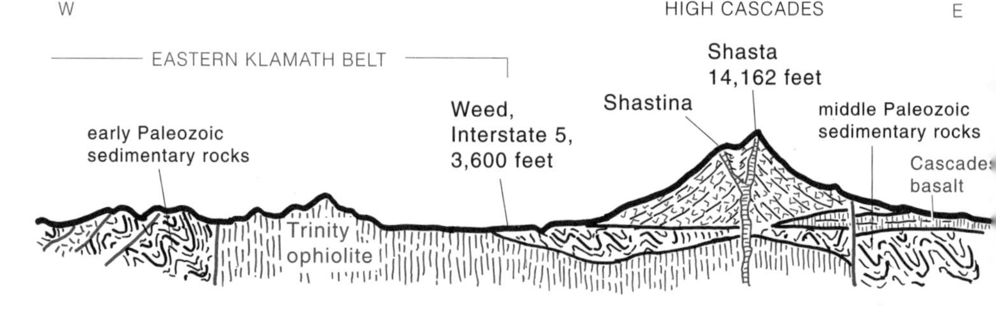

Geologic section across the line of Interstate 5 at Weed. Shasta stands on rocks of the Eastern Klamath belt. —Adapted from Wagner and Saucedo, 1987

Rocks in the hills west of the highway between Dunsmuir and Weed and on both sides of a short stretch of road at Yreka are part of the Trinity ophiolite. It was the floor of the ocean during Ordovician time, approximately 475 million years ago. Its black rocks are mainly peridotite, a generous slice off the earth's upper mantle, now largely altered to dark green serpentinite. Metamorphic rocks in the hills west of the road between Weed and Yreka were deposited as sediments on the Trinity ophiolite. They and the Trinity ophiolite belong to the Eastern Klamath terrane, the equivalent to the Shoo Fly terrane of the Sierra Nevada. They are among the oldest rocks in northern California.

Metamorphic rocks in the hills west of the road north of Yreka were deposited as sedimentary rocks during later Paleozoic time. They belong to the Western Paleozoic and Triassic belt of the Klamath block, the equivalent to the Calaveras complex of the northern Sierra Nevada.

Interstate 5 follows the valleys of Yreka Creek and the Klamath River in the 11 miles north of Yreka. Bedrock along the road is composed of

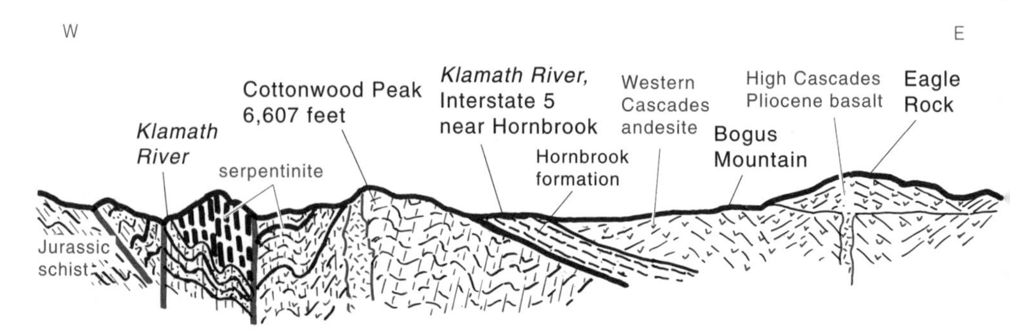

Geologic section across the line of Interstate 5 near the Oregon line. This is about as far south as the Western Cascades andesites extend.
—Adapted from Wagner and Saucedo, 1987

sedimentary rocks of the Hornbrook formation, laid down during late Cretaceous time. Some of them were deposited in the Modoc seaway. Their horizontal position shows that little tectonic movement has affected this part of California since late Cretaceous time.

Western Cascades Andesite

A large expanse of bedrock in the hills east of the northernmost segment of the road is andesite that erupted in the old Western Cascades volcanic chain. The chain went into business about 40 million years ago, erupted vigorously for more than 20 million years, then abruptly snuffed out from one end to the other, about 17 million years ago. The Western Cascades are a substantial mountain range in Oregon but just barely reach south into California. None of the original volcanic landforms survive. The range is a typical erosional landscape carved into volcanic rocks.

Shasta

Shasta is one of the largest active andesite volcanoes in the world. Its immense bulk of about 80 cubic miles cuts an enormous piece out of the skyline east of Interstate 5 along the way between Dunsmuir and Weed.

The French explorer La Perouse and his crew briefly glimpsed what they thought was a distant volcanic eruption while they were sailing along the coast of California at the latitude of Shasta in 1786. It was almost certainly the Hotlum Cone on the summit of Shasta—a minor eruption that scattered a thin layer of brown pumice across part of the mountain. The same eruption also sent an ash flow and several hot and cold mudflows about 7 miles down the valley of Ash Creek, on the east slope of Shasta, and sent a hot mudflow some 12 miles down the valley of Mud Creek.

Black Butte

Interstate 5 skirts the steep western flank of Black Butte about 5 miles south of Weed. It is a large rhyolite dome, which owes its dark color to the lichens that grow on the pink rock. Although it looks like a single volcano from Interstate 5, aerial photographs show that Black Butte is actually a line of rhyolite plugs. They rose out of the ground as extremely viscous masses of rhyolite magma, probably one after another. According to radiocarbon dates, that happened about 9,500 years ago, probably while Shastina was growing.

The outer shell of a rising rhyolite dome cools into a rigid carapace of solid rock, which then cracks into angular blocks as the molten rock within continues to move. Thus, the growing plug dome constantly sheds its outermost layers as steaming rubble that slides down its sides, clothing it in talus. Screes of angular blocks cover the steep sides of Black Butte

Black Butte, a series of lava domes, from the southeast.

so completely that no solid rock is exposed, except in a few craggy outcrops near the summit. Typical plug domes have no summit crater.

Like cinder cones, plug domes typically erupt during only one period of volcanism; the next volcanic eruption produces new ones. Plug domes are rather common in the High Cascades. Watch from Interstate 5 for several steep knobs that protrude like warts from the south slopes of Shasta.

Big Rockfall

Even at their best, andesite volcanoes are shabbily built mountains, great piles of loose rubble and ash with only a few solid lava flows to knit them weakly together. If there were a building code for mountains, they would never pass an inspection. And they tend to get weaker with time as hot water and steam circulating within them degrade their rocks into clay. Andesite volcanoes are a menace to those who live near them, whether or not they erupt.

Sometime around 300,000 years ago, the north side of Shasta collapsed in a great rockfall that spread northwest across the Shasta Valley, as much as 28 miles from the volcano. The rockfall contained about 6.5 cubic miles of rock. That debris must have been traveling incredibly fast to go so far across the nearly level surface of the Shasta Valley. Some geologists suggest that such rockfalls ride like hovercraft on a cushion of compressed air trapped beneath them. The mountain shows no sign of a scar where the great rockfall detached, no doubt because later eruptions repaired it.

Lumpy hills of rockfall in distance to west, viewed from Interstate 5 about 15.5 miles north of Weed.

Interstate 5 crosses the rockfall in much of the stretch between Weed and Grenada. Watch for lumpy hills that do not connect to make ridges and for marshy depressions that do not connect to make valleys. The landslide dump looks like an oversize glacial moraine, and its debris makes a chaotic deposit that looks like glacial till in roadcuts.

Mima Mounds

A roadside rest area on the big rockfall provides a good view of some Mima mounds, pronounced as in "lima bean." They are circular mounds, 30 feet or so in diameter and approximately 2 feet high. Many rise within a conspicuous outline of larger rocks nestled in the low spaces between them. They must have formed sometime since the rockfall moved about 300,000 years ago.

Mima mounds exist in many of the grassy areas of the Pacific Northwest, typically in herds of thousands. They are easiest to see when the low light of early morning or late evening casts them into shadow relief. Geologists and others have argued about them for decades, proposing all sorts of origins—from giant anthills to ice age gophers, even the vibrations of large earthquakes.

Hornbrook Formation

Interstate 5 crosses sedimentary rocks between the area about 1 mile north of the bridge across the Klamath River and the Oregon line. Most are greenish and gray sandstones, mudstones, and shale, the Hornbrook formation. Parts of the formation contain fossils of animals that lived in seawater during late Cretaceous time, about 70 million years ago.

The Hornbrook formation and the fossils it contains provide the best direct evidence that a Modoc seaway did indeed exist in northeastern California during late Cretaceous time. This area is one of the few places where these rocks peek out from under the edge of the volcanic rocks that cover most of the Modoc basin. The layers of late Cretaceous sedimentary rocks appear to extend eastward, beneath the much younger flood basalt flows and the still younger High Cascade volcanoes.

U.S. 97
Weed—Oregon
55 miles

All the rocks exposed along U.S. 97 are volcanic, none of them more than a few million years old and many no more than a few thousand. Most of the landscape is volcanic as well, the product of the same eruptions that made the rocks. The slow processes of erosion have not yet established an erosional landscape.

Weed is in the Shasta Valley, a low area between the Klamath block to the west and the Cascade volcanoes to the east. The hills north of the Shasta Valley are eroded in volcanic rocks that erupted from Western Cascades volcanoes, 17 or more million years ago. They extend south beneath much of the valley floor. Ice age glaciers spawned on Shasta reached the floor of the Shasta Valley, where they left large moraines in the area between Weed and the east side of Lake Shastina. U.S. 97 crosses lava flows

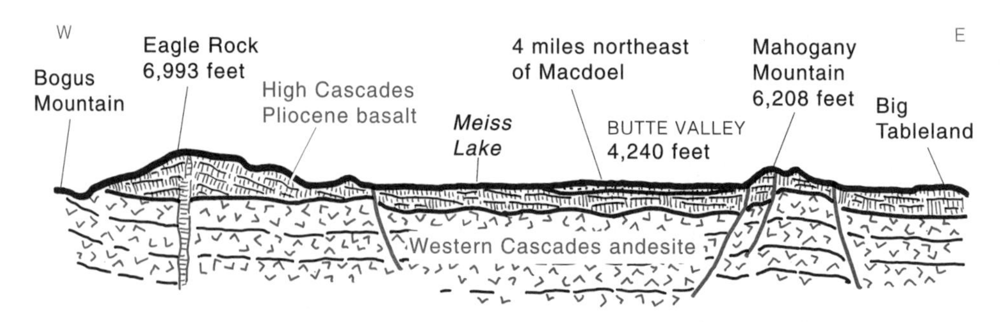

Geologic section across U.S. 97 north of Macdoel. The older volcanic rocks of the Western Cascades lie beneath the High Cascades rocks.
—Adapted from Wagner and Saucedo, 1987

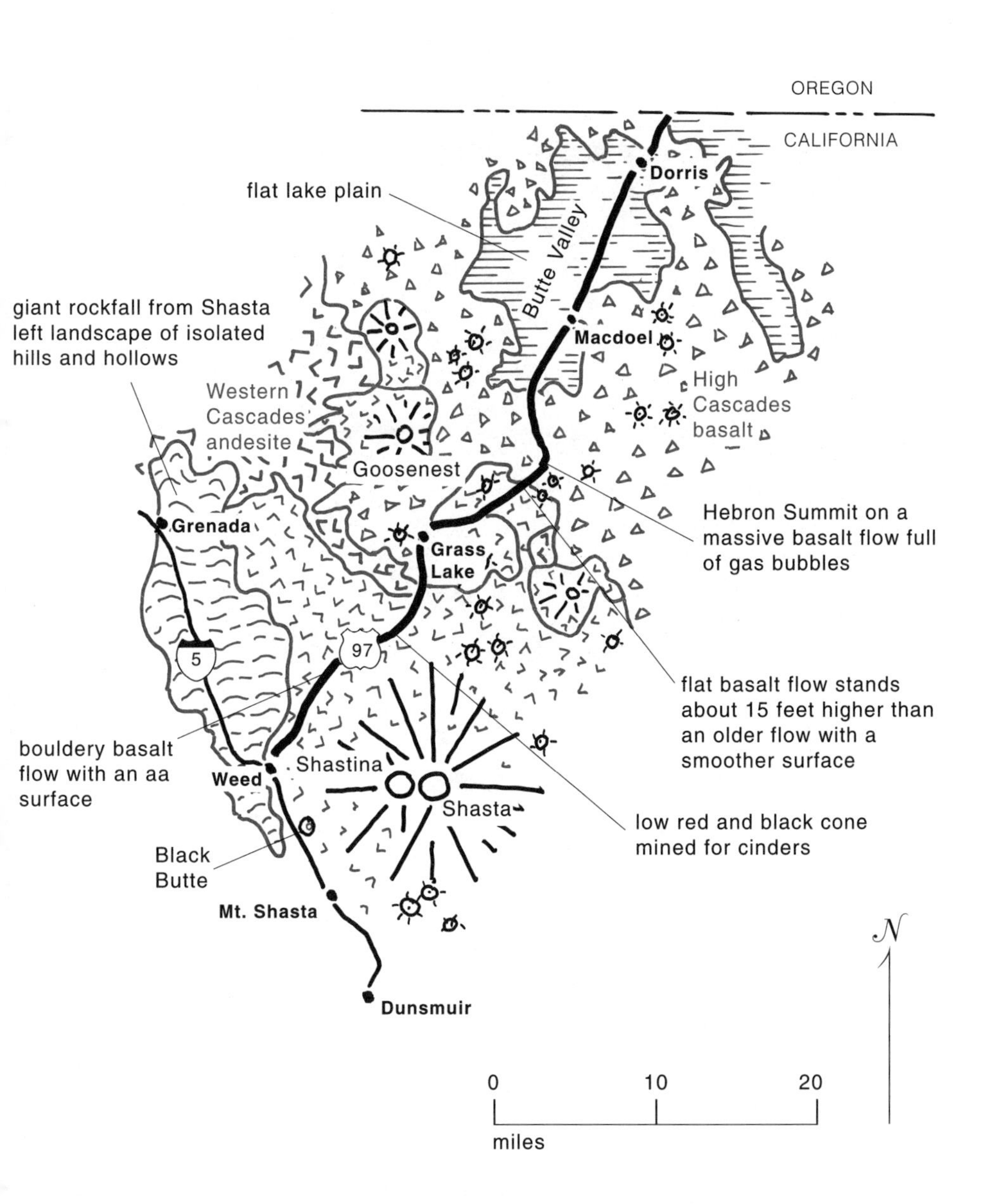

Rocks along U.S. 97 between Weed and the Oregon line.

Volcanic ash on low rolling landscape, 8 miles north of Weed.

and winds between volcanoes as it crosses the High Cascades between Weed and Macdoel. All belong to the High Cascades; some are as much as several million years old, some only a few thousand.

Lava Park Flow

U.S. 97 skirts the northern edge of the Lava Park flow between 5 and 10 miles north of Weed. The flow erupted from Shastina. Watch for its steep front, covered with rubbly blocks, a short distance south of the road. It still looks so fresh that many geologists guessed it was only a few hundred years old until radiocarbon dates showed that it erupted about 9,700 years ago. Even after all those years, hardly any plants grow on that raw rock.

Butte Valley

About 6 miles southwest of Macdoel, U.S. 97 crosses the boundary between the High Cascades to the southwest and the Modoc Plateau to the northeast. A steep slope marks the boundary.

Butte Valley is one of the dropped Basin and Range fault blocks of the Modoc Plateau. Its bedrock is mostly flood basalt flows erupted during middle Miocene time, sometime around 16 million years ago, from the crater basin volcano in southeastern Oregon.

Any surface as flat as the proverbial billiard table is almost certainly an old lakebed. Lake sediments, mostly clay, that lie beneath the flat valley floor between Macdoel and Dorris leave no doubt that it is indeed a lake plain. The lake existed in the wet climate of the last ice age and drained north into the Klamath River. All that remains is a marshy area called Meiss Lake, which has no outlet and fluctuates in size according to the weather.

California 36
Red Bluff—Susanville
112 miles

The western part of this route crosses volcanic rocks erupted from the Cascades in the past few million years. The eastern part follows so close to the boundary between the Cascades and the north end of the Sierra Nevada that it could fit equally well into either chapter.

Red Bluff is near the north end of the Great Valley, where its broad flat floor begins to narrow northward. East of Red Bluff, gravels that flank the northern Sacramento Valley underlie the lower slopes and lap onto the High Cascades volcanic rocks in the low hills to the east. Most of those are bouldery mudflows that geologists include in the Tuscan formation. They were deposited during Pleistocene time.

The road east of Mineral crosses many basalt lava flows that erupted at various times during the past several million years. Soil and brush cover the older ones; some of the younger ones still look fairly fresh. Mountains south of the road along the eastern half of the route are the broken northern end of the Sierra Nevada. They contain metamorphic rocks of the Calaveras complex and masses of granite. The High Cascades volcanic rocks lap onto them.

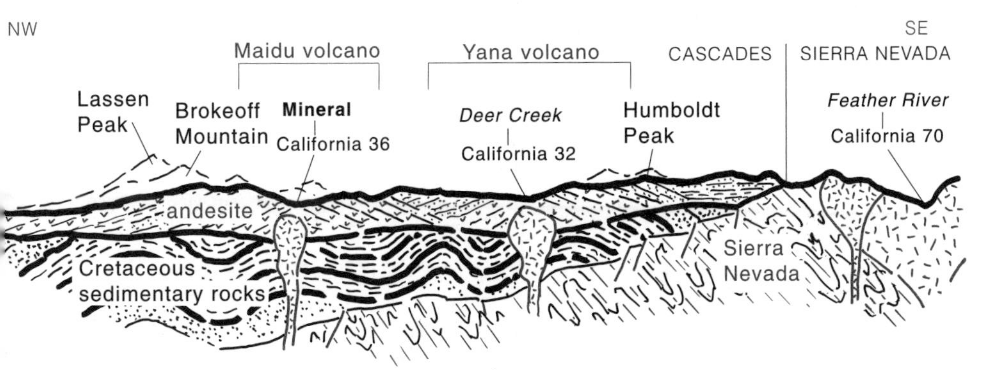

Geologic section across the line of California 36. The High Cascades volcanic rocks lie on much older Cretaceous sedimentary rocks deposited in the Sacramento Valley west of the Sierra Nevada.

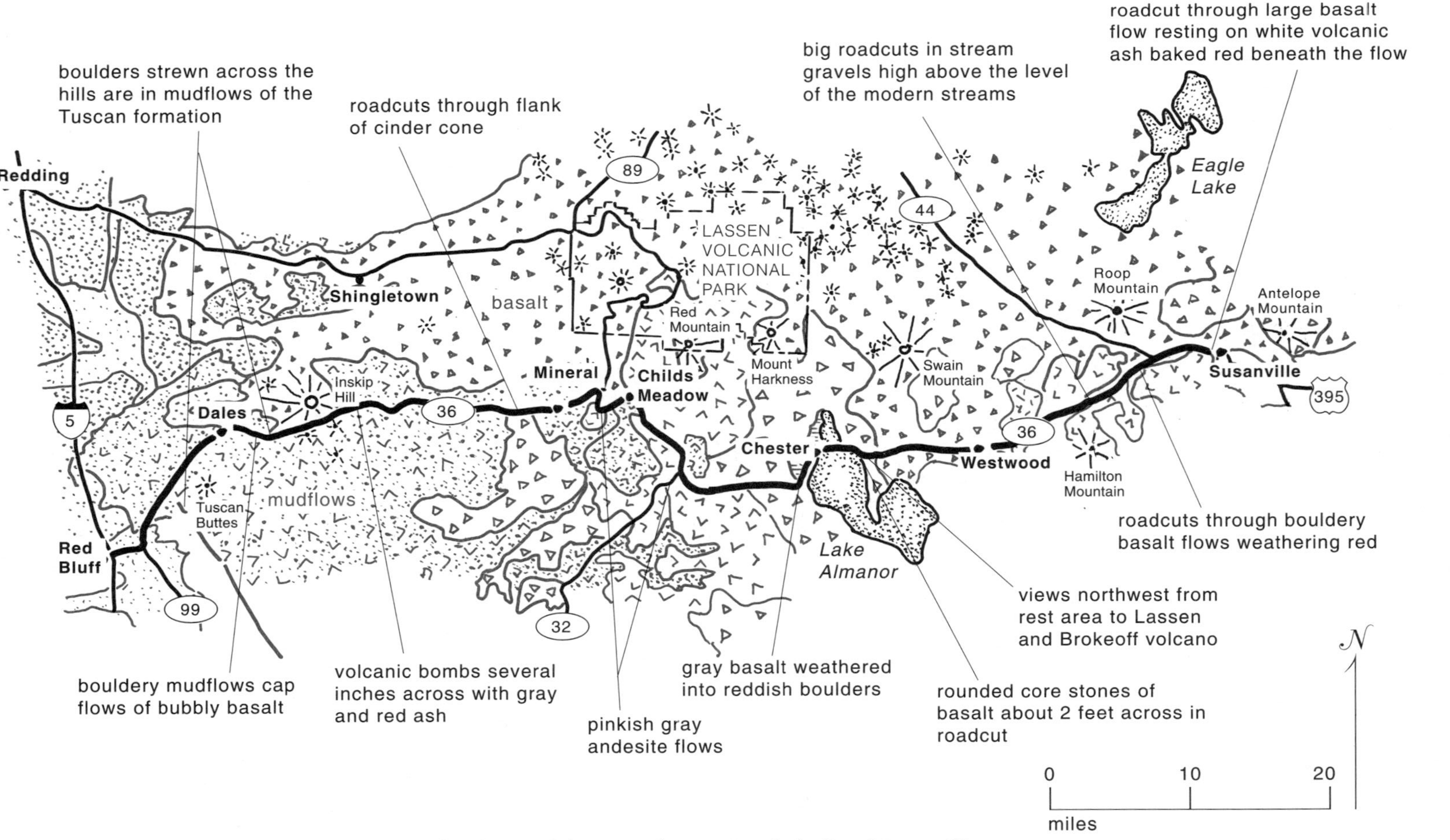

Rocks along California 36 between Red Bluff and Susanville.

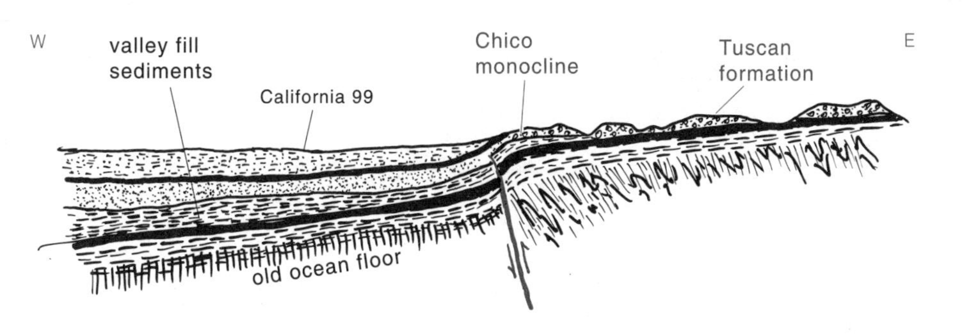

Section across the Chico monocline, a few miles south of California 36.

Chico Monocline

About 5 miles northeast of Interstate 5, the highway crosses the northward projection of the Chico monocline, which is not visible from the road. A monocline is a fold in which layers of sedimentary rocks simply bend as though they were a carpet draped over a step.

In this case, layers of volcanic sedimentary rocks in the Tuscan formation drape over a series of faults that trend northwest and raise the High Cascades with respect to the Sacramento Valley. The movement must have happened since the High Cascades volcanoes began to erupt, sometime within the past few million years. In one place the structure breaks a basalt lava flow that erupted 1.1 million years ago, so the faults have moved since then. Extremely precise measurements of the earth's magnetic and gravitational fields suggest that the fault zone follows the eastern boundary, at depth, of the old ocean floor under the Great Valley.

Mudflow Boulders of the Tuscan Formation

Boulders of dark andesite as much as 3 feet across litter the slopes of the hills between the area about 4 miles and 31 miles east of Red Bluff. The soil is mostly brown sand, once the matrix between the boulders. This picturesque landscape of scattered oak trees, grassy slopes, and dark boulders started as a series of huge volcanic mudflows.

The mudflows poured west from volcanoes south of Lassen, covering volcanic rocks on the lower slopes of the mountains. The lower ends of the mudflows interfinger with volcanic sands and gravels that were deposited on the floor of the Great Valley during Pliocene time, several million years ago. The mudflows picked up the boulders from older lava flows on the flanks of the mountains and carried them along.

Bouldery mudflow exposed in a roadcut about 12 miles east of Dales. The exposure is about 25 feet high.

Rocks feel lighter underwater than in the air because the water buoys them up—not enough to float them but enough to make them feel lighter. Mud is much denser than clear water, so it exerts a much greater buoyant effect on rocks. This explains why mudflows can carry boulders that clear water could not move. Millions of years of rains carried away much of the fine mud that once buried the boulders, leaving them littering the ground surface.

Age dates on volcanic ash in the Tuscan formation show that it is about 3 million years old. The mudflows are probably about the same age. The Tuscan formation, which is as much as 1,700 feet thick in some places, originally covered about 2,000 square miles before erosion removed parts of it. The mudflows originated on the flanks of two deeply eroded old volcanoes, Yana and Maidu.

Andesite volcanoes consist largely of volcanic ash and rubble, excellent raw material for a mudflow. They are especially likely to spawn mudflows during an eruption, when hot ash melts the snow on the mountain and the erupting plume of steam generates thunderstorms above it. Those mudflows are likely to come down the mountain boiling hot. Hot rocks within volcanoes, even dead volcanoes, commonly drive a circulation of hot water and steam that alters the original volcanic rocks to clay, which is even weaker. So old volcanoes often spawn mudflows long after their last eruption. Long seasons of wet weather, a rapidly melting snowpack, or an earthquake may trigger them into motion.

Mudflows commonly move at highway speeds, following valleys and filling low areas. Normal stream flow eventually cuts channels through them, leaving their bouldery remnants high on the stream banks.

The Tuscan Buttes rise about 500 feet above the surrounding slopes about 10 miles northeast of Red Bluff and 2 miles southeast of California 36. They are volcanic plugs made of gray to pink andesite that intrude the Tuscan formation. Some of the later mudflows in the Tuscan formation partly cover them, but none of those mudflows contain boulders of their distinctive rocks, so none of the mud came from them. They probably erupted in late Pliocene time, perhaps about 2 million years ago.

Basalt and Cinder Cones

Roadcuts about 3 miles east of Dales expose flows of black basalt that erupted during Pleistocene time, less than about 2 million years ago. Bouldery mudflows cap them. Dark brown and black basalt flows, some sandwiched between a few feet of equally dark cinders and ash, are exposed in cliffs and roadcuts about 7 miles east of Dales. These thin flows erupted from Inskip Hill, the small volcano that looms over the north side of the highway a few miles farther east. It erupted at some time within the past 2 million years.

The several cinder cones that dot the flanks of Inskip Hill erupted when rising basalt magma encountered rocks soaked with water. The water flashed into steam that coughed globs of molten basalt glowing into the air. They puffed into bubbly cinders and small volcanic bombs as they sailed into place on the growing cinder cone.

Between Dales and Mineral, California 36 crosses gray volcanic ash weathered in places to shades of red. It contains chunks of basalt that range in size from large cinders to lava bombs as much as a foot across. The chunks were blown out of nearby cinder cones and rained down on the surrounding landscape. Many were streamlined as they flew molten and soft through the air. Those were not occasions for a cloth umbrella.

The road passes just south of a cinder cone about 3 miles west of Mineral. Steam evidently permeated the cone for years after it erupted, altering the black cinders and leaving them stained with red iron oxide. A few roadcuts near the national forest boundary expose massive flows of gray andesite not related to the cinder cone.

Mount Lassen.

California 36 crosses andesite and some basalt between Mineral and the area about 3 miles east of the junction of California 89. An obscuring cover of trees hides most of it. The highway crosses a subdued landscape on pale volcanic ash and basalt lava flows between California 89 and the area about 8 miles east of Chester. The lava flows erupted during Pliocene time, several million years ago.

Lake Almanor

J. S. Diller, a pioneering geologist, noted in 1895 that the valley that Lake Almanor now floods was then called Big Meadows. The eastern part of the basin was swampy then but filling with sand, mud, and plants. Evidently, the wetland was the last remnant of a lake that once flooded the valley floor. The flat surface that California 36 crosses at Chester is a lake plain on sediments deposited in it. The old lake overflowed south until erosion of the spillway drained it. A dam built in the early 1900s essentially reconstructs the natural lake.

Lake Almanor straddles the irregular boundary between the High Cascades and the Sierra Nevada. Pliocene volcanic rocks are exposed along the north side of the lake and in the hills to the north. Metamorphic rocks of the Calaveras complex appear along its southern and eastern sides.

Sierra Nevada

The hills south of the highway, east of Westwood, are eroded in metamorphosed volcanic rocks that erupted during Triassic and Jurassic time, part of the Calaveras complex. They are near the broken northern edge of the Sierra Nevada, which is now partly buried under the much younger volcanic rocks of the Modoc Plateau. California 36 skirts that boundary between Lake Almanor and Susanville. The landscape south of the highway is typically erosional, the result of millions of years of erosion on old rocks. The landscape north of the highway is almost entirely the result of volcanic eruptions and of Basin and Range faulting. It owes little to erosion. Much of the faulting is hardly younger than its rocks.

California 36 follows a valley floored with continental sediments laid down during Eocene time, about 50 million years ago. They continue from the area about 4 miles east of Westwood to the junction with California 44. Hills above the road are eroded in basalts and andesites. Basalts of the Modoc Plateau are in the valley floor between the area about 3 miles west of the junction and Susanville.

Diamond Mountains

The north end of the Diamond Mountains rises on the southern horizon, in the area south of Susanville. The mountains are a large block of the eastern Sierra Nevada that detached and moved east during the Basin and Range faulting of about the past 16 million years. This northern end of the range is mostly Sierra Nevada granite. Farther south, the granite disappears under a cover of pale volcanic rocks that erupted since Basin and Range faulting began.

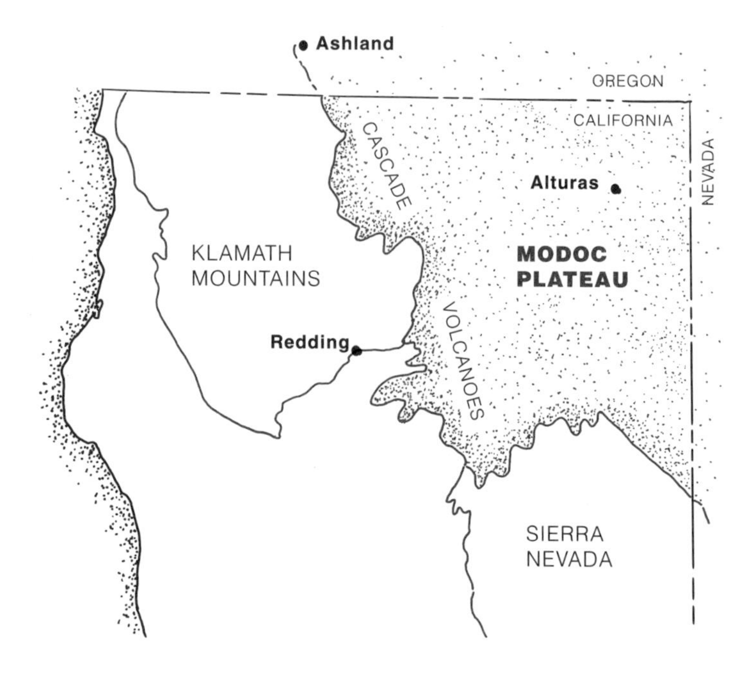

The Modoc Plateau.

Gas bubbles in basalt, 12 miles south of Ravendale.

8

Modoc Plateau

The steep rise of the High Cascade volcanoes defines the western margin of the Modoc Plateau. The torn northern edge of the Sierra Nevada defines its southern boundary. The Modoc Plateau continues north into Oregon and fades eastward into the Basin and Range of northern Nevada.

A local cover of young, sedimentary deposits fills the valley floors, but the hard bedrock in the Modoc Plateau is entirely volcanic. But the history of the Modoc Plateau began long before any volcanic eruptions. We divide that story into three periods: before, during, and after eruption of the great flood basalt lava flows of middle Miocene time.

BEFORE THE FLOOD BASALTS

Few exposed rocks tell of the Modoc Plateau before the flood basalts covered it. In the scarcity of direct evidence, we rely upon conjecture. The existence of some kind of Modoc seaway is not conjectural, but the details of its size, boundaries, and origin are. Too many younger volcanic rocks cover the evidence.

We imagine the Modoc seaway extending about 60 miles east of the Klamath block, the distance the block moved west. The seaway almost certainly lapped against the broken north edge of the Sierra Nevada and against some unknown coast in Oregon, now buried under volcanic rocks. It certainly lapped onto the eastern edge of the Klamath block. It seems reasonable to assume that the enclosing land rapidly shed large volumes of sediment into the rather small Modoc seaway. It filled, probably by the end of Cretaceous time, some 65 million years ago.

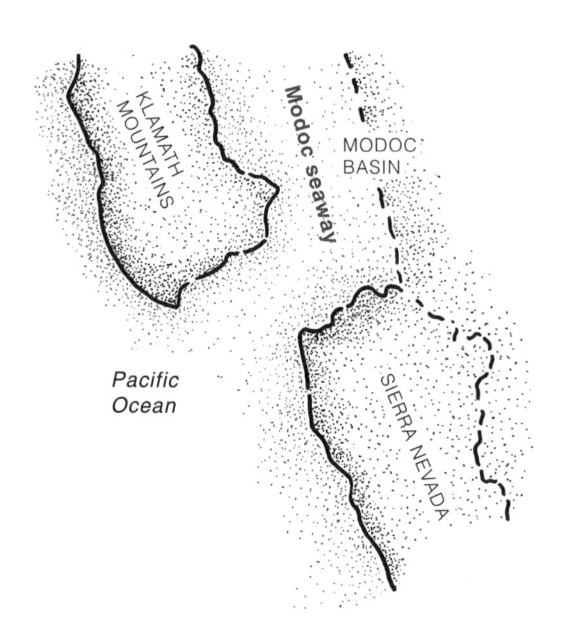

Northern California as it may have been during Cretaceous time.

Wherever they exist, deep accumulations of sedimentary rocks deposited in seawater during Cretaceous time are likely to contain oil and gas. The old Modoc seaway may be the last unleased, unexplored, and undrilled sedimentary basin in North America. And it may remain that way for a long time. Its cover of fairly young volcanic rocks looks forbidding and would greatly complicate exploration.

The Cretaceous sedimentary rocks of the Hornbrook formation disappear eastward beneath sedimentary rocks deposited on dry land during early Tertiary time. We imagine that the Modoc Plateau was then a broad interior plain, probably a desert, enclosed between the same mountains that enclosed the Modoc seaway. The picture abruptly changed when the flood basalt flows began to erupt about 17 million years ago.

FLOOD BASALTS

The coming of the flood basalts in middle Miocene time coincided with the establishment of most of the San Andreas fault, the rise of the Coast Range, and the beginning of crustal stretching in the Basin and Range. The ultimate cause of that great coincidence of events remains a matter of vigorous debate. Whatever their cause, the events of middle Miocene time were a major turning point in the geologic history of California.

Flood basalt flows differ from ordinary basalt flows in being enormous almost beyond comprehension. Ordinary basalt flows typically contain considerably less than 1 cubic mile of lava. Flood basalt flows typically contain tens of cubic miles of lava, considerably more than 100 in some

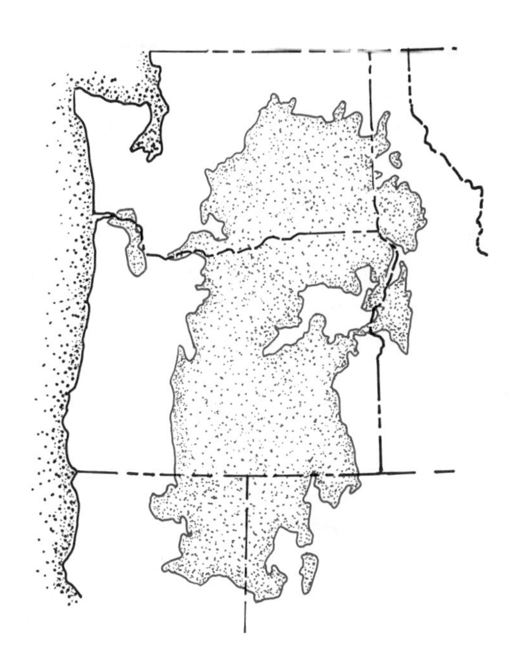

The flood basalt province in the Pacific Northwest.

cases. Ordinary basalt flows rarely cover more than a few dozen square miles; flood basalt flows pour across thousands or tens of thousands of square miles, probably within a few days. Every eruption of flood basalt was a major catastrophe. The number of flood basalt flows is unknown, but it is certainly in the hundreds.

The flood basalts of the Modoc Plateau are near the southern tip of a province that extends north into eastern Washington, east into western Idaho, and west to the western Cascades, the Willamette Valley, and in places, to the coast of Oregon. It is the youngest of more than a dozen large flood basalt provinces that have erupted here and there on the continents during the past 250 million years.

Geologists have paid too little attention to the flood basalts in the Modoc Plateau and know too little about them. Most appear to resemble the Steens basalts of southeastern Oregon, which are distinctive in containing more aluminum than ordinary basalt. The extra aluminum makes itself visible in abnormally large crystals of plagioclase feldspar, an aluminum mineral. Those large crystals show up as white flakes in the black basalt.

Although most of the enormous flood basalt province features only basalt, large volumes of rhyolite erupted with the basalt in southeastern Oregon. A good bit of that rhyolite erupted as ash, and part of that ash blew in great sheets into the Modoc Plateau. You see it in outcrops of light pink or sparkling white stuff interlayered with the dark basalt. All flood basalt provinces include a relatively small area in which rhyolite erupted along with the basalt.

Why flood basalts? We believe a large asteroid or comet struck southeastern Oregon a little more than 17 million years ago. Other geologists entertain quite different ideas. The matter is the subject of vigorous debate that seems unlikely to end in a general consensus anytime soon.

Asteroids and comets travel at tremendous speeds, generally in the range between 15 and 60 miles per second. Their tremendous speed gives them enough energy of motion to cause extremely violent explosions as the energy converts into heat, if they strike the earth. The large ones, those more than 1 mile across, explode with a release of energy far greater than that contained in all the world's nuclear arsenals combined.

The explosion of a large asteroid or comet, one several miles in diameter, would open a crater deep enough to expose the partially melted rocks in the asthenosphere portion of the earth's mantle, where the rocks are already partially molten. Those extremely hot rocks are too weak to support a deep crater. So the walls of the initial crater would immediately collapse, converting it into a much broader and much shallower crater basin.

Although much shallower than the initial crater, the new crater basin would be deep enough to considerably reduce the pressure on the already partly melted rocks in the asthenosphere. That would permit instantaneous, wholesale melting to generate enormous volumes of basalt magma. The magma would rise into the crater basin, converting it into an enormous lake of molten lava, probably within a matter of hours, or a few days at most.

We think the asteroid struck in southeastern Oregon because that is where the Basin and Range began, where the volcanic Snake River Plain began its long march across southern Idaho, where rhyolite erupted along with the flood basalts, and where the projected trends of several large swarms of basalt dikes intersect. All those convergences of quite different features of the same age are far too much to ask of coincidence.

The flood basalt eruptions poured into the Modoc Plateau because it was then a low area near the crater basin volcano. When the great eruptions finally ended, sometime around 15 million years ago, the region must have been an interior plain with a flat, almost level basalt surface—the ultimate in geologic and topographic simplicity.

AFTER THE FLOOD BASALTS

Movement along Basin and Range faults has since broken that flat surface into mountain ranges and broad valleys that trend north. Meanwhile, volcanoes associated with both the High Cascades and with the Basin and Range have covered much of the Modoc Plateau with volcanic rocks much younger than the flood basalts. Desert sediments filled the valley floors. The simple region of 15 million years ago is now complex, both geologically and topographically.

Crustal stretching probably began in the northern Basin and Range while the flood basalt flows were erupting. The stretching thinned the continental crust, just as stretching thins a rubber band. That stretching relieved pressure on the rocks in the partly melted asthenosphere portion of the mantle, permitting them to melt into basalt magma. As the basalt magma rose into the continental crust, it melted some of the rocks there to make rhyolite magma. Basalt and rhyolite erupted to build small volcanoes in the Basin and Range, especially in its western part. They covered large areas of the flood basalts in the Modoc Plateau with lava flows and rhyolite ash.

The younger volcanoes in the Modoc Plateau lie along the main trend of the High Cascades. Most are larger than the small basalt cinder cones typical of the Basin and Range. These volcanoes erupt andesite, along with some basalt and rhyolite. Their lavas also cover large areas of the flood basalts.

Basin and Range Valleys

Deep accumulations of desert valley sediments fill all the broad basins in the Modoc Plateau. The sediments are souvenirs of long periods of extremely dry climate, when too little rain fell to maintain a network of streams that could carry sediment to the ocean.

But the climate of the Modoc Plateau was much wetter during the ice ages. Lakes sparkled in the floors of the large valleys. Those ice age lakes overflowed into the Pit River. Then they shriveled in the dry years since the last ice age, once again losing their outlets. And the Pit River also shrank. The ice age lakes left sediments in their floors and shorelines around their margins. The lake sediments now make absolutely flat floors in the lower parts of the basins.

MEDICINE LAKE VOLCANO AND LAVA BEDS NATIONAL MONUMENT

Except for the old lakebed sediments and the rhyolite domes around Medicine Lake, the geologic map shows cinder cones and lava flows of basalt and basaltic andesite. The distinction between those two rock types is due to small differences in chemical composition that cause hardly any difference in the look of the rock. Most of the contacts on the map simply outline lava flows and groups of flows of various ages. A rash of basalt cinder cones dots much of the countryside.

The Medicine Lake volcano is a peculiar giant. It erupts basalt and rhyolite, the two ends of the compositional spectrum. The lack of ordinary,

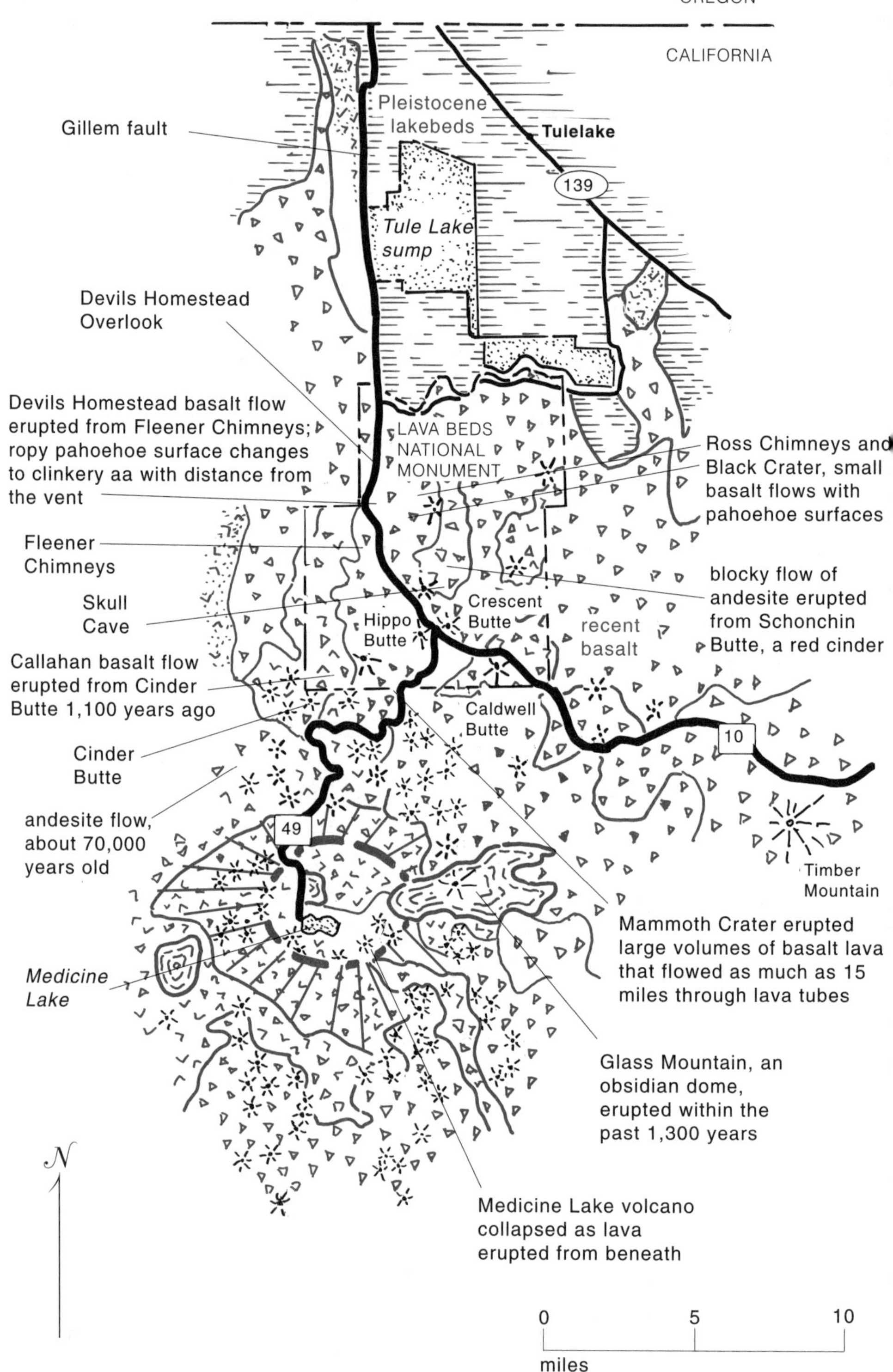

Medicine Lake volcano and Lava Beds National Monument.

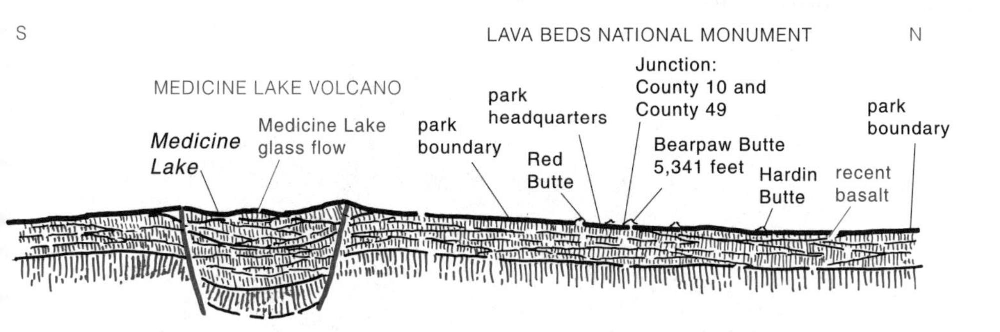

A geologic section through the Medicine Lake volcano, showing its deep stack of lava flows and the collapse caldera in its summit.

medium andesite distinguishes this volcano from the High Cascades, as does its position about 50 miles east of their main trend. It is much larger than any of the Basin and Range volcanoes. And it is much too young to associate with the flood basalt province.

Age dates suggest that basalt began to erupt in the Medicine Lake volcano about 1.9 million years ago. Continued eruptions built an enormous shield volcano. Then its summit collapsed to open an oval caldera about 6 miles long, 4 miles wide, and 500 feet deep. Later eruptions built a ring of eight smaller volcanoes around the rim of the caldera basin, nearly obscuring it. Medicine Lake floods what remains.

The volcano has erupted at least six times, from four different vents, in the past 2,000 years. These more recent eruptions produced a lot of rhyolite. Some of the rhyolite erupted as pale pumice and ash, much more as enormous lava flows of black obsidian glass. The youngest rocks may be less than 200 years old. Any volcano that has erupted that recently is likely to erupt again. Color this one active.

The Medicine Lake volcano is a marvelous place to see the results of fairly recent volcanic activity. Lava Beds National Monument, which includes a generous northern slice of the Medicine Lake volcano, provides the best access to most of them.

Some of the basalt flows in Lava Beds National Monument are old enough to support a good plant cover; others are so fresh they look as though they erupted last week. Looking at those is almost as good as watching an eruption in progress and is much safer. So it is surprising to find that those younger flows are more than 1,000 years old. It is a bit sobering to see how little the slow processes of soil formation have accomplished in all that time.

Ropy surface of a pahoehoe flow near Valentine Cave.

Basalt Eruptions

Molten basalt is normally fluid enough to make lava flows that, from the air, look like puddles of tar that run down streambeds. Steam bubbles freely from the fluid lava, ensuring that most basalt eruptions are quiet—by volcanic standards. It is steam trapped within the more viscous varieties of lava that incites violent eruptions.

Basalt lava pours over the countryside and down its stream valleys at a rate that depends on the volume of lava, its temperature, its steam content, and the steepness of the slope. An increase in any of those will speed the flow. But even on their speedier days, basalt flows rarely move faster than a vigorous person can walk.

Very mobile basalt develops a smooth surface with a ropy appearance that geologists call pahoehoe, a Hawaiian term. The lava runs mostly under the cooling surface of the flow, within an insulating shell of hot, but solid, basalt. The lava within the flow moves without significantly cooling and may continue for miles.

A lower temperature and lesser charge of steam tend to make basalt lava more viscous. The flow develops an outer zone full of gas bubbles that breaks into a surface of clinkery rubble as its sluggish movement shatters the crust and rafts the angular chunks along. Geologists call those flows

This aa surface developed on the Devils Homestead lava flow where it flowed slowly down a gentle slope.

aa, another Hawaiian term. A moving aa flow tips its clinkery cover down its front and then crawls across it like a caterpillar tractor moving on its tread. Cliff exposures of aa flows show a ledge of solid basalt between layers of rubble. The basalt typically breaks into a palisade of vertical columns.

Different as they look, pahoehoe and aa flows consist of exactly the same kind of rock, with the same chemical composition and composed of the same minerals. Both kinds of surface commonly exist on different parts of the same flow.

Lava Tubes

Lava Beds National Monument contains almost 300 lava tubes that formed as molten basalt within the flow drained from beneath the solid crust of pahoehoe lava. Fantastic little dripstone ornaments, lavacicles, decorate some of them. The ornaments formed as molten basalt dribbled off the ceiling and down the walls. Many lava tubes have horizontal ledges along their walls, which record periods when the surface of the flowing lava remained stationary long enough to cool and solidify a bit. The roofs of basalt tubes eventually collapse, opening rows of sinkholes and winding depressions that look like valleys in the flow. Remnants of the roof make natural bridges across those depressions.

Some of the deeper lava tubes have permanent ice in their lower parts. Caves fill with ice if their air circulation is almost entirely vertical. Air can sink vertically into a cave only if it is colder, and therefore denser, than the air already inside. Rocks are excellent insulators, so a large mass of cold air and ice can easily survive a long summer, provided no warm air can blow through.

Rhyolite

Some of the later eruptions in the Medicine Lake volcano produced rhyolite. If the molten rhyolite is heavily charged with steam, it will explode when it reaches the surface. Some of the rhyolite erupts as pumice, a frothy volcanic glass that may be light enough to float on water. The rest erupts as rhyolite ash, which is basically extremely small fragments of exploded pumice.

Look on the surfaces of basalt lava flows in Lava Beds National Monument to see that some have a decorative cover of bits of yellowish pumice about the size of garden peas. A rhyolite explosion somewhere in the Medicine Lake highlands must have blown those bits of pumice all over

Entrance to a lava tube through its broken roof.

the countryside. Age dates show that it erupted between about 1,600 and 1,100 years ago. No pumice sprinkles the younger lava flows.

Rhyolite magmas that do not contain steam erupt quietly to make bulging lava domes or swollen lava flows composed mostly of obsidian, black volcanic glass. A number of large obsidian lava flows exist in the Medicine Lake area, especially around Mount Hoffman, about 2 miles east of the lake. Obsidian is as shiny as any other kind of glass, and it breaks into slick fragments with edges as sharp as razor blades. Indians considered it the finest raw material for knife blades and arrowheads.

MODOC PLATEAU ROADGUIDES

Most roads in the Modoc Plateau follow the broad Basin and Range valleys, occasionally crossing a mountain range to get from one valley to the next. The roads cross desert valley fill and ice age lake sediments along most of their routes. They cross mountains on hard volcanic bedrock.

U.S. 395

Susanville—Oregon

142 miles

Susanville is at the southern edge of the Modoc Plateau. The Diamond Mountains on the southern horizon are granite detached from the northern Sierra Nevada along a Basin and Range fault. The road east of Susanville crosses the Honey Lake Valley. The road north of the Honey Lake Valley passes east of Shaffer Mountain, a small High Cascades volcano.

Every solid rock along the route is volcanic. Most are flows of black basalt; the others are white rhyolite ash. The route crosses the broad valleys on lakebed sediments deposited during the wet years of the last ice age. Those sediments turn to dust during dry weather, to the most awful muck when the weather is wet.

Some of the basalt is flood basalt that erupted about 16 million years ago, and some of the rhyolite may also be part of that picture. Most of the volcanic rocks erupted since then, from the small volcanoes that make most of the landscape on both sides of the road. Those that produced basalt or rhyolite probably belong to the Basin and Range. If they erupted andesite, they belong to the High Cascades.

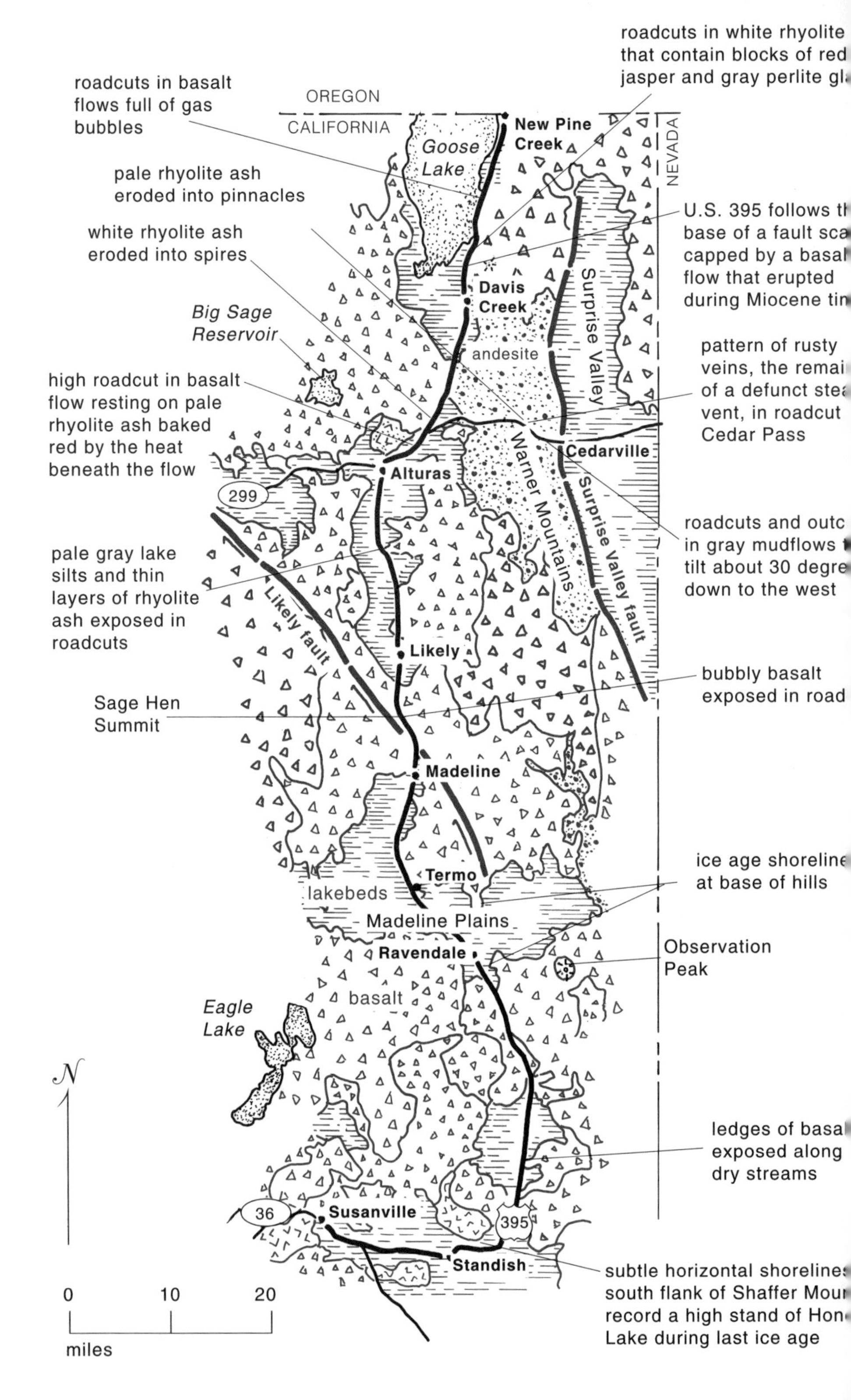

Rocks along U.S. 395 between Susanville and the Oregon line.

A basalt flow over rhyolite ash baked red beneath the flow, just west of Susanville.

Madeline Plains

The highway between the area south of Ravendale and Madeline crosses the Madeline Plains, a surface so flat and level it must be an old lakebed. Old shorelines low on the hills around the edge of the Madeline Plains leave no doubt that ice age lakes flooded them. This area was a western arm of Lake Lahontan, a large ice age lake that extended far into Nevada. The lake filled during the wet climate of the last ice age and evaporated in the much drier climate since.

Basalt

The highway north of the Madeline Plains crosses 10 miles of rugged hills eroded into basalt lava flows that erupted during the past few million years, probably Basin and Range rocks.

The highway between Likely and Alturas follows the long, straight valley of the South Fork of the Pit River. Volcanic and sedimentary rocks partially filled a basin that dropped as a block of the earth's crust moved along a

Dill Butte, about 3 miles northeast of Ravendale, was an island when ice age lakes flooded the Madeline Plains. The higher of the two old shorelines is the older.

White rhyolite ash north of Alturas.

fault, several million years ago. The Warner Mountains, visible on the skyline 25 miles to the east, may be the raised side of the same fault block. Volcanic rocks in the lower hills west of the road erupted while Basin and Range faults were moving. The road follows a valley now partially filled with younger volcanic and sedimentary rocks.

In the 12 miles north of Alturas, U.S. 395 follows the North Fork of the Pit River, passing exposures of black basalt, white rhyolite ash, and loose gravel, all part of the material that fills the valley floor. Here and there, the rhyolite ash weathers into distinctive conical shapes that look almost like giant tepees rising among the trees. It probably erupted in southeastern Oregon while the flood basalt flows were erupting. Farther north, the valley opens into the broad basin that holds Goose Lake.

About 5 miles north of Alturas, the highway passes spectacular roadcuts through a black basalt flow that poured across a bed of rhyolite ash. The ash was nearly white until the heat of the flow baked it to the color of brick, and almost to that consistency. A small fault offsets both. Watch between 6 and 12 miles north of Alturas for big roadcuts in white rhyolite ash, distinctly layered pale gray to pinkish rhyolite ash, and gray andesite. Farther north, a dark gray basalt flow makes the rimrock that caps the hills.

Goose Lake

Goose Lake is a ghost of its former self, rapidly becoming ghostlier. It floods a broad fault block basin, part of the Basin and Range. Look

Pinnacles eroded into rhyolite ash between Alturas and Goose Lake.

Basalt flow on rhyolite ash exposed in a roadcut north of Alturas.

for the straight fault scarp along its eastern side. Old shorelines low on the mountains show that it was much larger during the last ice age than now. Lake sediments lie beneath the broad expanse of flat ground south of the lake.

Despite the drier climate that has prevailed since the last ice age, Goose Lake retained its outlet through the Pit River into the early 1900s and remained fresh. Then progressive diversion of its water supply into irrigation systems dropped the lake level below its outlet. Now that it no longer overflows into the Pit River, Goose Lake is becoming salty and alkaline. Fish still survive and grow quite large, but fishermen complain about their flavor.

Surprise Valley Side Trip

Surprise Valley, east of the Warner Range, straddles the border of California and Nevada. It is about 55 miles long and 8 to 10 miles wide. California 299, the lonely road over the Warner Range from Alturas to Cedarville, is a good trip.

Surprise Valley contains three ephemeral lakes. Upper Lake is near its north end, Middle Alkali Lake is in the middle east of Cedarville, and Lower Lake is near the south end. All are playas, dry except during wet weather, when a sheet of salty, alkaline water floods them. During the wet times of the ice ages, the valley flooded to a maximum depth of 550 feet, the elevation of the highest old shoreline above the valley floor. Look for its level trace faintly etched on the mountainsides. The tract of sand dunes east of Middle Alkali Lake covers an area about 10 miles from north to south and several miles from east to west. It is one of the largest patches of desert dunes in northern and central California.

Warner Range

The Warner Range consists basically of a big block of volcanic rocks, mostly basalt lava flows, interlayered with some sedimentary rocks. They tilt down to the west at an average angle of about 25 degrees. The block is a large slice that tilted as it moved east along one of the curving faults of the Basin and Range.

Age dates show that the basalt lava flows erupted at various times between about 36 and 5 million years ago, from Oligocene to Pliocene time, so they are not a simple geologic unit. The age of the older flows suggests that they probably belong to the defunct Western Cascades. Those that erupted during middle Miocene time are probably flood basalts, and those erupted since then belong to the High Cascades. Basalt lava flows of any age look alike, so those distinctions are not apparent from the road.

Ragged pinnacles of agglomerate west of Cedar Pass.

Gray andesite mudflows east of Cedar Pass.

View south along the Surprise Valley fault scarp from Cedarville.

The Warner Range rises abruptly above the western edge of the Surprise Valley along a straight line, an obvious fault scarp—the Surprise Valley fault. It is on the northern projection of the Sierra Nevada fault and is probably its northern continuation.

The Surprise Valley fault chops all the lava flows in the Warner Range, so it must have started to move since the youngest of them erupted, within the past 6 million or so years. Those youngest flows, on top of the Warner Range, stand at an elevation of about 9,700 feet, about 5,200 feet above the floor of the Surprise Valley. It seems reasonable to suppose that the same rocks exist beneath the 7,000 feet of desert basin fill sediments in the valley floor. If so, the Surprise Valley fault must have moved about 12,200 feet within the past 6 million years, an average of approximately 2 vertical feet every thousand years. That is comparable to the rate of movement on the Sierra Nevada fault, extremely fast in the long perspective of geologic time.

Hot Springs

Surprise Valley contains quite a few hot springs, both on the broad valley floor and along the trace of the Surprise Valley fault. Most produce water with more dissolved solids than anyone would care to drink, but a few springs and wells produce drinkable water.

The most recent volcanic activity was the eruption of basalt flows in the southern part of the Warner Mountains, probably a few million years ago; that is too long ago to explain the existence of the hot springs. The hot springs tend to line up along the Surprise Valley fault or along trends parallel to it. Evidently, water sinks deep into the crust along fracture zones, absorbs heat from the rocks at depth, then rises to the surface. That is the commonest mechanism of all hot springs. Many people hope that the hot water in the Surprise Valley may someday provide geothermal energy.

Mud Volcanoes

Early March 1951. Loud noises and shaking ground awakened the people of Lake City, which is about 8 miles north of Cedarville. Steam eruptions about 2 miles northeast of town were blowing clouds of mud as much as 1 mile into the air, where they drifted southeast on the wind. Within hours, the eruptions had produced more than 300,000 tons of mud, building miniature volcanoes as much as 15 feet high. By the end of an exciting day, the activity had subsided to a group of bubbling mud pots.

Nothing volcanic happened. Hot springs were belching steam through deep deposits of mud laid down in the floor of Upper Lake. These were the first recorded mud eruptions in the area but certainly not the first. Aerial photographs taken before 1951 show the eroded remains of older mud volcanoes. The springs are still hot and will probably erupt more mud sometime.

California 299
Redding—Fall River Mills
75 miles

Much of Redding stands on gravels deposited in enormous alluvial fans that spread onto the floor of the Sacramento Valley during Pliocene time, when the climate was very dry. Fall River Mills stands on silts deposited on the flat floor of a former lake, a lake plain. The lake existed during the ice ages of Pleistocene time, when the climate was apparently wet. The road between Redding and Fall River Mills passes a wide variety of rocks in the Sacramento Valley, the southern edge of the Klamath block, and the High Cascades.

Klamath Block

The 20 miles of highway between Bella Vista and Round Mountain threads a course along the boundary between the Klamath block and the High Cascades. The hills north of the road are eroded in the old metamorphic rocks of the Eastern Klamath belt. Occasional roadcuts and

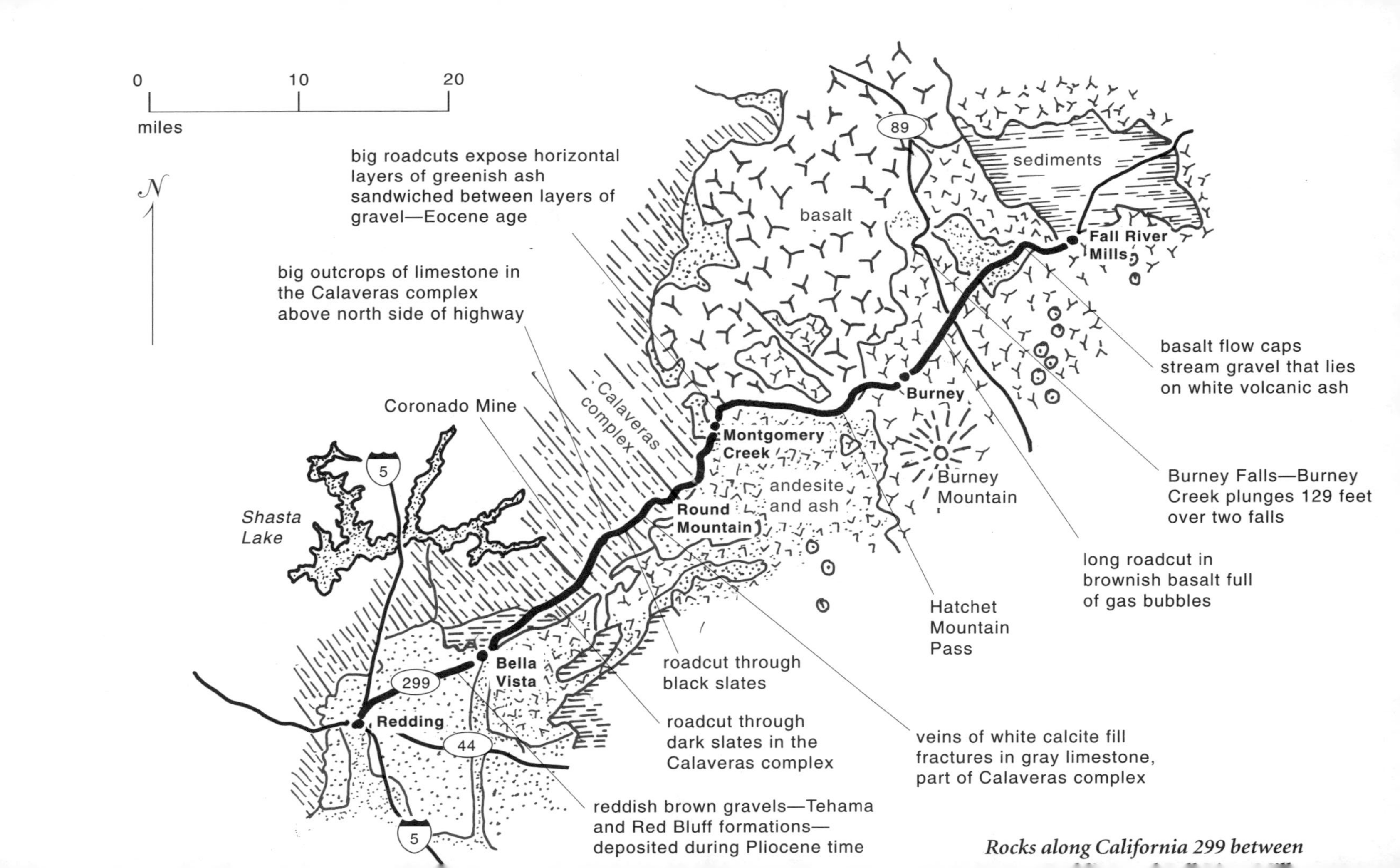

Rocks along California 299 between

outcrops expose gray to rusty rocks that look slabby because of their slaty fracture. Those are Triassic and Jurassic metamorphic rocks. The hills south of the road are a volcanic landscape that belongs to the High Cascades. Volcanic rocks visible from the road are mostly brownish ash erupted during Pliocene time.

High Cascades Volcanic Rocks

High Cascades volcanic rocks visible south of the road between Bella Vista and Round Mountain are mostly brownish volcanic ash that erupted during the past few million years. They lap onto and cover the southern edge of the Klamath block and presumably also the sedimentary rocks deposited along it while the Modoc seaway existed during late Cretaceous time. The road between Round Mountain and Montgomery Creek crosses dark gray andesite rubble and basalt lava flows. Both erupted in the High Cascades during Pliocene time.

The oldest rocks along the road east of Montgomery Creek are Miocene flood basalt flows and their associated rhyolite ash. They are exposed near Hatchet Mountain Pass and in a large area to the north. Elsewhere, most of the Miocene volcanic rocks are buried under younger volcanic rocks of the High Cascades, mostly andesite ash and basalt lava flows erupted during the past several million years. The older High Cascades rocks carry a fairly complete cover of soil and brush. Some of the younger ones are still almost bare.

Between Round Mountain and Fall River Mills, California 299 winds through a High Cascades volcanic landscape. The original volcanic

A dark ash flow covers a hill eroded on a pale rhyolite ash flow, 10 miles east of Montgomery Creek.

The dark streaks are obsidian, flattened blocks of pumice in a gray ash flow 10 miles east of Montgomery Creek.

Volcanic bombs in a red cinder cone 15 miles east of Burney are full of gas bubbles.

landforms survive almost intact, partly because they are young, but largely because their porosity invites the rain to soak into them, instead of running off and carving gullies.

A spectacular roadcut about 2.5 miles west of Hatchet Mountain Pass exposes a dark gray ash flow that covered a hill eroded in an older ash flow of pink to yellowish rhyolite. Although they look and are different, both of those ash flows have about the same composition. The difference between them is due mainly to their degree of welding. The pale ash is in the upper part of an ash flow, which was too cool to weld when it settled. It is full of tiny gas bubbles that reflect the light, giving it a pale color. An iron oxide stain makes the pink and yellowish hues. The dark gray ash flow is in the lower part of a later flow that was still hot enough to weld when it settled, and that compacted under the weight of the ash above, now mostly lost to erosion. The streaks of black obsidian formed where all the gas is squeezed out and no bubbles remain to reflect light.

Basin and Range faults broke the Modoc Plateau into broad valleys and mountain ridges that dominate the landscape east of Burney. Basalt lava flows that erupted during the past few million years of Pliocene and Pleistocene time floor the valleys. The small volcanoes and cinder cones that erupted them punctuate the general flatness. Many of the basalt flows exposed in roadcuts east of Burney are conspicuously full of open tubes that formed as bubbles of steam rose through the molten lava from wet ground below.

Watch a few miles east of the junction with California 89 for pale gray rhyolite ash exposed in roadcuts. The thin layers suggest that it fell like snow from a plume of airborne ash, instead of arriving in a glowing ash flow.

A big roadcut almost 3 miles west of Fall River Mills beautifully exposes the layering and volcanic cinders in a red cinder cone. The cinders are red because steam permeated the cone for a few years after it erupted.

The view southeast from the area just east of Fall River Mills reveals a group of volcanoes that erupted during Pliocene time. The foreground is a field of basalt flows erupted sometime during the past 3 million or so years. The rash of small cinder cones that produced them dots the view.

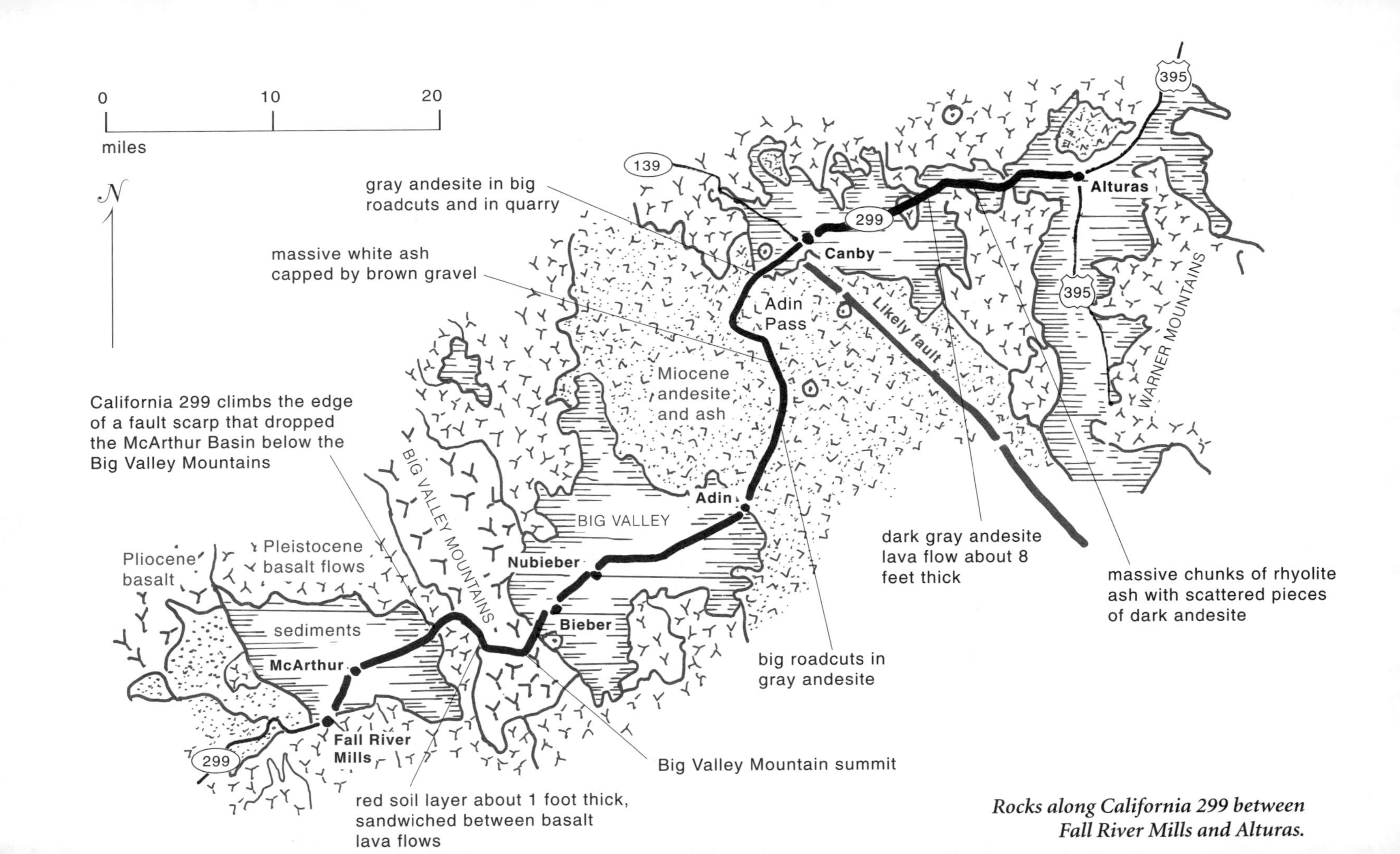

Rocks along California 299 between Fall River Mills and Alturas.

California 299
Fall River Mills—Alturas
106 miles

The boundary between the High Cascades and the Modoc Plateau lies just west of Fall River Mills. Between there and Alturas, California 299 crosses four broad basins with three ranges separating them. The ranges are blocks of the earth's crust that rose along curving faults; the basins dropped along the same faults.

Rocks exposed in the ranges are entirely volcanic—black basalt lava flows and occasional beds of volcanic ash. Rocks exposed in the valley floors are sediments, washed into the valleys from the surrounding mountains, and volcanic rocks that erupted as the ranges rose and the basins dropped. The geologic map shows that the segment of road across the Big Valley Mountains crosses Miocene flood basalt flows. These almost certainly lie beneath the entire Modoc Plateau, but younger volcanic rocks cover them in most places. Alturas stands on a flat lake plain, on sediments deposited in a lake that flooded its valley during the last ice age.

Fall River Valley

McArthur is in the middle of the Fall River valley. Some road maps show the rather fresh basalt flows that cover the northern part of the valley. They appear to have erupted from the Medicine Lake volcano, about 38 miles to the north.

Fall River Springs, about 7 miles north of Fall River Mills, is one of the largest in the country. It discharges some 1.25 million gallons of water per day, an amazing volume in such dry country. Some of the water probably comes from the Pit River, some from much farther north. Large springs are fairly common in volcanic areas, mainly because many basalt lava flows are open and permeable, as are the rubbly zones above and beneath them. Both the basalt and the rubbly zones store large volumes of water and provide good natural avenues for groundwater flow.

Big Valley Mountains

All of the rocks along the road across the Big Valley Mountains are volcanic. Most are basalt lava flows, but you also see occasional exposures of brown volcanic ash in thin layers.

If you are driving east, watch for the abrupt escarpment along the western base of the Big Valley Mountains. Any mountain wall that steep and that straight is almost certainly a fault scarp. The highway climbs about 50 feet onto a Pleistocene flood basalt flow weathered to shades of brown. Since the fault cuts that flow, the last major movement on the fault was less than 2 million years ago. About 1 mile farther east, the highway turns to climb right up the face of the main scarp.

Red soils about 1 foot thick mark the breaks between basalt flows. The red soils are laterites, or tropical soils, mementos of the wet and warm climate that prevailed during the period of flood basalt eruptions. It is typical to see a brown stain of iron oxide on the old fracture surfaces in flood basalt flows. That, apparently, is another souvenir of the climate of middle Miocene time.

Big Valley

The surfaces on both sides of the road between Nubieber and Bieber in Big Valley look suspiciously flat, like a lake plain, and the silty sediments beneath the surface leave no doubt. Those old lake sediments become extremely soggy when the weather is wet, then blow in the wind when it is dry.

The road between Bieber and Adin crosses low hills in the valley floor. The hills are eroded into deposits of mud, sand, and gravel that were swept into the valley from the surrounding mountains during Pliocene time. The climate was too dry then to maintain streams that might have carried them farther.

The road east of Adin curves through a narrow notch into Round Valley, then winds across a broad mountain block to Canby. All the rocks exposed in the mountains are volcanic: lava flows of black basalt and gray andesite, along with thick beds of volcanic ash in shades of pale gray.

Warm Springs Valley

The Likely fault defines the boundary between the eastern flank of the Big Valley Mountains and the western side of the Warm Springs Valley. The road crosses it about 0.75 mile west of Canby. The fault trends northwest and slips horizontally, the side to the west moving north. Basin and Range faults move vertically and trend almost north.

Canby and Alturas are at opposite sides of the broad Warm Springs Valley. Most of the route between them crosses low hills eroded into deposits of mud, sand, and gravel swept in from the surrounding mountains. The wetter climates of Pleistocene time eroded the deposits into the modern roadside landscape.

Dark gray basalt appears along the road about midway between Canby and Alturas. The basalt flows erupted in the valley floor during Pliocene time and, later, were mostly buried under the sediments that surround them. Watch for those pale gray sediments in the roadcuts. The low hills 1 mile or so north of most of this stretch of road are basalt flows erupted from farther north during Pleistocene time.

9

Basin and Range

The steep eastern front of the Sierra Nevada looks east toward the equally steep Wasatch Front in Utah, just east of Salt Lake City. Between the two fronts lies a swarm of mountain ranges with broad basins between them, all trending north. The mountains rose and the basins sank as the earth's crust between them stretched. Most of that happened during the past 17 or so million years. The distance between the sites of Reno and Salt Lake City is now at least twice what it was before the stretching began.

Basin and Range.

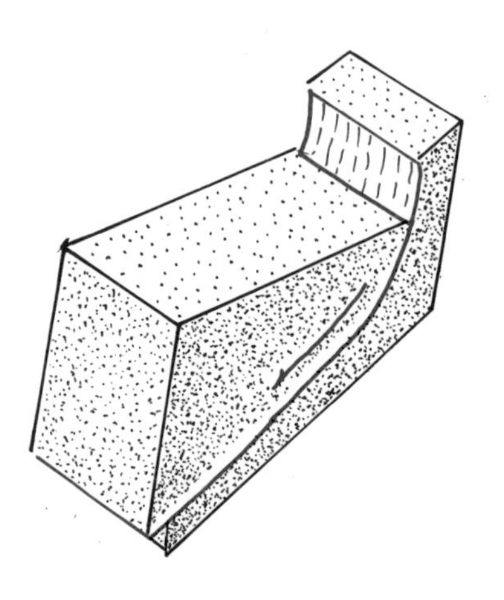

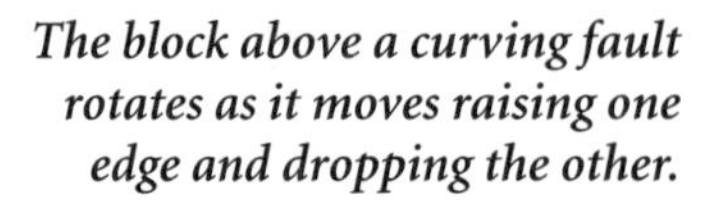

The block above a curving fault rotates as it moves raising one edge and dropping the other.

CURVING FAULTS

Until a few decades ago, most geologists thought of the mountains and valleys of the Basin and Range as blocks of the earth's crust that rose or sank along steeply inclined faults. Now, geologists think instead of slices of the upper continental crust moving along curving faults. The part of the fault exposed at the surface does indeed dip steeply into the earth, but it flattens with depth to make a surface that is concave upward. The moving slice rotates on that curving surface, so one side rises to become a mountain range, while the other sinks to become a basin.

Most of the fault slices in the western part of the northern Basin and Range appear to have moved east, while those in the eastern part moved west. All moved toward the center, toward the projected line of a structure called the Nevada rift. Most of the age dates on dikes of igneous rocks in the Nevada rift range between 20 and 15 million years. That, it now seems, is the time when the crust began to stretch in the northern Basin and Range. The exact time may well be 17 million years ago, when the volcanoes of the Western Cascades snuffed out and flood basalt lava flows began to erupt.

The broad center of the Basin and Range appears to have stopped moving long ago. Only the faults along the eastern and western margins of the province are still moving. Sporadic volcanic activity concentrates along the western margin, most spectacularly in the Long Valley caldera.

LONG VALLEY CALDERA

The Long Valley caldera does not at all resemble the towering volcanic cones on travel posters. Those are typical andesite volcanoes. The Long Valley caldera looks like a large rhyolite volcano, a resurgent caldera. The Yellowstone volcano in Yellowstone National Park is the only other active resurgent caldera in the United States.

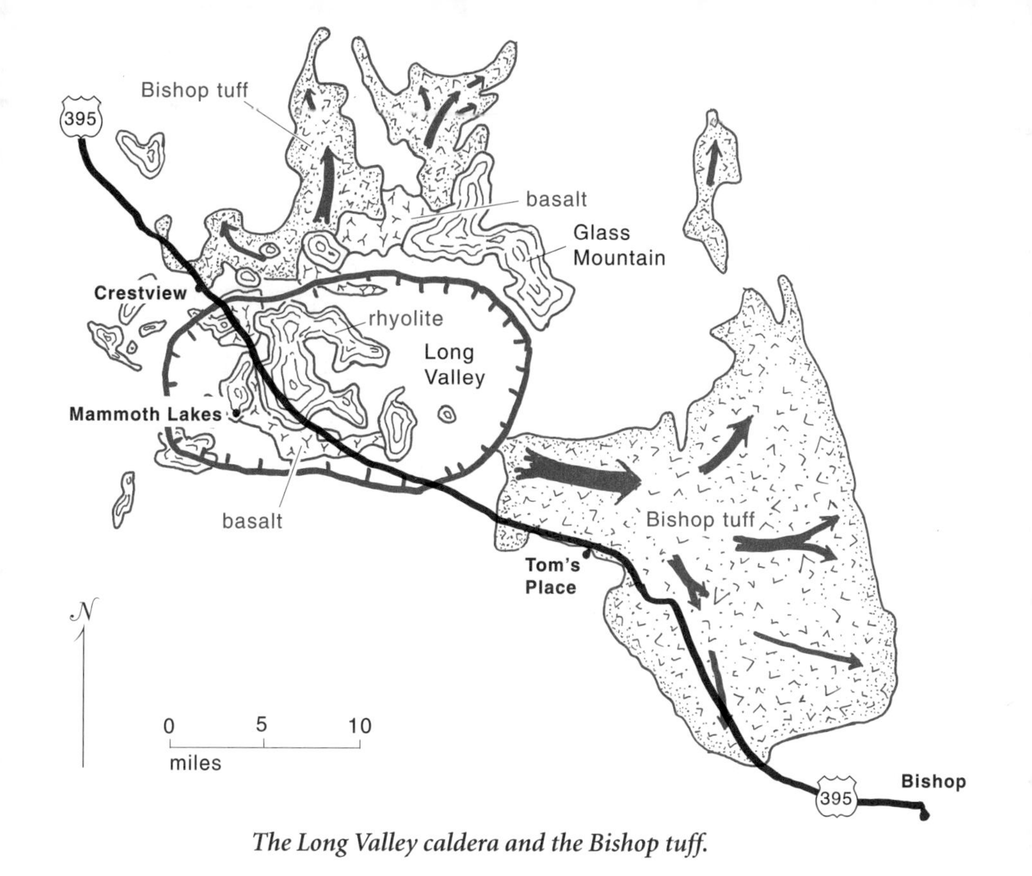

The Long Valley caldera and the Bishop tuff.

The northwest rim of the Long Valley caldera.

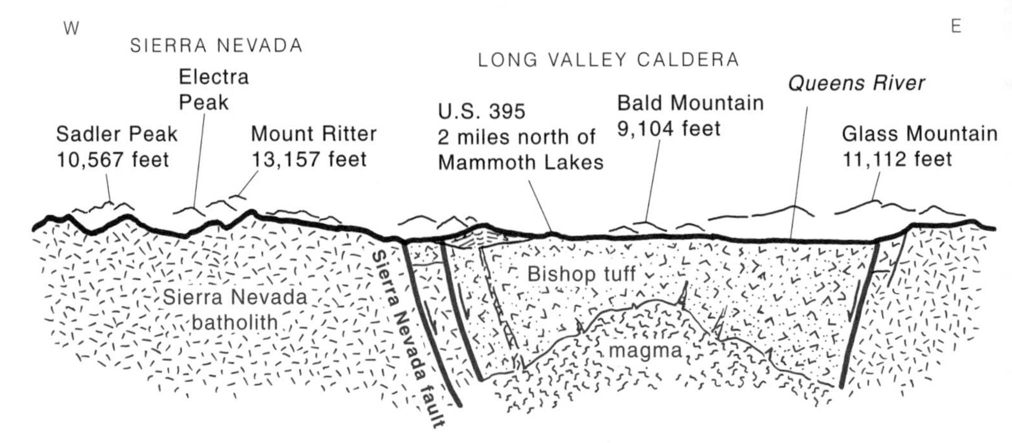

Geologic section through the Long Valley caldera.

Most resurgent calderas stage several major eruptions, at intervals of hundreds of thousands of years. None have erupted since long before people learned to write, so we have only the rocks to read. They tell of enormous clouds of red-hot steam so heavily laden with rhyolite ash that they are denser than air and pour across the ground surface at highway speeds. Geologists call them rhyolite ash flows. When a flow finally settles, the particles of ash in its hot interior commonly fuse together into solid rock, welded ash. Meanwhile, a towering cloud of steam, less heavily laden with ash, rises and drifts downwind, dropping volcanic ash like snow. Many tens of cubic miles of rhyolite ash erupt in a single event.

Meanwhile, the ground collapses into the emptying magma chamber beneath to make the caldera, a broad basin as much as 25 miles across. It sinks along a set of curving faults that align into a crude ring. The continuing eruptions fill the caldera so completely that it is hardly visible as part of the landscape. Then the column of buoyant magma rising beneath the filled caldera slowly bulges its center into a broad highland, the resurgent dome.

The most recent major eruption in the Long Valley caldera happened about 760,000 years ago. The Volcanic Tableland north of Bishop is a huge dump of rhyolite ash, the Bishop tuff. It erupted as an ash flow within a few days, perhaps within a few hours. The cloud of airborne ash dropped rhyolite as far east as the Great Plains, at least 1,000 miles away. An exact match in chemical composition fingerprints the distant ash.

More recent and much smaller eruptions produced numerous large rhyolite domes, masses of extremely viscous magma nearly without steam. They rose quietly, like spring mushrooms after a long rain. Most erupted along the ring of faults around the edge of the caldera; others erupted within or beyond the caldera, making the landscape break into a rash of rhyolite domes. Basalt lava flows cover large areas of the western end of the caldera and small areas elsewhere.

Although no one can know for sure whether the Long Valley caldera will stage another catastrophic eruption, most geologists consider that a distinct possibility. It is the source of numerous small earthquakes and swarms of small earthquakes, the kinds of tremors typical of an active volcano. Small eruptions have happened within the past 1,000 years. Studies of seismographic records show that a large volume of magma exists within a few miles of the surface. The question of whether any future eruptions will be violent hinges on the amount of water that the magma at depth contains, and that is beyond knowing.

DESERT LANDSCAPE

All of the northern Basin and Range is dry enough to qualify as a desert. The landscapes of deserts look different from those of wetter regions because different processes of erosion operate.

The sparse plant cover permits splattering raindrops to splash the soil and surface runoff to wash it away. That extremely effective combination makes soil erosion far more rapid than in humid regions.

Desert mountains rise abruptly from a broad desert plain that spreads away from their flanks.

The horizontal line on the hill is the shoreline of an ice age lake near Madeline Plains north of Susanville on U.S. 395.

Water running off some desert mountains erodes their lower parts to make broad bedrock benches, called pediments. With others, the debris that washes off the rocky hill slopes spreads across the lower flanks of the mountains as alluvial fans. Pediments and fans combine to make broad desert alluvial plains that slope gently down from the mountain front and across the valley floor.

All the water that enters the northern Basin and Range evaporates within it. The drying streams leave the sediment eroded from the mountains in the valleys, filling them to great depths, typically several thousand feet. The landscapes of the Basin and Range consist mainly of the raised edges of the fault blocks standing as mountains above the sea of sediment that floods the basins.

Undrained basins hold water. Shallow playa lakes flood their lowest parts after a rain, then dry up with the dry weather. A few playas hold water all the time. During the wet years of the ice ages, the undrained desert basins filled to make much larger freshwater lakes that lasted as long as the ice age. Watch low on the mountainsides for their horizontal shoreline benches, like rings around a giant bathtub.

Too many movies and too much television have trained too many people to associate sand dunes with deserts. In fact, dunes cover only a few percent of the desert surface in California.

BASIN AND RANGE ROADGUIDES

Most main roads in the Basin and Range follow the broad valleys, with mountain ranges rising on either side. Side roads offer the best, and in many areas the only, close views of the hard bedrock.

U.S. 395

Bishop—Nevada

136 miles

This segment of U.S. 395 passes through one of the most spectacular parts of California. The glacially carved peaks of the high Sierra Nevada rise on the western horizon. The road crosses the Long Valley caldera and passes through a marvelous outdoor museum of volcanic features, many of them erupted within the past thousand or so years. A veneer of young volcanic rocks covers much of the older Sierran granite and metamorphic rocks that were sliced off the Sierra Nevada and moved east during the past few million years.

Any people who may have followed the future route of U.S. 395 during the last ice age would have seen a glittering wall of ice and snow rising in the west. The high snowfields of the Sierra Nevada spawned enormous glaciers that filled all the large valleys and in places flowed east to the line of the highway. When all that ice melted, it left a magnificent alpine landscape at high elevations where the glaciers gnawed on hard bedrock. It also left enormous glacial deposits at lower elevations—the other side of the erosional coin. Deposits along the road include moraines made of bouldery glacial till and spreading alluvial fans of glacial outwash.

Sierra Nevada Front and the Owens Valley

The viewpoint a few miles south of Sherwin Summit provides a scenic view of the Sierra Nevada front, here a towering fault scarp. Farther south it becomes huge fracture surfaces that dip steeply down to the east. Other fractures that trend east break the mountain front into vertical splinters.

The Owens Valley and the White Mountains beyond it are both parts of the same fault slice. The White Mountains sheared off the Sierra Nevada and moved east, opening the Owens Valley behind it. Meanwhile, the Sierra Nevada, freed of the burden of the detached slice, rose.

Tungsten Hills

The Tungsten Hills are the low middle ground about 8 miles straight west of Bishop. The granite in them weathers to tan outcrops that support sagebrush but few trees.

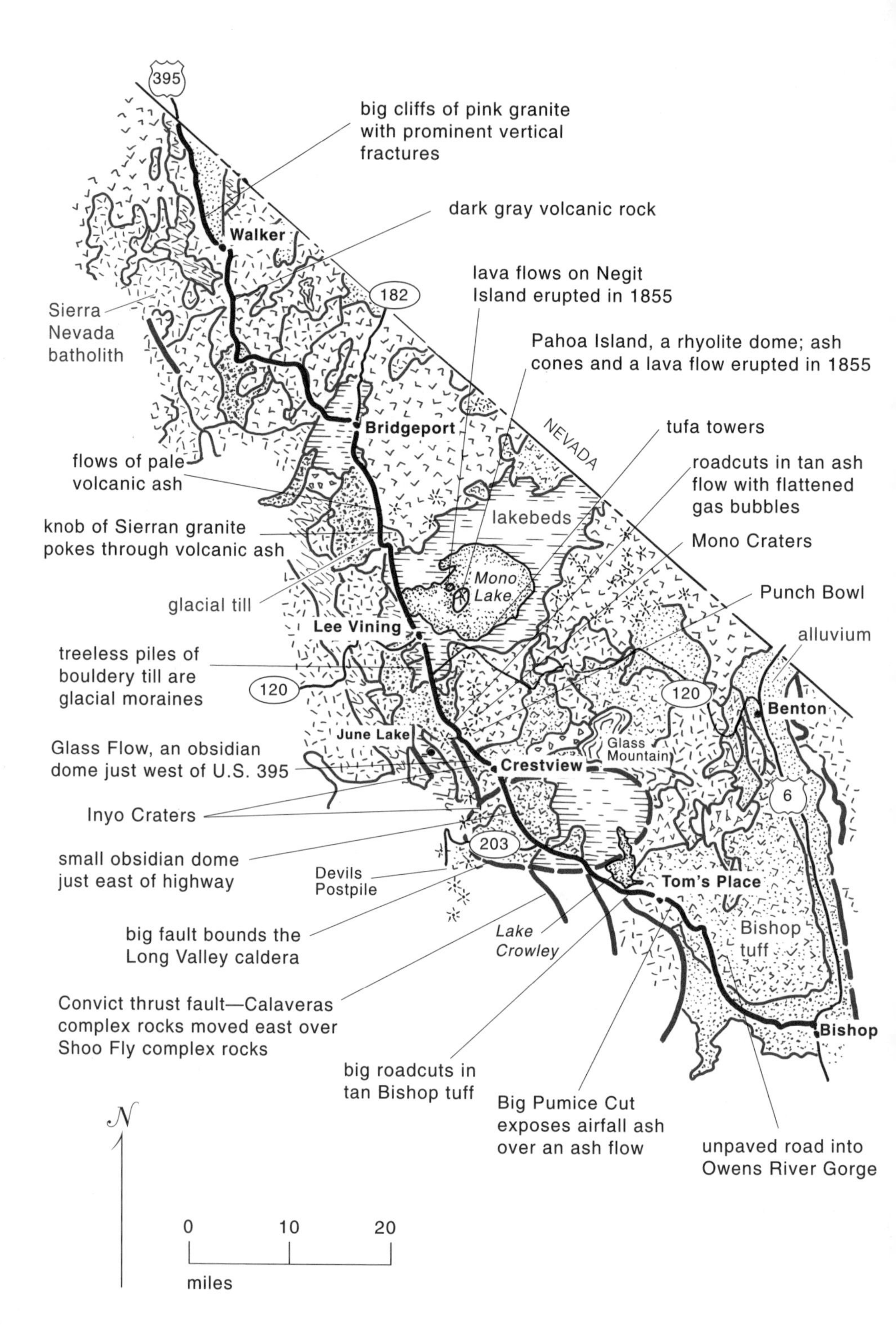

Rocks along U.S. 395 between Bishop and Nevada.

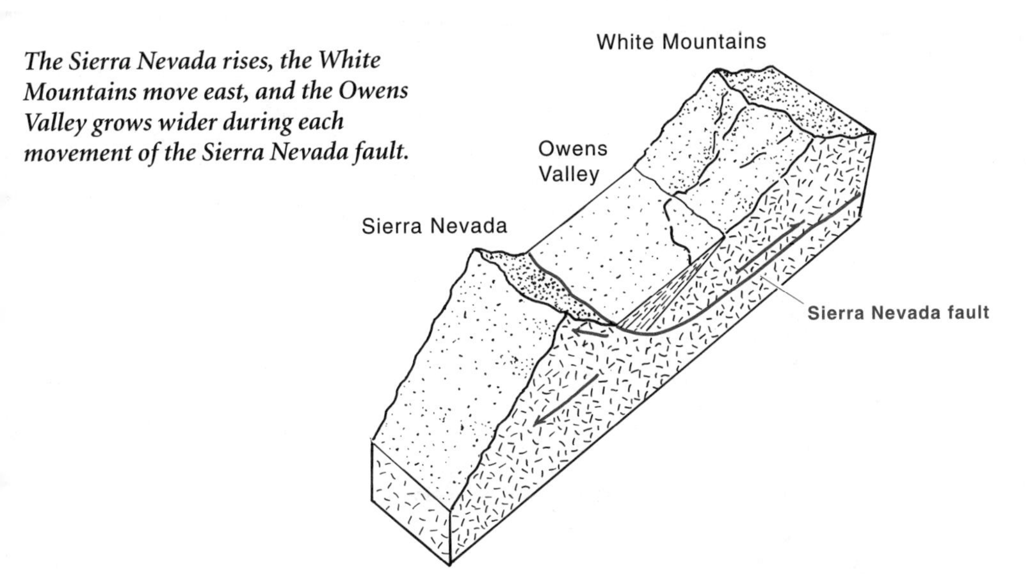

The Sierra Nevada rises, the White Mountains move east, and the Owens Valley grows wider during each movement of the Sierra Nevada fault.

The molten granite magma invaded large masses of marble, with dramatic results. Steam escaping from the magma permeated the marble, carrying with it large amounts of dissolved silica, as well as lesser amounts of many other substances. They reacted with the marble, converting it to a spectacular rock called skarn, which consists largely of garnet along with a menagerie of other minerals, many of which are quite rare. Most skarns consist mainly of large and colorful crystals, nicely shaped.

Like many skarns, those in the Tungsten Hills contain scheelite, the mineral form of calcium tungstate. Scheelite typically occurs in small, nondescript grains, in a dishwater shade of gray. They are hard to see except when they glow fiercely under ultraviolet light. Tungsten prospectors sally forth in the dark of night armed with a black light.

Tungsten is a vitally important metal used in making electric light filaments and a wide variety of superhard alloys and industrial abrasives. No industrial society can long survive without it. The Pine Creek Mine began producing large quantities of tungsten from the Tungston Hills in 1918 and was, during its many active years, the world's major source.

U.S. 6 Side Trip

U.S. 6 follows several streams along the valley floor between Bishop and Benton. Everything along the road is modern stream sediment. The White Mountains cut a jagged skyline profile east of the highway. The low Volcanic Tableland rises west of the highway in the southern part of the route, and several ranges of low hills stand farther north.

The White Mountains are a slice of the Sierra Nevada that detached and moved east on a fault during the past few million years. The rocks,

The east front of the Sierra Nevada—view south from U.S. 395, 17 miles northwest of Bishop.

like those in the eastern Sierra Nevada, include granite and associated metamorphic rocks, with a partial cover of volcanic rocks that erupted during Tertiary time. Similar rocks exist in a number of isolated hills between the White Mountains and the Sierra Nevada. All are parts of a formerly continuous mass that broke into big slices and moved east, slice by slice. The metamorphosed sedimentary rocks in the White Mountains include the formations of the eastern Sierra Nevada, as well as some that were laid down in Precambrian time, before animals appeared.

Almost all of the rocks north and west of U.S. 6 are Bishop tuff, the stuff of the Volcanic Tableland. The Benton Range, west and southwest of Benton, is the top of a slice of Sierran granite that rises like an island above the Volcanic Tableland. More of the same granite is exposed in the hills on both sides of California 120, between Benton and the southern end of the Adobe Valley.

Volcanic Tableland—the Bishop Tuff

The Volcanic Tableland is the large area with a flat upper surface straight north of Bishop. It is the Bishop tuff, a great flow of rhyolite ash that covers about 320 square miles. U.S. 395 skirts its western and southern margins between Bishop and Lake Crowley.

One of the best places to see the Bishop tuff from U.S. 395 is a big roadcut just north of Sherwin Summit, about 3 miles southeast of Lake

The Bishop tuff, view east from U.S. 395, 5 miles north of Bishop. The White Mountains rise in the background.

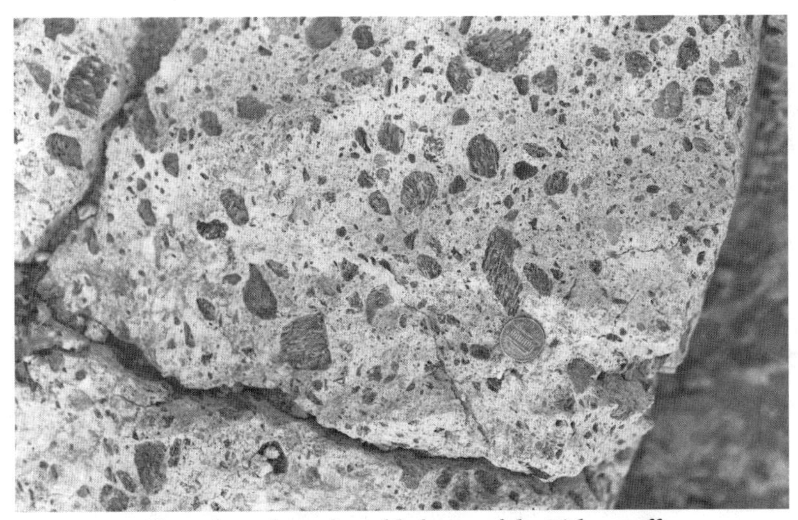

Close view of poorly welded part of the Bishop tuff. The dark chunks are pumice. View is 18 inches wide.

Crowley. Many of the roadcuts south of the summit are almost as good. The bouldery stuff in the lower part of the cut is ancient glacial till buried under the ash. The road east and north into the Owens River Gorge from about 14 miles north of Bishop provides good views, as does the road to Long Valley Dam. North of Sherwin Summit, the tan and rusty Bishop tuff east of the highway contrasts with bouldery outcrops of light gray granite to the west.

All of the Bishop tuff is rhyolite ash, but in several slightly different forms. The fresh rock is pale gray. It weathers to various shades of gray, pink, red, brown, and purple. Water dissolving and redepositing mineral matter partly solidified the upper part into a weakly consolidated rock. The lower part really hard rock; ash that was still partly melted when it settled, fused together into solidly welded tuff. Exposures in the Owens River Gorge below the Long Valley Dam show magnificent columns that stand vertically in some places, make radiating patterns in others. The soft, upper part of the Bishop tuff contains evenly distributed chunks of pumice as much as a few inches across. Deep in the ash flow, the larger chunks of pumice are flattened and fused into dark glass.

A number of side roads that cross the top of the Volcanic Tableland reveal other aspects of the Bishop tuff. The straight cliffs as much as 30 feet high that trend north across the surface are fault scarps. The low mounds, about 50 to 100 feet high and dotted across the Tableland's surface, are the

Crooked Creek, near Long Valley Dam site, meanders along the flat bottom of its valley, entrenched within the Volcanic Tableland.
—W. T. Lee photo, circa 1904, U.S. Geological Survey

Welded lower part of the Bishop tuff.

Columns in the Bishop tuff in Owens River Gorge, 12 miles north of Bishop.

sites of fumaroles. These vented hot volcanic gases, mostly steam, from the interior of the deposit, probably for years after the eruption. The gases altered the rhyolite ash around the vent, making it much harder and more resistant to erosion. Now, that altered ash stands in erosional relief.

Long Valley Caldera

The highway crosses the Long Valley caldera near the Mammoth Lakes junction. Its map outline is a broad oval that measures about 18 miles from east to west, about 10 miles from north to south, and has a surface area of approximately 175 square miles. Lake Crowley straddles its southeastern margin.

The highway traces part of the southern edge of the Long Valley caldera between Lake Crowley and its intersection with California 203. Glass Mountain Ridge, 10 miles to the northeast, is the northeastern margin. All the low ground between is in the eastern part of the caldera. The highway north of the intersection with California 203 crosses the western part of the resurgent dome, in the center of the caldera. A fault dropped a slice of the dome to create the low route the highway follows.

The small dome east of the highway a few miles north of Mammoth Lakes junction is a mass of obsidian, the glassy form of rhyolite. Lookout Mountain is east of the highway, about 7 miles north of Mammoth Lakes junction, near the northern edge of the caldera. Lookout Mountain is a large rhyolite dome that erupted from the north rim of the caldera sometime around 130,000 years ago, about 630,000 years after the monstrous eruption that produced the Bishop tuff. The rock is black obsidian.

Numerous hot springs within the Long Valley caldera provide good evidence that the rocks at depth are still hot. Casa Diablo Hot Springs features a variety of springs, bubbling mud pots, and fumaroles that blow hot gases. Some of the hot springs occasionally erupt as geysers. Hot Creek is north of the highway, several miles east of Casa Diablo Hot Springs. A number of springs enter the creek, including some that boil. They give the creek a fairly complete range of temperatures at all seasons, warm enough in places for people to swim comfortably in the winter.

The Once and Present Lake

Lake Crowley is a reservoir that was impounded behind the Long Valley Dam in 1941 to supply water to Los Angeles, causing great and continuing consternation in the northern Owens Valley. The reservoir partly replaces a vanished natural lake.

Long ago, the Owens River filled a dropped area to make a lake that once flooded much of the Long Valley caldera. The lake's surface reached an elevation of approximately 7,800 feet about 600,000 years ago, when

The big moraine that bulges from the valley of McGee Creek, in the east front of the Sierra Nevada.

it was more than 300 feet deep. Then the lake started draining across the Volcanic Tableland. The spillway eroded its channel into the Owens River Gorge and finally drained the lake sometime during the past 100,000 years. The dam in the gorge approximately restores the original lake as it was in its later years. The bulging resurgent dome has warped the old shorelines enough to make it hard to reconstruct the exact outlines of the original lake.

Mountain Glaciers

The Sierra Nevada snatched enough snow out of the moist winds of the ice ages to build an ice cap. It spawned glaciers that filled the valleys in the Sierra Nevada front, in some places as far as U.S. 395. One scoured the large valley of McGee Creek, southwest of Lake Crowley, leaving the big lateral moraine that flanks the lower part of the valley. More glacial moraines appear west of the highway, farther north.

Mammoth Lakes and Beyond

California 203 ventures west from U.S. 395 to pass through the western part of the Long Valley caldera. It provides a much closer view of many volcanic features than U.S. 395. Follow California 203 to reach Mammoth Lakes, Mammoth Mountain, Earthquake Fault, and Devils Postpile National Monument.

Mammoth Mountain is a volcano made of a series of at least ten lava domes and thick lava flows that erupted between about 180,000 and 50,000 years ago. The lavas are rhyolite and similar rocks, extremely viscous stuff.

They erupted on the southwestern edge of the caldera, no doubt from the ring faults that define its margin.

Between Mammoth Lakes and Mammoth Mountain, California 203 passes Earthquake Fault. It is an open crack in the ground, several feet wide and as much as 65 feet deep, that trends north for a distance of at least several hundred yards. It probably extends much farther under a fill of debris. It is not clear that the Earthquake Fault ever caused an earthquake, and it may not be a fault.

Faults slip. The opposite sides of the fracture move past each other, offsetting the rocks. The Earthquake Fault is certainly a fracture, but it is hard to tell whether it has slipped because it cuts homogeneous volcanic rocks in which any offset would be almost impossible to see. Some geologists have suggested that it is just a big shrinkage crack that opened as the rocks beneath cooled. But it does seem suspicious that the Earthquake Fault parallels a number of active faults in the region, including the one that controls the alignment of the Inyo and Mono Craters. Whatever else it may be, the opening of the crack certainly required some crustal stretching.

Devils Postpile National Monument

Anyone can recognize a basalt lava flow from a mile away. A dark ledge that breaks into rows of vertical columns is almost certainly basalt. A cliff exposure looks like a log stockade. The fractures that outline the columns

An elegant display of basalt columns at Devils Postpile.
—F. E. Matthes photo, U.S. Geological Survey

e tops of basalt nns exposed on ıciated surface, Devils Postpile.

make a typical shrinkage pattern. Geologists who have watched basalt lava flows in Hawaii report seeing the polygonal fracture pattern develop on their surfaces as soon as they solidify, long before they cool. It seems that the rock shrinks mainly when it crystallizes, then a bit more as it cools.

The Devils Postpile is notable among basalt lava flows because its columns are so beautifully developed and elegantly displayed. They are a foot or so across and as much as 60 feet high. The flow probably erupted sometime around 700,000 years ago, somewhere near Mammoth Pass. In the manner of lava flows, it poured down a valley, which it filled to a depth of about 600 feet. The ice age glacier that scraped the top off the Devils Postpile flow left a polished bedrock surface with a pattern of parallel scratches. The polygonal tops of the basalt columns look almost like the pattern of a tiled floor. They come in several versions, with four, five, six, or even seven sides. Six is the most common number.

Mammoth Gold Mine

The old town of Mammoth started as a gold mining camp. The original town site was 2 miles south of Mammoth Lakes, facing the mine on Gold Mountain. The ore is in gold quartz veins around the margins of a granite intrusion. An aerial tramway carried ore from Gold Mountain to a large stamp mill at Mill City between 1878 and 1880. Then the company flopped because the mine could not produce enough gold to pay for the mill—one of the usual reasons. Another operator built yet another stamp mill during the 1890s, then promptly went under. People have fooled around with the deposit off and on ever since, with no good luck worth mentioning.

Lost Cement Mine

U.S. 395 crosses Deadman Creek just south of Crestview. The creek got its name from one of those wild, nineteenth-century incidents involving rumors of gold that finally led to murder. In this case, the ore was described as a lava that looked like reddish cement studded with gold nuggets. The idea of a lava flow studded with gold nuggets is even more ridiculous than the notion that anyone could actually lose a mine.

Most of the fabulous lost mines of the nineteenth century were myths concocted as fronts for fencing operations. Most mines that produced gold from quartz veins ran into occasional bonanzas of extremely rich ore. The miners sneaked pieces of bonanza ore home, which then became stolen property, hot as a poker fresh out of the fire. Every district had its fence who spun a tale about a secret mining claim and occasionally stole off by night to work it. Then he would reappear loaded with bonanza ore that looked just like that from the real mine. The mythical mine that no one but the fence could find is all that remains of some old gold districts.

Inyo and Mono Craters

The Inyo and Mono Craters are a long, almost straight line of little volcanoes that starts within the Long Valley caldera and extends north to Mono Lake. The Inyo Craters are the southern part of that line, west of U.S. 395. The Mono Craters are the northern part, east of the highway.

The two southernmost craters are open holes in the ground more than 600 feet across, shaped like funnels with little ponds in their floors. The blanket of debris that surrounds them consists entirely of broken fragments of volcanic rocks, with no fresh ash. Most geologists interpret them as steam explosion craters, blasted where magma rising to shallow depth boiled groundwater. When the steam pressure exceeded that of the overlying rocks, it suddenly exploded with a crashing bang. A log buried in the debris blanket around the southern crater gave a radiocarbon date of 650 years.

Farther north are five volcanoes, rhyolite domes. Their rocks include pale gray rhyolite, frothy pumice, and black obsidian glass, all varieties of rhyolite. The Obsidian Dome Road, about 2 miles north of Crestview, leads 1.5 miles west to Obsidian Dome. The northernmost Inyo Crater is Wilson Butte, the rocky dome with steep sides just west of U.S. 395, about 2 miles south of the turnoff to June Lake. Radiocarbon dates on wood associated with the five rhyolite domes range between 1,000 and 500 years.

The Mono Craters continue north from Wilson Butte to Mono Lake. They include 20 of the world's most elegant little rhyolite volcanoes. They are aligned along the east edge of Pumice Valley, a big caldera with its western edge at the foot of the Sierra Nevada, west of U.S. 395. Radiocarbon dates range from about 40,000 to 550 years.

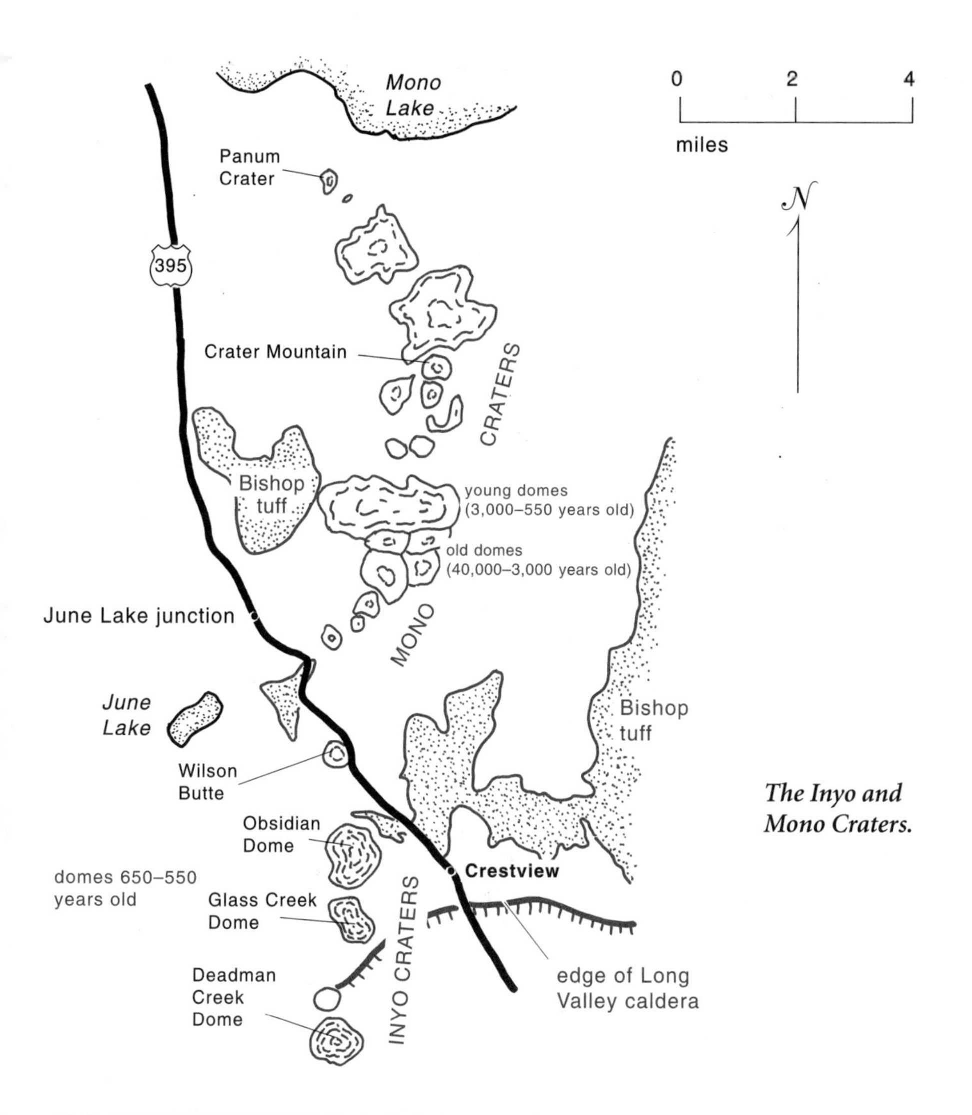

The Inyo and Mono Craters.

Mono Craters from the west.

Imagine a rising mass of granite magma encountering groundwater as it approaches the surface. The magma absorbs water, which turns to steam and rises to the top of the molten mass. That sets the stage. When the steaming top of the magma breaks the surface, it erupts rhyolite ash and blocks of pumice to make a small circular volcano with a crater in the center, a tuff ring. That disposes of the steam. If the dry magma beneath rises a bit more, it pokes up through the tuff ring to make a rhyolite dome. If it rises even more, the dome becomes a ponderously thick lava flow that pours across one side of the tuff ring. People in the Owens Valley call the domes "coulees." The Mono Craters illustrate every part of that progression, from tuff ring to lava coulee.

Big tan, rusty roadcuts in the Bishop tuff 1.5 miles south of the junction to June Lake contain chunks of pumice with flattened gas bubbles. Evidently the bubbly pumice compacted before it solidified. A road that turns north from the highway, almost 1 mile south of the turnoff to June Lake, leads about 1 mile east to the Punch Bowl, at the south end of the Mono Craters. It is a steam explosion crater within a ring of ash and with a rhyolite dome poking out of it. Most of the rhyolite domes and lava flows north of the Punch Bowl are less accessible.

California 120 turns east from U.S. 395 and passes between two of the Mono Craters on its way to Mono Lake. Look for Panum Crater, just south of Mono Lake. It is the north end of the chain, north of California 120, and 2.5 miles from U.S. 395. Panum Crater is a handsome tuff ring, a perfectly intact circle with a rhyolite dome poking up through its crater like a cherry in a breakfast tart. Radiocarbon dates show that it formed in two eruptions about 700 and 1,200 years ago. Northwest Coulee, just south of California 120, is a lovely lava flow that demolished the north side of its tuff ring and spread across the nearby countryside. The steep, high front of the flow expresses the extreme viscosity of rhyolite magma.

Mono Lake

Mark Twain made Mono Lake and its weird tufa towers famous in 1872. The lake began to shrink just a few years later as farmers diverted water for irrigation. It really began to shrink fast in 1940, when more water was diverted to supply Los Angeles. A court order issued in 1994 requires Los Angeles to raise the level of the lake to an elevation of 6,390 feet and stabilize it there. That might save the lake but will not restore it to the level that Mark Twain saw.

Mono Lake is a tiny remnant of Russell Lake, one of the large lakes that flooded parts of the Great Basin during the wet years of the ice ages. Old shorelines on the surrounding mountains show that its surface was as much as 750 feet above the level of Mono Lake. Russell Lake drained southeast through the Adobe Valley, then into the Owens River.

Negit, the dark island in Mono Lake, is an andesite cone that erupted during Pleistocene time.

Tufa towers at the south edge of Mono Lake.

Mono Lake is a chemical sump of the desert. Salt and other soluble substances enter with the surface runoff that occasionally pours into the lake but do not drain out because the lake has no outlet stream. The water in Mono Lake is about three times as salty as seawater and so alkaline that it makes your skin feel soapy, leaving it thirsting for a lotion. The water is getting saltier as the lake shrivels; the two islands grow steadily larger as they merge into the shore to become peninsulas. That was one of the sad spectacles that drove the lawsuit leading to the court order of 1994.

A side trip of a few miles on California 120 leads between Panum Crater and Northwest Coulee to the famous tufa towers on the south shore of Mono Lake. Tufa towers exist in a number of other desert lakes, but these are probably the best and certainly the most famous. They grow where submerged springs discharge slightly acidic fresh water into the strongly alkaline water of the lake. The fresh water contains dissolved calcium carbonate, which promptly precipitates as the mineral calcite when it meets the alkaline lake water. The towers form along the rising column of fresh water, apparently with some help from algae.

Bridgeport Valley

North of Conway Summit, U.S. 395 enters the south end of the Bridgeport Valley, with its flat cover of glacial outwash gravels. Volcanic rocks erupted a few million years ago make up the hills east of the valley. Buckeye Ridge, west of the south end of the valley, is mostly Sierran granite. Hills farther east and north are mostly volcanic rocks.

The road crosses the Bridgeport Valley at Bridgeport, just south of the Bridgeport Reservoir. Rocks in the 20 miles of hills that separate the Bridgeport Valley from the Antelope Valley to the north are mostly Sierran granite with a patchy cover of volcanic rocks, most of them fairly pale. Volcanic rocks also make up the ridge crest east of the highway, 3 to 6 miles

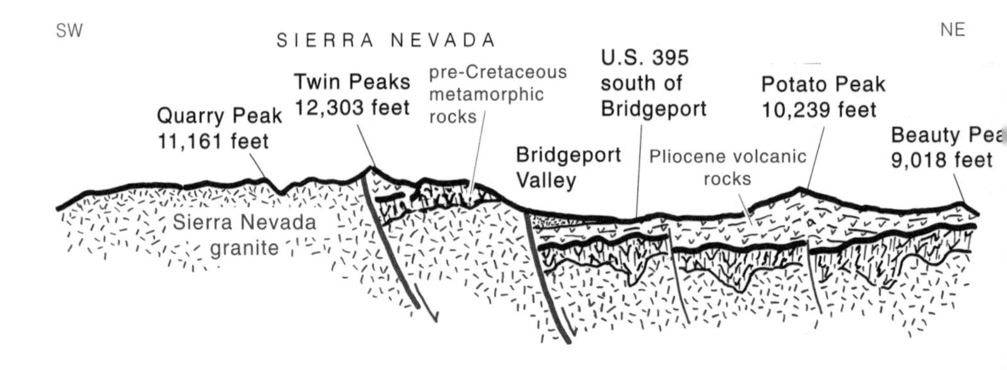

Geologic section across the Sierra Nevada, the Bridgeport Valley, and the volcanoes to the east.

north of the junction with California 108. Basin and Range faults detached the granite from the main mass of the Sierra Nevada while the volcanic rocks erupted. Younger faults broke the whole conglomeration into blocks.

Antelope Valley

The road enters the Antelope Valley at Walker and follows its west side into Nevada. Like the Bridgeport Valley, the Antelope Valley is a basin on the dropped side of a slice of the Sierra Nevada that moved east on a fault. The fault is on the west side of the valley, which is also the lower side—the usual pattern.

Most of the rocks along the edge of the valley are granite. Those in the big, pinkish cliffs just north of Walker break along a nearly vertical fracture, so the cliff looks like a rack of giant pencils. The darker rocks, such as those near the south end of Topaz Lake, are older, metamorphosed sedimentary rocks.

U.S. 395

Reno—Susanville

81 miles

The route between Reno and Susanville passes through Basin and Range country, where the ranges rise and the valleys sink as great blocks of the land move east along faults. The high crest of the Sierra Nevada rises on the western skyline, above the steep scarp of the Sierra Nevada fault. Deep deposits of sediment washed in from the nearby mountains lie beneath the flat floors of some of the valleys. Volcanic fields, which are not flat, spread across other valley floors. The older bedrock in the hills is granite, along with some small areas of metamorphic rocks north of Reno. The volcanic rocks erupted as the Basin and Range opened.

Slices of the Sierra Nevada

U.S. 395 enters California 15 miles northwest of Reno. It follows Long Valley Creek through a valley that dropped along the Last Chance fault as the mountains to the west rose. Stream sand, stream gravel, and lake silts flank both sides of the valley. Older rocks in the mountains are metamorphosed volcanic rocks that erupted during Jurassic time. The highway north and south of Hallelujah Junction is on valley fill, mostly of Pliocene age. The valley floor probably owes its tilt down to the west to rotation on the concave surface of the Last Chance fault.

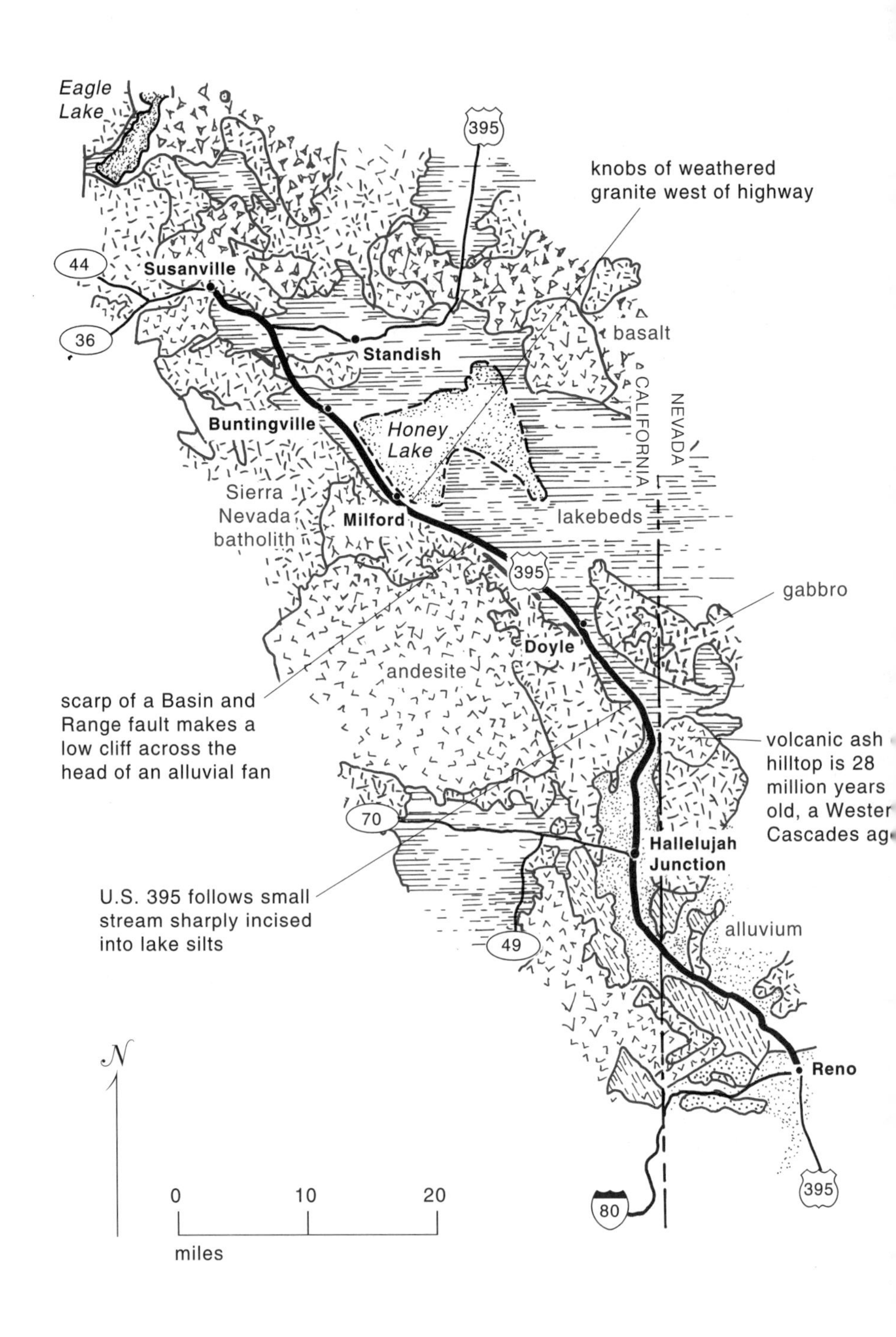

Geology along U.S. 395 between Reno and Susanville.

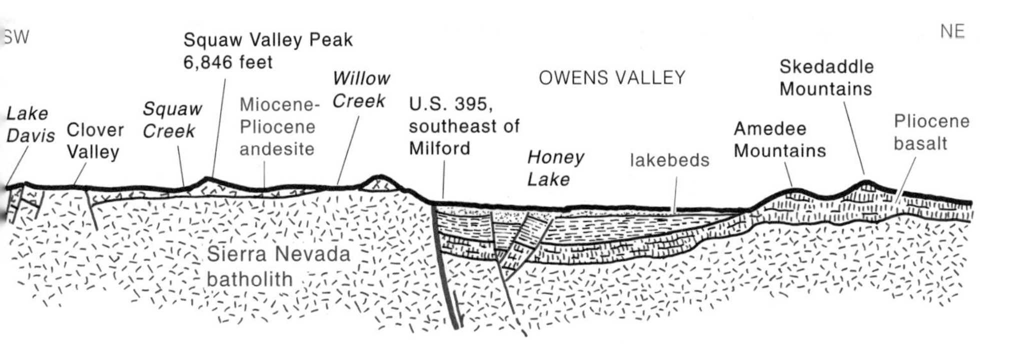

Geologic section across the Sierra Nevada and Owens Valley at Honey Lake. The Pliocene basalt erupted before the Sierra Nevada began to rise and is now broken across the great fault.

About 11 miles north of Hallelujah Junction, the highway follows Long Valley Creek through a narrow notch between mountains eroded in Cretaceous granite. Long Valley Creek began flowing on soft valley fill sediments that covered the granite. Then it entrenched its bed onto the hard granite and eroded its narrow valley into it.

The Diamond Mountains rise west of the highway north of Hallelujah Junction, all the way to Susanville. They are a big block of the Sierra Nevada that detached along a Basin and Range fault and moved east. Rocks in the Diamond Mountains are like those in the nearby Sierra Nevada: a foundation of Mesozoic granite with a partial cover of much younger volcanic rocks. Granite is exposed in most of the northern part of the range; volcanic rocks completely cover it south of Honey Lake.

The Fort Sage Mountains lie directly east of the highway in the area south of Honey Lake, east of Long Valley. They are another detached block of the Sierra Nevada that moved much farther east than the Diamond Mountains. They consist mainly of Mesozoic granite partly buried under a patchy cover of pale andesite and rhyolite.

The Diamond and Fort Sage Mountains contain the northernmost masses of granite along this road. Shaffer Mountain and the Skedaddle Mountains, north and northeast of Honey Lake, are volcanoes and so are the mountains farther north. That change marks the boundary between the northern edge of the Sierra Nevada and the High Cascades.

Honey Lake

Travelers look across the expanse of Honey Lake along a distance of nearly 15 miles in the area just southeast of Susanville. It is an oversize playa—very much a part of the desert and certainly no oasis. In wet seasons, Honey Lake is a broad sheet of shallow water, briny and alkaline. When the weather dries, it shrivels to soupy mudflats and sunbaked mud crusty with white salts.

During the wet years of the ice ages, Honey Lake was an arm of Lake Lahontan, which flooded a large area of western Nevada. It now receives much of its water from the Sierra Nevada by way of the Susan River. Long Valley Creek, which the highway follows along half the route between Reno and Susanville, also supplies some water.

A long spell of wet weather in 1868 filled Honey Lake to a depth of about 25 feet, flooding wide expanses of the flat ground around it. All that rain in the desert spread consternation and despair among farmers who had been working some of that land. Then dry weather returned, Honey Lake dwindled to its usual insignificance, and joy returned to the flooded farmers.

This straight mountain front along U.S. 395 south of Honey Lake is a fault scarp.

View south along the Sierran front south of Susanville.

Glossary

agglomerate. A rock that consists of variously shaped fragments of volcanic rocks embedded in a matrix of volcanic ash.

amphibole. A family of silicate minerals that abound in many kinds of igneous and metamorphic rocks, most commonly in those with a pale color. The commonest amphibole is hornblende, which crystallizes into glossy, black needles. Rare blue amphiboles exist in the blueschists of the California Coast Range.

andesite. A gray volcanic rock intermediate in composition between basalt and rhyolite.

anticline. A fold that bends layered rocks into an arch.

ash. Small shreds of lava blown from a volcano by escaping steam. Ash particles are light enough to drift on the wind.

ash flow. A dense cloud of volcanic ash and steam that pours across the ground surface. Many ash flows are so hot that the particles of ash weld into solid rock as they stop moving.

asthenosphere. A zone of partially melted rock within the earth's mantle, the lower boundary of the lithosphere.

basalt. A common volcanic rock composed mainly of the black mineral augite and pale plagioclase feldspar. It may or may not contain olivine. Basalt is hard and black.

Basin and Range. A geologic province of western North America in which the earth's crust stretched and broke into a pattern of mountain ranges and valleys that trend generally north.

batholith. A large mass of granite exposed over an area of more than 40 square miles. Some cover tens of thousands of square miles.

basement. The fundamental rocks of the continental crust, mainly granite, schist, and gneiss.

blueschist. A rare kind of metamorphic rock that forms when sedimentary rocks recrystallize under conditions of high pressure and low temperature, typically in the depths of an oceanic trench. It contains crystals of blue amphibole.

caldera. A volcanic basin that forms as the ground surface collapses into an emptying magma chamber during a large eruption.

chert. A sedimentary rock composed mostly of microscopic grains of quartz. Stone Age peoples commonly chipped arrowheads and knife blades out of chert.

cinder cone. A small basalt volcano that consists of a pile of loose rubble blown from the vent by escaping steam. Most cinder cones finish their careers by erupting one or two lava flows, then never erupt again.

continental crust. The raft about 25 miles thick that makes the foundation of all continents. It consists mostly of the common rocks granite, gneiss, and schist, and floats high because these rocks are lighter than those of the underlying mantle.

crust. The upper surface of the lithosphere. Continental crust consists mostly of granite, gneiss, and schist; oceanic crust consists of basalt.

digger pine. A species of pine tree, common in parts of California, that has long, bluish needles and extremely large cones. Also known as coulter pine.

dike. A sheet of igneous rock emplaced as molten magma filled a fracture.

dome. The term usually refers to an extrusion of extremely viscous rhyolite magma. It also refers to landforms eroded in granite.

eclogite. A rare metamorphic rock that forms when basalt recrystallizes under conditions of high pressure and low temperature, typically in the depths of an oceanic trench. It consists mostly of green pyroxene and red garnet.

fault. A fracture in the earth's crust along which the rocks on either side have shifted.

flood basalt. An enormous basalt flow that may contain more than 100 cubic miles of lava. Flood basalt flows that erupted during middle Miocene time exist in the Modoc Plateau of northeastern California.

Franciscan complex. The name commonly applied to the extremely deformed sedimentary rocks in the California Coast Range.

fumarole. A vent that blows hot gas, typically associated with volcanoes.

gabbro. A dark igneous rock composed primarily of black pyroxene and greenish plagioclase, along with variable amounts of green olivine. Gabbro has the same composition as basalt, but differs from it in containing much larger mineral grains.

garnet. A family of silicate minerals that abound in many kinds of igneous and metamorphic rocks. Most garnets are red, but a few are green or white.

geyser. A hot spring that from time to time erupts a column of hot water and steam.

gneiss. A coarsely granular metamorphic rock with a streaky grain due to parallel alignment of mineral grains. Most gneisses also show distinct color banding. Pronounce it "nice."

granite. An igneous rock composed mostly of orthoclase feldspar and quartz, in grains large enough to see without using a magnifier. Most granites also contain mica or amphibole.

grus. Coarse sand formed when granite weathers and disintegrates.

hornblende. An iron and calcium silicate mineral, the common form of amphibole. Hornblende is common in rocks of the continental crust. It typically crystallizes into glossy, black needles.

igneous rock. Any rock, volcanic or intrusive, that crystallized from a molten magma.

laterite. The typical red soil of regions that have a warm and humid climate. Laterite consists mainly of iron oxide, aluminum oxide, and kaolinite clay, in no particular order of abundance.

lava. Molten rock on the earth's surface; erupted magma.

limestone. A sedimentary rock composed mostly of calcite, the mineral form of calcium carbonate.

lithosphere. The relatively cold and rigid outer rind of the earth, which extends down to a depth of about 60 to 100 miles. The lithosphere consists of the outermost part of the mantle, as well as its cover of continental or oceanic crust.

magma. Molten rock. The term *magma* is generally applied to molten rock below the earth's surface. If it erupts, it becomes lava.

mantle. The part of the earth between the crust on the outside and the core within—the largest part of the planet. The mantle consists mainly of peridotite.

marble. Metamorphosed limestone, composed of the mineral calcite in grains larger than those in the original limestone.

marine terrace. The surface of a strip of shallow seafloor, commonly a wavecut bench, raised above sea level.

mélange. A chaotic mess of rocks swept together within an ocean trench.

metamorphic rock. Rocks that recrystallized at high temperature, and in many cases under high pressure.

metamorphism. Recrystallization of a rock to form a new rock composed of new mineral grains. Metamorphism typically occurs at high temperature and most commonly under high pressure as well.

mica. A family of silicate minerals that crystallize into thin flakes. The most common varieties are the black mica, biotite, and the white mica, muscovite. Micas abound in many kinds of igneous and metamorphic rocks, most commonly in those with a pale color.

mudflow. A thick slurry of mud. Mudflows can carry much larger rocks than ordinary streams because mud is much denser than clear water, and therefore exerts a much greater buoyant effect.

mudstone. A sedimentary rock formed from mud.

natural levee. A slightly elevated stream bank deposited as floodwater dumps its load of sediment in the vegetation beside the stream.

obsidian. A glassy form of rhyolite. Obsidian is typically black, and may contain gray or brown streaks. It forms from rhyolite magma that contains no steam.

oceanic crust. The bedrock ocean floor: basalt lava flows on top, basalt dikes below them, and gabbro at the base. It is generally about 5 miles thick.

oceanic ridge. The line along which two tectonic plates move away from each other. Basalt erupts from the gap between them to make new oceanic crust.

oceanic trench. A depression on the ocean floor that develops where the floor bends down to begin its descent into the mantle at a collision plate boundary.

olivine. An iron and magnesium silicate mineral that typically forms glassy green crystals. Look for olivine in black igneous rocks.

ophiolite. A slab of oceanic crust now on land.

orthoclase. A variety of the mineral feldspar that is rich in potassium. Orthoclase typically makes white or pink crystals with a blocky shape. Abundant in granite and in many kinds of schist or gneiss.

peridotite. A dense, black igneous rock composed mostly of black pyroxene, along with green olivine and any of several other minerals. The earth's mantle probably consists mainly of peridotite, so it is the most abundant rock in the planet, even though you rarely see it exposed at the surface.

placer. A segregation of heavy minerals, typically in stream or beach sands. Pronounce the *a* as in "placid."

plagioclase. A variety of the mineral feldspar that is rich in sodium and calcium, in various proportions. Plagioclase makes white or pale greenish crystals with a blocky shape. It is abundant in many kinds of igneous and metamorphic rocks, especially in those with a dark color.

plate. A segment of the earth's outer rind, the lithosphere. A dozen or more plates of widely various sizes move about on the earth's surface.

pumice. A form of rhyolite so full of minute gas bubbles that it is essentially a glass foam.

pyroxene. A family of silicate minerals that occur mostly in dark igneous and metamorphic rocks. The commonest and most abundant pyroxene is augite, which occurs mainly in black igneous rocks such as basalt, gabbro, or peridotite. It forms stubby black crystals. The rare green pyroxene, jadeite, occurs in eclogite.

quartz. A mineral form of silica, silicon dioxide. Quartz is one of the most abundant and widely distributed minerals in continental rocks. It comes in a wide variety of forms, including clear crystals, sand grains, and chert.

resurgent caldera. A type of volcano that erupts enormous volumes of rhyolite while developing a large caldera basin, fills the caldera with rhyolite, then erupts again hundreds of thousands of years later.

resurgent dome. A bulge that develops in the floor of a large caldera as a mass of molten rhyolite magma rises beneath. Typically forms in large rhyolite volcanoes.

rhyolite. A volcanic rock, the eruptive equivalent of granite. Rhyolite is typically pale, and most commonly occurs as sheets of white volcanic ash. It also forms large domes and thick lava flows.

rhyolite dome. A mass of viscous rhyolite that slowly erupts to make a volcanic mountain.

ribbon chert. Thinly layered and colorful chert abundant in the Franciscan complex. Typically occurs in shades of red, green, yellow, and white.

right lateral. If a fault moves horizontally so that the side on your right comes toward you as you look along it, the sense of movement is right lateral.

sandstone. A sedimentary rock made primarily of sand.

schist. A metamorphic rock that contains enough mica to confer a flaky texture, or enough amphibole to make it splintery.

sea stack. A resistant mass of rock that stands in the surf as a small island.

seismograph. The instrumental record of an earthquake.

seismometer. An instrument used to observe and record earthquakes.

serpentinite. A soft, slippery green rock that forms when hot mantle peridotite reacts with water at the crest of an oceanic ridge or deep beneath the oceanic trench.

shield volcano. A large pile of thin basalt lava flows with gently sloping sides.

slate. Slightly metamorphosed shale or mudstone that breaks easily along surfaces parallel to the imaginary surfaces that would bisect folds into mirror images.

stand. Water level at which a lake or sea stands for a period of time.

syncline. A fold that bends layered rocks down into a trough.

terrane. An assemblage of rocks that share a more or less common origin and history.

thrust fault. A type of fault that typically moves older rocks on top of younger rocks.

till. A disorderly mixture of debris of all sizes deposited directly from glacial ice.

transform boundary. A plate boundary along which lithosphere plates slide past each other without creating or destroying oceanic crust.

tuff. Rock made of volcanic ash.

tuff ring. A volcano that consists mainly of a doughnut ring of volcanic ash with a crater or a rhyolite dome in its center.

turbidite. Muddy sandstone in which the particles in the individual layers grade from coarse sand or pebbles at the base to mud at the top. The typical sedimentary rock of the deep ocean floor.

vein. A deposit of minerals that fills a fracture.

wavecut bench. A bedrock surface that slopes gently seaward from a sea cliff. Most wavecut benches are submerged except at extreme low tide.

welded ash. A rock composed of volcanic ash that was still partially molten when it settled to the ground, and fused into solid rock. An ash flow deposit.

Selected Reading

It is hard for the layperson to find much useful literature about the geology of California, or of nearly any other place. Most of the literature is in rather baffling technical language and available only in large research libraries. In northern and central California, those include the various university libraries, the library of the California Division of Mines and Geology, and the U.S. Geological Survey library in Menlo Park.

The best readily available resource is the state geologic map, which comes in sheets that cover areas of 1x2 degrees. About twenty-seven of them cover the area of this book. You can buy these from the California Division of Mines and Geology, 801 K Street, Sacramento 95814, or the regional office, 185 Berry Street, Suite 3600, San Francisco 94107.

Aalto, K. R., R. J. McLaughlin, G. A. Carver, J. A. Barron, W. V. Sliter, and K. McDougall. 1995. Uplifted Neogene margin, southernmost Cascadia–Mendocino triple junction region, California. *Tectonics* 14:1104–16.

Alt, D., J. W. Sears, and D. W. Hyndman. 1988. Terrestrial maria: The origins of large basalt plateaus, hotspot tracks, and spreading ridges. *Journal of Geology* 96:647–62.

Atwater, T. 1989. Plate tectonic history of the northeast Pacific and western North America. In *The Geology of North America.* Vol. 4, *The Eastern Pacific Ocean and Hawaii.* Ed. E. L. Winterer, D. M. Hussorg, and R. W. Decker, 21–72. Boulder, Colo.: Geological Society of America.

Aydin, A., and B. M. Page. 1984. Diverse Pliocene-Quaternary tectonics in a transform environment, San Francisco Bay region, California. *Geological Society of America Bulletin* 95:1303–17.

Bailey, R. A. 1987. Long Valley caldera, eastern California. In *Geological Society of America Centennial Field Guide.* Vol. 1, *Cordilleran Section.* Ed. Mason L. Hill, 163–68. Boulder, Colo.: Geological Society of America.

———. 1989. *Geologic map of the Long Valley Caldera, Mono-Inyo Craters Volcanic Chain, and vicinity, eastern California.* U.S. Geological Survey Misc. Invest. Map I–1933.

Bally, A. W., and A. R. Palmer, eds. 1989. *The Geology of North America.* Vol. A, *An Overview.* Boulder, Colo.: Geological Society of America.

Bateman, P. C. 1978. *Map and cross section of the Sierra Nevada from Madera to the White Mountains, central California.* Geological Society of America Map and Chart Series MC–28E.

Beard, J. S., and H. W. Day. 1987. The Smartville intrusive complex, Sierra Nevada, California: The core of a rifted volcanic arc. *Geological Society of America Bulletin* 99:779–91.

Blake, M. C., Jr., ed. 1985. *Franciscan Geology of Northern California.* Los Angeles: Society of Economic Paleontologists and Mineralogists, Pacific Section.

Blake, M. C., Jr., R. L. Bruhn, E. L. Miller, E. M. Moores, S. B. Smithson, and R. C. Speed. 1985. *C–1 Mendocino Triple Junction to North American Craton: Centennial Continent/Ocean Transect #12.* Boulder, Colo.: Geological Society of America.

Blake, M. C., Jr., A. S. Jayko, R. J. McLaughlin, and M. B. Underwood. 1988. Metamorphic and tectonic evolution of the Franciscan Complex, Northern California. In *Metamorphism and Crustal Evolution of the Western United States.* Ed. W. G. Ernst, 1035–60. Englewood Cliffs, N.J.: Prentice-Hall.

Blake, M. C., Jr., and others. 1991. Mendocino triple junction to Wyoming. In *Tectonic Section Displays, Centennial Continent-Ocean Transects.* Comp. R. C. Speed. Boulder, Colo.: Geological Society of America.

Borcherdt, R. D., J. F. Gibbs, and K. R. Lajoi. 1975. *Maps Showing Maximum Earthquake Intensity Predicted in the Southern San Francisco Bay Region.* U.S. Geological Survey Misc. Field Studies Map MF–709.

Burchfiel, B. C., and G. A. Davis. 1981. Triassic and Jurassic tectonic evolution of the Klamath Mountains–Sierra Nevada geologic terrane. In *The Geotectonic Development of California.* Ed. W. G. Ernst, 50–70. Englewood Cliffs, N.J.: Prentice-Hall.

Burchfiel, B. C., P. W. Lipman, and M. L. Zoback, eds. 1992. *The Cordilleran Orogen, Conterminous United States.* Boulder, Colo.: Geological Society of America.

Burford, R. W., and R. V. Sharp. 1982. Slip on the Hayward and Calaveras faults determined from offset power lines. In *Special Publication of the California Division of Mines and Geology, no. 62.* Ed. E. W. Hart, S. E. Hirschfeld, and S. S. Schulz, 261–69.

Christiansen, R. L., and R. S. Yeats. 1992. Post-Laramide geology of the U.S. Cordilleran region. In *The Cordilleran Orogen, Conterminous United States, The Geology of North America.* Vol. G–3. Ed. B. C. Burchfiel, P. W. Lipman, and M. L. Zoback, 261–406. Boulder, Colo.: Geological Society of America.

Coleman, R. G., and others. 1988. Tectonic and regional metamorphic framework of the Klamath Mountains and adjacent Coast Ranges, California and Oregon. In *Metamorphism and Crustal Evolution of the Western United States.* Ed. W. G. Ernst, 1061–97. Englewood Cliffs, N.J.: Prentice-Hall.

Courtillot, V., H. Feinberg, J. P. Ragarn, R. Kerguelen, M. McWilliams, and A. Cox. 1985. Franciscan Complex limestone deposited at 24°N. *Geology* 13:107–10.

Cowan, D. S., and R. L. Bruhn. 1992. Late Jurassic to early Late Cretaceous geology of the U.S. Cordillera. In *The Cordilleran Orogen, Conterminous United States, The Geology of North America.* Vol. G–3. Ed. B. C. Burchfiel, P. W. Lipman, and M. L. Zoback, 169–203. Boulder, Colo.: Geological Society of America.

Curtis, G. H. 1954. Mode of origin of pyroclastic debris in the Mehrten Formation of the Sierra Nevada. *University of California Publications in Geological Sciences* 29:453–502.

Davis, G. A., and G. S. Lister. 1988. *Detachment and Faulting in Continental Extension: Perspectives from the Southwestern U.S. Cordillera.* Geological Society of America Special Paper 218, 133–59.

Day, H. W., P. Schiffman, and E. M. Moores. 1988. Metamorphism and tectonics of the northern Sierra Nevada. In *Metamorphism and Crustal Evolution of the Western United States.* Ed. W. G. Ernst, 737–63. Englewood Cliffs, N.J.: Prentice-Hall.

Dickinson, W. R. 1995. Forearc basins. In *Tectonics of Sedimentary Basins.* Ed. C. J. Busby and R. V. Ingersoll, 221–61. Cambridge, Mass.: Blackwell Science.

Dickinson, W. R., and E. I. Rich. 1972. Petrologic intervals and petrofacies in the Great Valley Sequence, Sacramento Valley, California. *Geological Society of America Bulletin* 83:3007–24.

Dickinson, W. R., C. A. Hopson, and J. B. Saleeby. 1996. Alternate origins of the Coast Range Ophiolite (California): Introduction and implications. *GSA Today* 6:1–10.

Dilek, Y., P. Thy, and E. M. Moores. 1991. Episodic dike intrusions in the northwestern Sierra Nevada, California: Implications for multistage evolution of a Jurassic arc terrane. *Geology* 19:180–84.

Donnelly-Nolan, J. M. 1992. Medicine Lake volcano and Lava Beds National Monument, Siskiyou and Modoc counties. *California Geology* 45:145–53.

Edelman, S. H., H. W. Day, E. M. Moores, S. M. Zigan, T. P. Murphy, and B. R. Hacker. 1989. *Structure Across a Mesozoic Ocean-Continent Suture Zone in the Northern Sierra Nevada, California.* Geological Society of America Special Paper 224.

Effimov, I., and A. R. Pinezich. 1986. *Tertiary Structural Development of Selected Basins: Basin and Range Province, Northeastern Nevada.* Geological Society of America Special Paper 208, 31–42.

Eppler, D. B. 1987. The May 1915 eruptions of Lassen Peak: II. May 22 volcanic blast effects, sedimentology and stratigraphy of deposits, and characteristics of the blast cloud. *Journal of Volcanology and Geothermal Research* 31:65–85.

Ernst, W. G. 1984. Californian blueschists, subduction, and the significance of tectonostratigraphic terranes. *Geology* 12:436–40.

Etter, S. D., D. M. Fritz, P. R. Gucwa, M. A. Jordan, J. R. Kleist, D. H. Lehman, J. A. Raney, and D. M. Worrall. 1981. *Geologic Cross Sections, Northern California Coast Ranges to Northern Sierra Nevada.* Geological Society of America Map and Chart Series MC–28N.

Foland, S. S., and K. J. Enzor. 1994. Reinterpretation of the northern terminus of the San Andreas transform system: Implications for basin development and hydrocarbon exploration. *American Association of Petroleum Geologists Bulletin* 78:1142.

Fox, K. F. 1983. *Tectonic Setting of Late Miocene, Pliocene, and Pleistocene Rocks in Part of the Coast Ranges North of San Francisco, California.* U.S. Geological Survey Professional Paper 1239.

Galloway, A. J. 1977. *Geology of the Point Reyes Peninsula, Marin County, California.* California Division of Mines and Geology Bulletin 202.

Garrison, R. E., and R. G. Douglas, eds. 1981. *Monterey Symposium: Monterey Formation and Related Siliceous Rocks of California.* Los Angeles: Society of Economic Paleontologists and Mineralogists, Pacific Section.

Geologic Map of California. 27 sheets, 1957–1967. California Division of Mines and Geology. Scale 1:250,000.

Hamilton, W. W. 1987. Crustal extension in the Basin and Range province, western United States. In *Continental Extensional Tectonics.* Ed. M. P. Coward, J. F. Dewey, and P. L. Hancock. *Geological Society of London Special Publication* 28:155–76.

Harper, G. D. 1984. Josephine ophiolite, northwestern California. *Geological Society of America Bulletin* 95:1009–26.

———. 1989. *Geologic Evolution of the Northernmost Coast Ranges and Western Klamath Mountains, California.* American Geophysical Union, International Geological Congress Field Guide T308.

Harper, G. D., J. B. Saleeby, and M. Heizler. 1994. Formation and emplacement of the Josephine ophiolite and the Nevadan orogeny in the Klamath Mountains, California-Oregon: U/Pb and $^{40}Ar/^{39}Ar$ geochronology. *Journal of Geophysical Research* 99:4293–321.

Harris, S. P. 1988. *Fire Mountains of the West: The Cascade and Mono Lake Volcanoes.* Missoula, Mont.: Mountain Press Publishing.

Hearn, B. C., Jr., J. M. Donnelly-Nolan, and F. E. Goff. 1981. *The Clear Lake volcanics: Tectonic Setting and Magma Sources.* U.S. Geological Survey Professional Paper 1141, 24–45.

Hearn, B. C., Jr., R. J. McLaughlin, and J. M. Donnelly-Nolan. 1988. *Tectonic Framework of the Clear Lake Basin, California.* Geological Society of America Special Paper 214, 9–20.

Hill, M. L., ed. 1987. *Geological Society of America Centennial Field Guide.* Vol. 1, *Cordilleran Section.* Boulder, Colo.: Geological Society of America.

Hopson, C. A., J. M. Mattinson, B. P. Luyendyk, and E. A. Pessagno, Jr. 1991. California Coast Range Ophiolite (CRO): Middle Jurassic/central Tethyan and latest Jurassic/southern Boreal episodes of ocean-ridge magmatism. *EOS, Transactions of the American Geophysical Union* 72:443.

Ingersoll, R. V. 1983. Petrofacies and provenance of late Mesozoic forearc basin, northern and central California. *American Association of Petroleum Geologists Bulletin* 67:1125–42.

Ingersoll, R. V., and T. H. Nilsen, eds. 1990. *Sacramento Valley Symposium and Guidebook* [Book 65]. Los Angeles: Society of Economic Paleontologists and Mineralogists, Pacific Section.

Irwin, W. P. 1981. Tectonic accretion of the Klamath Mountains. In *The Geotectonic Development of California.* Ed. W. G. Ernst, 29–49. Englewood Cliffs, N.J.: Prentice-Hall.

Irwin, W. P., and M. D. Dennis. 1979. *Geologic Structure Section across Southern Klamath Mountains, Coast Ranges, and Seaward of Point Delgada, California.* Geological Society of America Map and Chart Series MC 28–D. Scale 1:250,000.

Jayko, A. S., M. C. Blake, and T. Harms. 1987. Attenuation of the Coast Range ophiolite by extensional faulting, and the nature of the Coast Range "thrust," California. *Tectonics* 6:475–88.

Jennings, C. W., comp. 1977. *Geologic Map of California.* California Geologic Data Map Series, Map No. 2. California Division of Mines and Geology. Scale 1:750,000.

Jones, D. L., and W. P. Irwin. 1971. Structural implications of an offset Early Cretaceous shoreline in northern California. *Geological Society of America Bulletin* 82:815–22.

Kilbourne, R. T., and C. L. Anderson. 1981. Volcanic history and active volcanism in California. *California Geology* 34:159–68.

Lawson, A. C., chairman of the commission. 1908. *The California Earthquake of April 18, 1906.* Report of the State Earthquake Investigation Commission: Carnegie Institution, Washington Publication 87, 3 volumes, 1 atlas. Reprinted, 1969.

Lindgren, W. 1911. *The Tertiary Gravels of the Sierra Nevada of California.* U.S. Geological Survey Professional Paper 73.

Luedke, R. G., and R. L. Smith. 1981. *Map Showing Distribution, Composition, and Age of Late Cenozoic Volcanic Centers in California and Nevada.* U.S. Geological Survey Miscellaneous Investigations Series Map I–1091C. Scale 1:500,000.

McLaughlin, R. J., W. V. Sliter, D. H. Sorg, P. C. Russell, and A. M. Sarna-Wojcicki. 1996. Large-scale right-slip displacement on the East San Francisco Bay region fault system, California: Implications for location of late Miocene to Pliocene Pacific plate boundary. *Tectonics* 15:1–18.

Merritts, D. J., and A. C. Brustolon. 1995. The Mendocino triple junction: Active faults, episodic coastal emergence, and rapid uplift. In *Proceedings of the Workshop on Paleoseismology.* Convener, C. S. Prentice. U.S. Geological Survey Open File Report, 121–23.

Muffler, L. J. P., C. R. Bacon, R. L. Christiansen, A. Michael-Cline, J. M. Donnelly-Nolan, C. D. Miller, D. R. Sherrod, and J. G. Smith. 1989. Excursion 12B: South Cascades arc volcanism, California and southern Oregon. In *Field Excursions to Volcanic Terranes in the Western United States.* Vol. 2, *Cascades and Intermountain West.* Ed. C. E. Chapin and J. Zidek, 183–225. New Mexico Bureau of Mines and Mineral Resources, Memoir 47.

Nilsen, T. H., and S. H. Clarke, Jr. 1989. Late Cenozoic basins of northern California. *Tectonics* 8:1137–58.

Page, B. M., and T. M. Brocher. 1993. Thrusting of the central California margin over the edge of the Pacific plate during the transform regime. *Geology* 21:635–38.

Page, B. M., H. C. Wagner, D. S. McCulloch, E. A. Silver, and J. H. Spotts. 1979. *Geologic Cross Section of the Continental Margin off San Luis Obispo, the Southern Coast Ranges, and the San Joaquin Valley, California.* Geological Society of America Map and Chart Series MC–28G.

Pessagno, E. A., Jr., C. A. Hopson, J. M. Mattinson, C. D. Blome, B. P. Luyendyk, D. M. Hull, and W. Beebe. 1997. Coast Range ophiolite and its sedimentary cover (California Coast Ranges): Jurassic stratigraphy and northward tectonic transport. *Tectonics* manuscript.

Plafker, G., and J. P. Galloway, eds. 1989. *Lessons learned from the Loma Prieta, California, earthquake of October 17, 1989.* U.S. Geological Survey Circular 1045.

Powell, R. E., R. J. Welson II, and J. C. Matti. 1992. *San Andreas Fault System: Displacement, Palinspastic Reconstruction, and Geologic Evolution.* Geological Society of America Memoir 178.

Prims, J., and K. P. Furlong. 1995. Subsidence of San Francisco Bay: Blame it on Salinia. *Geology* 23:559–62.

Regional Geologic Map Series. 1982–1990. California Division of Mines and Geology. Scale 1:250,000.

Revenaugh, J., and C. Reasoner. 1997. Cumulative offset of the San Andreas fault in central California: A seismic approach. *Geology* 25:123–26.

Ring, U., and M. T. Brandon. 1994. Kinematic data for the Coast Range fault and implications for exhumation of the Franciscan subduction complex. *Geology* 22:735–38.

Ross, D. C., and D. S. McCulloch. 1979. *Cross Section of the Southern Coast Ranges and San Joaquin Valley from Offshore Point Sur to Madera, California.* Geological Society of America Map and Chart Series MD–28H.

Saleeby, J. B., and others. 1991. Central California offshore to Colorado Plateau. In *Tectonic Section Displays, Centennial Continent-Ocean Transects.* Comp. R. C. Speed. Boulder, Colo.: Geological Society of America.

Schweikert, R. A., N. L. Bogen, and C. Merguerian. 1988. Deformational and metamorphic history of Paleozoic and Mesozoic basement terranes in the western Sierra Nevada metamorphic belt. In *Metamorphism and Crustal Evolution of the Western United States.* Ed. W. G. Ernst, 789–822. Englewood Cliffs, N.J.: Prentice-Hall.

Sharp, W. D. 1988. Pre-Cretaceous crustal evolution in the Sierra Nevada region, California. In *Metamorphism and Crustal Evolution of the Western United States.* Ed. W. G. Ernst, 823–64. Englewood Cliffs, N.J.: Prentice-Hall.

Sloan, D., and D. L. Wagner, eds. 1991. *Geological Excursions in Northern California: San Francisco to the Sierra Nevada.* California Division of Mines and Geology Special Publication 109.

Speed, T., M. W. Elison, and F. R. Heck. 1988. Phanerozoic tectonic evolution of the Great Basin. In *Metamorphism and Crustal Evolution of the Western United States.* Ed. W. G. Ernst, 572–606. Englewood Cliffs, N.J.: Prentice-Hall.

Suchecki, R. K. 1984. Facies history of the Upper Jurassic–Lower Cretaceous Great Valley sequence: Response to structural development of an outer-arc basin. *Journal of Sedimentary Petrology* 54:170–91.

Suppe, J. 1979. *Southern Part of Northern Coast Ranges and Sacramento Valley, California.* Geological Society of America Map and Chart Series MC–28B.

Trehu, A. M. 1995. Pulling the rug out from under California: Seismic images of the Mendocino triple junction region. *EOS, Transactions of the American Geophysical Union* 76:369, 380–81.

Unruh, J. R. 1991. The uplift of the Sierra Nevada and implications for late Cenozoic epeirogeny in the western Cordillera. *Geological Society of America Bulletin* 103:1395–1404.

Venum, W. 1994. Castle Crags. *California Geology* 47:31–38.

Wagner, D. L., and E. J. Bortugno. 1982. *Geologic Map of the Santa Rosa Quadrangle, California.* California Division of Mines and Geology Map 2A. Scale 1:250,000.

Wagner, D. L., E. J. Bortugno, and R. D. McJunkin. 1991. *Geologic Map of the San Francisco–San Jose Quadrangle, California.* California Division of Mines and Geology Map 5A. Scale 1:250,000.

Wagner, D. L., C. W. Jennings, T. L. Bedrossian, and E. J. Bortugno. 1981. *Geologic Map of the Sacramento Quadrangle, California.* California Division of Mines and Geology Map 1A. Scale 1:250,000.

Wagner, D. L., and G. J. Saucedo. 1987. *Geologic Map of the Weed Quadrangle.* California Division of Mines and Geology Regional Geologic Map Series Map 4A. Scale 1:250,000.

Wahrhaftig, C., coordinator. 1989. *Geology of San Francisco and Vicinity.* 28th International Geological Congress Field Trip Guidebook T105, American Geophysical Union.

Wallace, R. E., ed. 1990. *The San Andreas Fault System, California.* U.S. Geological Survey Professional Paper 1515.

Waters, A. C. 1992. Captain Jack's Stronghold (Lava Beds National Monument). *California Geology* 45:135–44.

Wernicke, B. 1981. Low-angle normal faults in the Basin and Range Province: Nappe tectonics in an extending orogen. *Nature* 291:645–48.

Williams, H. 1929. The volcanic domes of Lassen Peak and vicinity, California. *American Journal of Science* 18:313–30.

———. 1932. Geology of the Lassen Volcanic National Park, California. *University of California Publications in Geological Sciences* 21:195–385.

———. 1969. Geology of the Lassen Volcanic National Park, California. In *Geologic Guide to the Lassen Peak, Burney Falls, and Lake Shasta Area,* 1–42. Sacramento, Calif.: Geological Society of Sacramento.

Williams, H., and G. H. Curtis. 1977. The Sutter Buttes of California, a study of Plio-Pleistocene Volcanism. *University of California Publications in Geological Sciences* 116:1–56.

Wolf, M. B., and J. B. Saleeby. 1995. Late Jurassic dike swarms, southwest Sierra Nevada Foothills terrane. In *Jurassic Magmatism and Tectonics of the North American Cordillera.* Ed. D. B. Miller and C. Busby. Geological Society of America Special Paper 299.

Wright, J. E., and S. J. Wyld. 1994. The Rattlesnake Creek terrane, Klamath Mountains, California: An early Mesozoic volcanic arc and its basement of tectonically disrupted oceanic crust. *Geological Society of America Bulletin* 106:1033–56.

Index

About the Authors

David Alt and **Donald W. Hyndman** are dedicated to bringing geology to the general public. They founded the Roadside Geology series, have written several of its books, and help edit others. When they are not working on books, Alt and Hyndman teach geology at the University of Montana in Missoula.